# Mineralogy of Arizona

Stope in oxide ore, Copper Queen mine, Bisbee. These old workings were probably mined in the 1890's and were subsequently partially filled with waste rock. Peter Kresam and Richard Graeme.

# Mineralogy
## of
# Arizona

John W. Anthony,
Sidney A. Williams,
and Richard A. Bideaux

THE UNIVERSITY OF ARIZONA PRESS
TUCSON, ARIZONA

## About the Authors . . .

JOHN W. ANTHONY received his B.S. and M.S. degrees in geology from the University of Arizona and his Ph.D. from Harvard University. Following several years as mineralogist with the Arizona Bureau of Mines, he joined the faculty of the University of Arizona, where he rose to the positions of professor and curator of the Mineralogical Museum. His studies in mineral synthesis, crystal structure analysis, and classical descriptive mineralogy have resulted in the publication of a number of technical papers. He has served as consultant to the minerals industries and was founding editor of the *Digest,* the bulletin of the Arizona Geological Society. Anthony is a fellow of the Geological Society of America, the Mineralogical Society of America, and the American Association for the Advancement of Science. In 1972 he served as general chairman of the first joint mineralogical meeting of amateurs and professionals, sponsored by the Mineralogical Society of America, the Friends of Mineralogy, the Tucson Gem and Mineral Society, and the University of Arizona.

SIDNEY A. WILLIAMS has authored or co-authored some thirty-five scientific papers and discovered sixteen new mineral species, naming two of them after his co-authors. A graduate of Michigan College of Mining and Technology (now Michigan Technological University), he received his Ph.D. in geology from the University of Arizona. He was a member of the faculty of Michigan College of Mining and Technology for three years, leaving that position to become mineralogist for Silver King Mines, Inc., in Nevada. In 1965 he joined Phelps Dodge Corporation as mineralogist/petrographer for the Western Exploration Office and director of the Geological Laboratory. Williams is a fellow of the Mineralogical Society of America.

RICHARD A. BIDEAUX, a University of Arizona graduate, received his M.A. degree in geology from Harvard University. He has discovered and helped to describe several new Arizona minerals which appear in this volume. For many years he was associated with the Tucson Gem and Mineral Society Show and also served as assistant editor and columnist for the *Mineralogical Record*. Although well known in mineralogical circles as a prominent collector of fine examples of Arizona minerals, Bideaux is president of the Tucson-based international consulting firm, Computing Associates, noted for computer applications in geology and mining engineering.

The authors and publisher express grateful appreciation to the Phelps Dodge Corporation for its special grant that made possible the inclusion of color reproductions in this volume.

THE UNIVERSITY OF ARIZONA PRESS

I. S. B. N. 0-8165-0601-9 cloth
I. S. B. N. 0-8165-0471-7 paper
L. C. No. 75-44670

# Acknowledgments

The authors have naturally accrued a number of debts during the compilation of this volume. We have noted throughout the catalog the names of those who provided information on specific mineral occurrences, and we are very grateful for their assistance. We are also happy to express our thanks to the several other people and organizations who helped in various ways.

Richard T. Moore, H. Wesley Peirce, and Robert T. O'Haire, of the Arizona Bureau of Mines, cheerfully provided counsel where sought. We are also indebted to the Bureau for permission to reproduce the Map and Index of Arizona Mining Districts. Carleton B. Moore, Peter R. Busek, and Charles F. Lewis, of Arizona State University, kindly made available information on the meteorites. For allowing access to their respective collections and catalogs we are indebted to the following: Paul E. Desautels, U.S. National Museum of Natural History; Clifford Frondel, Harvard University; Vincent Manson, American Museum of Natural History; Horace Winchell, Yale University; and Max H. Hey and Peter Embrey, British Museum (Natural History). Spencer R. Titley and John M. Guilbert, University of Arizona, read the section on porphyry copper deposits and made constructive comments.

Susan H. Hunt carried out a significant share of the literature research while an undergraduate in geology at the University of Arizona, conducting her sometimes tiring bibliographic investigations with an exceptional maturity of scholarship. Without the many contributions of Therese Murchison, formerly Acting Curator of the Mineralogical Museum, University of Arizona, the appearance of this book would have been long delayed; we are pleased to acknowledge that assistance as well as her unflagging enthusiasm.

Rose Samardzich typed the manuscript with her customary care.

The Arizona Bureau of Mines and the Tucson Gem and Mineral Society kindly provided financial assistance during preparation of the manuscript. The senior author wishes to thank the Department of Geosciences and the University of Arizona for the granting of a sabbatical leave during which this project was begun.

We are pleased to acknowledge the contribution made to the quality of the color illustrations by Messrs. William Hunt, Robert Mudra, and Marvin Deshler, who generously provided many fine micromount specimens, particularly of the zeolites.

The authors were fortunate to have enjoyed the closest of associations with two of the most distinguished of mineral photographers: Julius Weber, Associate in the Department of Mineralogy of the American Museum of Natural History, New York City; and Jeffery Kurtzeman, award-winning photographer, of Phoenix. These gentlemen, whose photographic representations of minerals are appreciated as much for their sensitivity of rendition as for their mastery of technique, were most generous of their time and materials, and we are grateful for their splendid efforts to "light up" this volume. We also thank those other individuals who provided specific photographs as acknowledged in the photograph captions.

The authors and publisher express grateful appreciation to the Phelps Dodge Corporation for its special grant that made possible the inclusion of color reproductions in this volume.

Finally, the authors wish to thank the staff of the University of Arizona Press for basic publication financing and for professional transformation of the manuscript into this book.

JOHN W. ANTHONY
SIDNEY A. WILLIAMS
RICHARD A. BIDEAUX

# Contents

# Illustrations

# CONTENTS

## TABLES

PART ONE

# INTRODUCTION

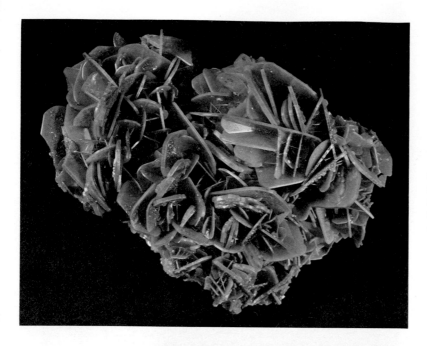

**Wulfenite.**
Glove mine, Cottonwood Canyon,
Santa Rita Mountains, Santa Cruz
County. Bandy collection.
Jeff Kurtzeman.

**Cerussite on aurichalcite.**
79 mine, Gila County. Wayne
Darby collection. Jeff Kurtzeman.

**Wulfenite on mottramite.**
79 mine, Gila County. Wayne
Thompson collection.
Jeff Kurtzeman.

# Introduction

Since the inception of modern mining activity in the Arizona Territory in the latter half of the nineteenth century, Arizona has been known as a copper-mining state. The statistics recounting the commercial end-product of a century of continuously accelerating effort expended in digging out the mineral wealth of this enormous area are impressive: over 40 billion pounds of copper

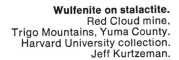

**Wulfenite on stalactite.**
Red Cloud mine,
Trigo Mountains, Yuma County.
Harvard University collection.
Jeff Kurtzeman.

**Aurichalcite and hemimorphite.**
79 mine, Gila County.
Wayne Thompson collection.
Jeff Kurtzeman.

were mined during this period. If there were any doubt that this is a lot of copper, the blasé should be impressed by the image of a solid copper pyramid, square-based, two football fields on an edge and two football fields high. By comparison, the largest of the pyramids of ancient Egypt, the Great Pyramid of Khufu, originally measured about 756 feet along the square base and was about 481 feet high. Although copper has constituted the great part of Arizona's metallic mineral wealth through the years, exploitation of the ores of gold, silver, lead, zinc, molybdenum, uranium, mercury, and tungsten has not been neglected; nor has development of a host of nonmetallic mineral resources which are becoming increasingly important as the population grows.

It is inevitable that this century of mining leave its imprint on any extensive collection of mineral data. The most cursory glance through the mineral occurrences listed in this volume will make it obvious that the "average" mineral reported is in or near a mine or mineral prospect.

The pace of geological and mineralogical activity in Arizona has accelerated during the 1960s and 1970s. The efforts of the many skilled geologists working in the state, augmented by the contributions of an ever-increasing number of skilled amateur collectors, have unearthed a host of new occurrences and brought to light a surprising number of mineral species not previously known to occur in the state.

We had several goals in undertaking this venture. We wished to collect into one source much of the Arizona mineralogical information that has become available since the 1959 edition of Galbraith's and Brennan's *Minerals of Arizona*. We also attempted to document as many as we could of the sources of the older, unreferenced, locality citations in that volume. Although our list of references is by no means exhaustive, the majority of significant papers is included.

We believe that this information will be of assistance to the geologist whose primary concern is with scientific aspects of mineralogy, and to the applied geologist searching for those natural mineral commodities so rapidly being depleted. We believe, too, that this book will appeal to those sharing with us aesthetic appreciation of the beauties inherent in the magnificent minerals so long a source of pride to Arizona. On the basis of geological and economic inference, it appears that Arizona, and for that matter the world, will seldom if ever again provide

mineral specimens to equal the quality of those from the distinguished Arizona mining camps.

The first comprehensive collections of information on the minerals of the state appeared in 1909 and 1910 when two pioneers of Arizona geology, William Phipps Blake (1826–1910), Territorial Geologist and Director of the School of Mines, University of Arizona, and Frank Nelson Guild (1870–1939), Professor of Chemistry and Mineralogy, University of Arizona, independently published summaries of the minerals of the Territory. Blake's effort appeared in 1909 as a report directed to the Honorable J. H. Kibbey, Governor of the Territory; it discussed approximately 100 minerals. Guild's book, *The Mineralogy of Arizona*, privately printed in 1910, described about 120 species. Blake, incidentally, also compiled the first California mineral list, and the centennial of that occasion was commemorated in the appearance of *Minerals of California*, by Murdoch and Webb (1966).

No further compilation was undertaken until 1941, when Frederic William Galbraith, Professor of Geology, University of Arizona, compiled *Minerals of Arizona*. By that time the number of species reported had grown to about 293. First issued as Bulletin 149 by the Arizona Bureau of Mines, the work progressed through two revisions, the second being co-authored with Daniel J. Brennan. This work enjoyed considerable popularity with amateur and professional geologists and mineralogists alike. With the exception of a few popular works aimed at the amateur mineral collector, *Minerals of Arizona* remained for years the only convenient single source of information on mineral occurrences in the state. The number of species reported in the 1959 revision had grown to about 403, of which twenty subsequently were dropped from the mineral list.

The body of data assembled in the works of our predecessors constitutes the nucleus of this book, and the reader will observe that many entries are unchanged from their earlier form. Quite a number of mineral occurrences were gathered by Galbraith from word-of-mouth sources either unknown or no longer available to us. Although in some instances the information is rather vague and not in the detail which could be desired, it is retained. In other instances, entries have been modified to accord with more recent data on specific occurrences. We have, of course, rejected a number of mineral citations in the earlier works where subsequent study has shown material to have been incorrectly identified.

Some mineral names have been changed to conform to the most widely accepted nomenclature. As examples we cite the preference for the name vesuvianite over idocrase; the substitution of acanthite for argentite to designate the silver sulfide phase stable under ordinary conditions of temperature, and the revision of the nomenclature of the silver halogen minerals to follow systematic current usage. Some minerals have been reduced from species to varietal status. For example, the mineral formerly called smaltite is now regarded as an arsenic-deficient variety of skutterudite.

New detail has been found in the mineralogies of some of the older well-known occurrences, primarily as the result of the widespread use of x-ray diffraction methods of mineral identification. A good example is seen in the studies of manganese oxide deposits. Several workers, among them Hewett and Fleischer, unraveled the complexities of some of these deposits whose mineralogies had been lumped under the catch-all terms "wad" and "psilomelane," and presented us with a number of mineral species previously not known to occur in Arizona.

Another unexpected complexity has arisen from studies of copper sulfides. New species such as djurleite have been discovered in copper sulfide ores formerly thought to consist only of chalcocite. One can anticipate that this and other related species will prove to be even more widespread than is known at this writing.

The largest single group of species newly reported here, however, was encountered in the oxidized portions of base metal deposits. The blessing (and oft-times we must add, bane) of the professional mineralogist who undertakes to identify exotic minerals brought to him, is the keen eye of the well-informed amateur collector who may approach the encyclopedic in his knowledge of the rare, colorful, and well-crystallized secondary oxidized minerals found in the near-surface portions of the mines of the southwest. Arizona has its share of such avid devotees, and we are most happy to acknowledge their contributions to whatever claim this book can lay to completeness. Their excursions (occasionally nocturnal and surreptitious) into mine workings inaccessible to most of us have provided a wealth of study material.

The means of pinpointing the geographical locations of mineral occurrences is an inheritance of Arizona's mining bias. In addition to being geopolitically divided into 14 counties, Arizona is partitioned into some 240 mining districts. These rather vaguely delineated regions were early established as quasi-political regulatory entities to promulgate rules to be followed in local mining practice; individual districts, like as not, followed different mining mores. With the advent of the county governmental structure, however, the need for this regulatory function diminished until the mining district remains largely as an anachronism. But the habit is hard to break, and it is still common practice in some western states to discuss mines, prospects, and even geology in terms of the old mining district names. Because so many of the locality references in the literature are to mining districts, we have found it expedient to perpetuate this well-ingrained custom. Maps are included showing the locations of the mining districts in the various counties.

An alternative would be to situate each mineral occurrence in a coordinate system such as latitude and longitude, or by means of the Standard Land Survey, which fixes a location by a statement of section, township, and range, referred to either the Gila and Salt River or the Navajo Base and Meridian. The latter system is used for a few localities listed in this book, but the requisite information often is lacking. Commonly, the only means of obtaining the required data is to scale them from maps. Many of the localities reported are so indefinite that a series of formal symbols designating them would impart to the reader an erroneous sense of precision. No one will be more aware of the vagueness of many of the localities here given than the authors.

In selecting the format of this book we have taken the position that inclusion of basic descriptive data for each mineral species would be needless duplication of information to which the reader has ready access in many excellent reference works. We have, however, continued the practice of stating chemical formulas; written versions of the chemical compositions have been added to assist those who may not have facility with chemical symbology. Brief statements of the geological modes of occurrence of each mineral species are also included.

The verbal sketches of the two kinds of mineral occurrences for which Arizona is particularly noted — the porphyry copper deposits and

the Colorado Plateau-type uranium and vanadium deposits — are designed to help the reader
appreciate the geological nature of these remarkable mineral assemblages to which reference is
repeatedly made throughout the systematic compilation.

The detailed descriptions of the mineralogies
at the Mammoth–St. Anthony mine at Tiger and
of the mines at Bisbee, which surely rank among
the mineralogical wonders of the world, are intended to partially fill a long-standing void in
the mineralogical literature; for although many
excellent geological studies of these mines have
been reported, there have appeared no inclusive
summaries of their mineralogies incorporating
recent work. We intend that these descriptions
also serve as types to which the many other
similar mineral occurrences throughout the state
can be compared.

As a convenience to the reader, mineral entries are arranged in alphabetical order rather
than according to a conventional chemical-struc-
tural classification.

The names of individuals who provided information on mineral occurrences follow each citation.

We have noted catalog numbers of some of
the Arizona specimens held in the collections of
some of the museums with whose holdings we
are familiar. These entries follow the mineral

TABLE 1

**Codes for Museums Containing
Mineral Specimens**

| Code | Museum |
|------|--------|
| AEC | Energy Research and Development Commission, Grand Junction, Colorado (formerly the Atomic Energy Commission) |
| AM | American Museum of Natural History |
| ASU | Arizona State University, Center for Meteorite Studies |
| BM | British Museum (Natural History) |
| H | Harvard University |
| S | National Museum of Natural History, Smithsonian Institution |
| UA | University of Arizona |

locality citations and are coded as shown in
Table 1.

The soil-forming clay minerals and the common rock-forming silicate minerals have been
accorded what may appear to be somewhat
cavalier treatment. Each of these groups is discussed in an extensive literature: the former in
the journals of soil and agricultural science, and
civil engineering; the latter in the geological

**Quartz twinned after the Japanese
law, lightly coated by hematite.**
Washington Camp, Santa Cruz
County. Davis collection.
Jeff Kurtzeman.

## TABLE 2
### Minerals First Discovered in Arizona

| Mineral | Citation Source | County |
|---------|-----------------|--------|
| Ajoite | Schaller and Vlisidis, 1958 | Pima |
| Andersonite | Axelrod et al., 1951 | Yavapai |
| Antlerite | Hillebrand, 1889 | Mohave |
| Arizonite (?) | Palmer, 1909 | Mohave |
| Bayleyite | Axelrod et al., 1951 | Yavapai |
| Bermanite | Hurlbut, 1936 | Yavapai |
| Bideauxite | Williams, 1970 | Pinal |
| Bisbeeite (?) | Schaller, 1915 | Cochise |
| *Brezinaite | Bunch and Fuchs, 1969 | Pima |
| Buetschliite | Milton and Axelrod, 1947 | Coconino |
| Butlerite | Lausen, 1928 | Yavapai |
| Chalcoalumite | Larsen and Vassar, 1925 | Cochise |
| *Cliftonite (?) | El Goresy, 1965 | Coconino |
| Coconinoite | Young et al., 1966 | Coconino |
| Coesite | Chao et al., 1960 | Coconino |
| Coronadite | Lindgren and Hillebrand, 1904 | Greenlee |
| Creaseyite | Williams and Bideaux, 1975 | Pinal |
| Cryptomelane | Richmond and Fleischer, 1942 | Cochise |
| Emmonsite | Hillebrand, 1885 | Cochise |
| Fairchildite | Milton and Axelrod, 1947 | Coconino |
| Flagstaffite | Guild, 1920 | Coconino |
| Gerhardtite | Wells and Penfield, 1885 | Yavapai |
| Graemite | Williams and Matter, 1975 | Cochise |

| Mineral | Citation Source | County |
|---------|-----------------|--------|
| Guildite | Lausen, 1928 | Yavapai |
| Hemihedrite | Williams and Anthony, 1970 | Pinal |
| Jeromite (?) | Lausen, 1928 | Yavapai |
| Junitoite | Williams, 1976 | Pinal |
| Jurbanite | Anthony and McLean, 1976 | Pinal |
| Kinoite | Anthony and Laughon, 1970 | Pima |
| *Krinovite | Olsen and Fuchs, 1968 | Coconino |
| Lausenite | Lausen, 1928; Butler, 1928 | Yavapai |
| *Lonsdaleite | Frondel and Marvin, 1967 | Coconino |
| Luetheite | Williams, 1976 | Santa Cruz |
| Moissanite | Kunz, 1905 | Coconino |
| Murdochite | Fahey, 1955 | Pinal |
| Navajoite | Weeks et al., 1954, 1955 | Apache |
| Papagoite | Hutton and Vlisidis, 1960 | Pima |
| Paramelaconite | Koenig, 1891 | Cochise |
| Ransomite | Lausen, 1928 | Yavapai |
| Selenium | Palache, 1934 | Yavapai |
| Shattuckite | Schaller, 1915 | Cochise |
| Spangolite | Penfield, 1890 | Cochise |
| Stishovite | Chao et al., 1962 | Coconino |
| Swartzite | Axelrod et al., 1951 | Yavapai |
| Wherryite | Fahey et al., 1950 | Pinal |
| Wickenburgite | Williams, 1968 | Maricopa |
| Yavapaiite | Hutton, 1959 | Yavapai |
| Yedlinite | McLean et al., 1974 | Pinal |

*Found in meteorites

journals. It is not practicable to include a systematic listing of the many occurrences of the soil clay minerals or of the common rock formers, and we recognize that persons interested in these will consult the extensive specialized literature sources. Some minerals (quartz, for example), in addition to their occurrences as essential constituents of rocks, form under a wide variety of other geological conditions, and, in general, the rock associations of such minerals are not systematically treated in this compilation.

Forty-eight minerals first described from Arizona localities are listed in Table 2. Four minerals of questioned validity as species — arizonite, bisbeeite, cliftonite, and jeromite — are included. Yavapai County contributed twelve of these new minerals, followed by Coconino (10), Cochise (8), Pinal (8), Pima (4), Mohave (2), Gila (2), and one each from Apache, Greenlee, Maricopa, and Santa Cruz Counties.

Two hundred and twenty-four mineral species are reported here which did not appear in the 1959 edition of *Minerals of Arizona*. There are no published accounts of Arizona occurrences of 77 of these minerals. With the addition of these 224 species, the Arizona list totals about 605. The newly reported minerals are listed in Table 3.

## TABLE 3
## Species Added to the Arizona Mineral List
## 1959–76

| | | | |
|---|---|---|---|
| Acmite | Corkite | Langite | Riebeckite |
| Adamite | Cornetite | Leucite | Ripidolite |
| Aegerine-augite | Cornubite | Leucophosphite | Robertsite |
| Akaganéite | Cornuite (?) | Litharge | Roedderite |
| Alamosite | Cornwallite | Lithiophorite | Saléeite |
| Allophane | Cowlesite | Lonsdaleite | Salmonsite |
| Amesite | Creaseyite | Ludwigite | Saponite |
| Anthonyite | Creedite | Luetheite | Sapphirine |
| Apophyllite | Danalite | Mackayite | Schoepite |
| Arizonite (?) | Datolite | Maghemite | Sherwoodite |
| Arsenolite | Delessite | Manandonite | Starkeyite |
| Austinite | Devilline | Manganaxinite | Stellerite |
| Axinite | Diabantite | Mesolite | Sternbergite |
| Babingtonite | Diadochite | Meta-autunite | Stetefeldtite |
| Basaluminite | Diaspore | Metahalloysite | Stevensite |
| Bassanite | Dickite | Metanovacekite | Stilbite |
| Bayldonite | Djurleite | Metasideroni- | Stishovite |
| Bementite | Duftite | trite | Stringhamite |
| Berlinite | Eglestonite | Meta- | Strontianite |
| Beta- | Erionite | uranocircite | Svanbergite |
| uranophane | Eucryptite | Metavoltine | Sylvite |
| Bideauxite | Fairchildite | Miersite | Szomolnokite |
| Bieberite | Flagstaffite | Millerite | Teineite |
| Birnessite | Fornacite | Mixite | Tephroite |
| Bixbyite | Forsterite | Mordenite | Thorite |
| Boltwoodite | Francolite | Nacrite | Thuringite |
| Botryogen | Gahnite | Natroalunite | Tilasite |
| Brannerite | Gearksutite | Natrolite | Tinzenite |
| Bravoite | Gismondine | Nekoite | Todorokite |
| Brezinaite | Gmelinite | Neotocite | Tsumebite |
| Brunsvigite | Graemite | Nepheline | Tyrolite |
| Buetschliite | Harmotome | Offretite | Umohoite |
| Buttgenbachite | Hausmannite | Palygorskite | Unnamed spe- |
| Cacoxenite | Helvine | Papagoite | cies (sodium |
| Calomel | Hemihedrite | Paratacamite | analogue of |
| Cannizzarite | Hercynite | Parsonsite | zippeite) |
| Carbonate- | Herschelite | Penninite | Uranocircite |
| cyanotrichite | Heulandite | Perite | Uranospinite |
| Carnallite | Hexahydrite | Perovskite | Ureyite |
| Celadonite | Hidalgoite | Pharmacosi- | Uvarovite |
| Cesarolite | Hinsdalite | derite | Vaesite |
| Chabazite | Hisingerite | Phillipsite | Valleriite |
| Chalcophanite | Hollandite | Phlogopite | Vandendriess- |
| Chenevixite | Hydrobasalumi- | Phosphosiderite | cheite |
| Chloritoid | nite | Phosphuranylite | Variscite |
| Chondrodite | Hydronium | Picrochromite | Vesignieite |
| Chrysoberyl | jarosite | Plumbogummite | Weeksite |
| Clausthalite | Illite | Plumbo- | Weissite |
| Cliftonite (?) | Ilsemannite | nacrite (?) | Whewellite |
| Clinochlore | Ilvaite | Polyhalite | Wickenburgite |
| Clinochrysotile | Jadeite | Posnjakite | Wittichenite |
| Clinoclase | Jeromite (?) | Prehnite | Xenotime |
| Clinohumite | Johannsenite | Prosopite | Xonotlite |
| Clinoptilolite | Jordisite | Pseudoboleite | Yavapaiite |
| Clintonite | Junitoite | Pseudobrookite | Yedlinite |
| Coconinoite | Jurbanite | Pumpellyite | Yttrotantalite |
| Coesite | Kinoite | Ramsdellite | Zeunerite |
| Colusite | Krinovite | Rhodonite | Zincite |
| Cordierite | Ktenasite | Richterite | |

[ 8 ]

PART TWO

# ARIZONA MINERAL DEPOSITS

# 1. The Porphyry Copper Deposits

Arizona produced over 20 million tons of copper in the first 100 years or so of its mining history. In 1972 alone, the value of the total mineral production of the state slightly exceeded one billion dollars, 92 percent of which was from nonferrous metals. Copper accounted for over 94 percent of this production and made up 54 percent of the total United States copper production. Gold and silver in amounts of about 100,000 ounces and 6.5 million ounces, respectively, were produced during the same year, substantially all as by-products of copper mining.

Lest the mineral collector mourn the loss of incalculable quantities of fine mineral specimen material that he would imagine must have been demolished in the course of extraction of these enormous amounts of copper, he should be aware of the nature of the remarkable concentrations of this metal in the earth's crust. For reasons that will emerge, the mineral-specimen characteristics of the ore in these large deposits are quite unprepossessing. In fact, the skilled eye of a trained observer often is required to recognize that the ore contains copper minerals at all. There are, happily, many exceptions to this generality.

The minerals from which the great bulk of this copper production came actually are few in number; chalcopyrite, chalcocite-like minerals and, locally, bornite are the only abundant sulfides; chrysocolla, malachite, and cuprite are the principal "oxide" minerals.

Substantially all this metal has been produced from virtually unique copper concentrations termed "porphyry" or "disseminated" copper deposits. Because so many of the minerals listed in this book are associated with this type of occurrence, and since the deposits have been responsible for impressive statistics of production, a brief review of their geology and mineralogy is presented.

For reasons that are geologically complex (and not fully understood, although clearly related to large-scale movements of the earth's crust), many of these deposits are in the southern half of Arizona, localized in the Basin and Range physiographic province, and overflowing into adjacent parts of Sonora and New Mexico.

The following features are typical of these copper occurrences:

1. The deposits are characteristically large, measuring in size from a few tens and several hundreds of millions to billions of tons of ore reserves. Their sizes depend in large part on the grade of ore which prevailing metallurgical practice can profitably process. Mineralized areas of noncommercial grade may extend well beyond the economically-dictated boundaries of the mines.

2. The deposits are low grade, the ore invariably averaging less than one percent (20 pounds of copper per ton of rock). Modern mines of this type produce copper ores whose average tenor is between about 0.4 and 0.6 percent copper.

3. Ore mineralization is always associated with an intrusive calc-alkalic porphyritic rock of intermediate composition, typically quartz monzonite, tonalite, or granodiorite. An intrusive coarse-grained rock of granitoid texture of similar composition may have preceded the porphyry intrusion. The geological ages of these rocks range from late Mesozoic through middle Tertiary, for the most part, although an earlier (Jurassic) age is ascribed to the intrusives at Bisbee. The term "porphyry copper" derives from the common presence of rock of this type.

4. The host, or country, rock into which the igneous rocks intruded may be of almost any type or geologic age; limestones, shales, sandstones, volcanics, schists, and gneisses are representative.

5. Structural movements, probably assisted

by shrinkage of the intrusive rock masses during cooling, fractured their fabrics and those of the intruded rocks and prepared them for the subsequent entrance of aqueous fluids. The rocks of these deposits are typically intensely shattered and may contain stockworks and brecciated areas as manifestations of post-consolidation structural events.

6. Waves of hot, aqueous, hydrothermal fluids moved upward from a source thought to be genetically related to the origin of the igneous intrusives themselves and permeated the fractured rocks, chemically attacking and mineralogically altering them. Since the hydrothermal fluids were not in chemical equilibrium with the invaded rocks, the fluids reacted with them in a corrosive manner. The mineralogical products of this chemical activity are varied and complex and frequently so pervasive that in many instances the original rocks have been so made over as to be nearly unrecognizable.

In part, the hydrothermal alteration process may have overlapped in time the late stages of cooling and consolidation of the porphyritic intrusive. Important and abundant minerals that formed during this phase in the developmental history of the porphyry deposits include sericite (muscovite), biotite, hydromica, chlorite, epidote and clinozoisite, calcite, clay minerals (especially the kaolin and montmorillonite groups), potassic feldspar (usually orthoclase), albite, and quartz. All these minerals will not necessarily be present in one deposit. Large masses of rock in some deposits have been so completely altered as to become aggregates of essentially nothing but sericite, quartz, and pyrite. The large amounts of sericite in some of the deposits could probably have formed only if potassium had been introduced with the hydrothermal fluids. The common development of abundant quartz argues for the introduction of silica, and iron must also have been brought into the deposits to account in part for the tremendous quantities of pyrite formed. Copper, of course, is another element introduced in abundance.

7. Associated with the intrusion of the igneous rocks, and probably in part also related to the hydrothermal alteration, contact metamorphic mineral assemblages, termed skarn or tactite, were developed in the country rocks of some of these deposits. The mineralogies developed depend largely on the kinds of country rock present: carbonate sediments give rise to calc-silicates such as diopside-hedenbergite, tremolite, and garnet, as may shales, impure sandstones, and volcanic rocks; shales may become silicified; carbonate rocks are frequently marblized by recrystallization of calcite; and, locally, bodies of magnetite may form, especially in limestones.

8. During and after hydrothermal alteration, pyrite and other sulfur-bearing minerals were introduced along with copper sulfides, especially chalcopyrite. The copper-bearing sulfides probably continued to form after the initial hydrothermal alteration.

The major primary (hypogene) ore mineral, chalcopyrite, and, much less importantly, bornite, seem to have been deposited more or less simultaneously. Pyrite is typically the most abundant sulfide and often encloses blebs of chalcopyrite or bornite within it. Pyrite also was usually the first sulfide to form and often continued to crystallize throughout these primary phases of the development of the deposit. Small amounts of molybdenite, sphalerite, galena, or other sulfides may be present. The primary sulfide minerals occur as disseminated grains (whence the name "disseminated" deposit), as veinlets filling fractures in the host rocks, and as replacements of other minerals. In places, the disseminated sulfide grains are seen to cluster markedly in the vicinity of the dark ferromagnesian minerals of the porphyry — biotite, pyroxenes, or amphiboles. Molybdenum, gold, and silver, while they are in small concentrations in the copper ores, may be important byproducts of the extractive processes.

9. With the cessation of alteration and primary sulfide mineralization, the deposits remained quiescent through periods of geologic time during which their upper exposed portions were subjected to weathering. Erosion stripped away overlying rock, exposing the primary sulfides (protore) to further weathering. Oxygen contained in the atmosphere and in meteoric and underground waters chemically attacked and dissociated pyrite and other sulfides in the upper reaches of the deposits. Sulfuric acid and iron and copper sulfates formed. In the presence of acidic waters, copper and iron were in solution, but whether or not the solutions could percolate downward depended on the chemical nature and permeability of the rocks themselves. If they were strongly reactive to the acid solutions (limestone, for example), the copper and iron in solution would tend to precipitate as carbonate minerals, but if the rocks were chemically nonreactive or neutral in behavior (for example, sandstone or quartzite), the metal-bearing solutions could migrate downward into

the deposit. It is perhaps fortuitous that the porphyritic rocks which are always present in these deposits behave neutrally to these acidic solutions and do not inhibit their downward passage into the upper parts of the mineralized rocks.

As the downward percolating solutions encountered neutralizing or chemically reducing environments, copper and iron ions precipitated as new sulfides or replaced pre-existing minerals (especially the primary chalcopyrite) to form secondary sulfides. Chalcocite is by far the most abundant of these secondary (supergene) copper sulfides, although subsidiary covellite, bornite, djurleite, and even chalcopyrite, are known to form in this manner.

This complex process, which removes copper from the upper portion of the deposit and transports it in solution into the unoxidized rock below, depends on a number of circumstances, paramount among which is a supply of pyrite sufficient in quantity to provide the necessary acid.

It is generally held that the reducing conditions encountered below the groundwater table in the vicinity of the deposit were an important factor in localizing precipitation of secondary sulfide minerals. The relatively simple undulating surface of the groundwater table in many places controlled the distribution of chalcocite so that a more or less blanket-like configuration resulted. The chalcocite blanket is typically situated between an upper oxidized zone and a lower primary ore zone and commonly overlaps both. In many deposits the supergene chalcocite mass became sufficiently enriched in grade and of such size as to constitute the major ore body

of the mine. The presence of the chalcocite-enriched zone has transformed many deposits, which otherwise would have been too low in grade, into commercially feasible ventures. The deposit at Morenci, for example, would not be of workable grade were it not for immense tonnages of supergene chalcocite. In some deposits local complexities in the behavior of the water table resulted in zonations which are not as simple as suggested, and some ores are complex mixtures of primary and secondary sulfides and oxides of copper; the Castle Dome, Copper Cities, and Bisbee deposits are examples. With improved extractive techniques and deeper levels of mining, the industry is finding that it need rely less and less on the presence of secondarily enriched ores.

10. The residua left near the surface after the oxidation of sulfides and the ensuing leaching processes are termed leached or oxidized capping or gossan. The presence of these distinctive and interesting mineralogical assemblages has led to the discovery of large ore bodies at depth. Gossans are typically siliceous microcavernous rocks consisting of the altered and weathered original rock impregnated with mixtures of secondary iron oxides and sulfates, the aggregate of which is loosely termed "limonite." The most common minerals in the limonitic cappings are hematite, lepidocrocite, goethite, and jarosite. Alunite is sometimes present, as is turquoise. Copper solutions that could not migrate out of the oxidized zone because of the presence of reactive gangue formed new mineral phases at varying distances from the point of origin of the solutions. Malachite, azurite, chrysocolla, cuprite, native copper, and a vari-

**Banded malachite and azurite.** Bisbee, Cochise County. Robert W. Jones collection. Jeff Kurtzeman.

**Copper.**
San Manuel mine, Pinal County.
Wayne Thompson collection.
Jeff Kurtzeman.

**Silver.**
Tombstone, Cochise County.
Les Presmyck collection.
Jeff Kurtzeman.

ety of less common secondary minerals can be formed in the oxidized zone of the deposits. Indeed, the copper-bearing portions of the oxidized zones of some of the porphyry copper mines have provided large amounts of ore. This is true, for example, of the Ray, Inspiration, and Twin Buttes deposits.

Man-made leaching systems, reminiscent of this secondary leaching process, are in common use to dissolve copper from some of these oxidized zones, ore from which cannot be metallurgically processed in the manner of the sulfide ores. *In situ* leaching is practiced at some mines; water and sulfuric acid are spread on the surface, and the copper-charged liquids which percolate downward through the oxide ores are tapped underground. Other mines "heap-leach" broken ores that have been spread on the surface.

A distinctive feature of the surface outcrops overlying many of the large, low-grade, porphyry copper ore bodies is their coloration. A variety of tones and hues in reds, browns, yellows, and grays results from the abundance of secondary iron minerals. Sufficient amounts of malachite and chrysocolla may impart greenish tinges to the surface exposures.

Spectacular mineral specimens are not, as a rule, found in the low-grade sulfide deposits. The open pit mines from which most such ores are produced are usually thought to be mineralogically unexciting and rather drab in appearance, as those who have stood high on the lip of one of these enormous excavations will likely attest. Again, there are exceptions because the mineralization is not always as monotonous as the foregoing may have suggested. The beautifully symmetrical New Cornelia open pit mine at Ajo, for example, which is unusual in that the ore consists largely of primary chalcopyrite and bornite with little supergene mineralization, contained a large mass of coarse orthoclase-rich pegmatite associated with the quartz monzonite porphyry.

On the other hand, colorful and exotic minerals have come in gratifying abundance from the oxidized portions of many of the same deposits. A few noteworthy examples might include such occurrences as the rare phosphate-sulfate, tsumebite, found in the capping at Morenci; the copper silicates ajoite and papagoite, first described from the New Cornelia mine, and crystallized native copper and magnificent azurite crystals from the same mine; the copper molybdate, lindgrenite, in the oxidized ores of the Live Oak pit of the Inspiration mine near Globe, and from the Esperanza mine in the foothills of the Sierrita Mountains; the delicate, gypsum-enclosed chalcotrichites and native copper found in the Sierrita pit near Tucson.

From the point of view of the mineral collector, perhaps most important of all as a fount of mineral specimen wealth in Arizona have been those regions peripheral to a few of the porphyry copper deposits. Not all fine mineral specimens from the state have come from such areas, of course, but the importance of several such localities which can be construed as being genetically related to porphyry copper mineralization makes this group of deposits worthy of special mention.

Little attention has been given in the preceding discussion to these marginal regions enveloping the mineralized central porphyry. The contacts between the intrusive rocks and the invaded country rocks are often complex; the structural deformations which shattered the marginal host rocks into which the hydrothermal fluids seeped also gave rise to systems of faults and shattered zones which often extend for considerable distances outward from the central intrusives. Solutions related to those responsible for the disseminated mineralization also moved outward from the porphyry mass into the permeable surrounding rocks and altered and mineralized them in and adjacent to the peripheral faults and sheared zones. Sulfide minerals were deposited in open fissures to create veins, and, locally, the hydrothermal solutions chemically attacked and replaced susceptible host rocks to give rise to replacement deposits. Limestones and dolomites were particularly receptive to this replacement process. Although they are smaller than the porphyry

**Chalcotrichite in calcite.**
Ajo, Pima County. Jeff Kurtzeman.

**Chrysocolla, with quartz
and malachite.**
Live Oak pit, Miami, Gila County.
Manuel Fraijo collection.
Jeff Kurtzeman.

deposits proper, the concentration of ore minerals in the vein and replacement deposits is much more pronounced, and the minerals formed in them tend to be more coarsely crystallized.

Where open spaces were prevalent in the host rocks, the conditions may have been locally ripe for the growth of large, occasionally spectacular, crystalline masses and single crystals of many kinds of primary minerals. Oxidation and secondary enrichment processes functioned in these peripheral ore bodies much as in the mineralized porphyry masses, even though the sulfide mineralization was of higher grade and confined to smaller areas. Porosity and open-space conditions developing in the oxidized upper portions of these deposits permitted the growth of large crystals of many kinds of secondary minerals. It is from this provenance that some of the world's great specimens have been won.

Prospectors who scoured the mountains of southern Arizona during the latter half of the nineteenth century, often attracted by colorful secondary copper minerals, stumbled upon outcrops of these peripheral ore deposits but were hardly aware of the potentially greater wealth that lay beneath the unspectacular oxidized cappings. Some of their discoveries, of course, gained renown in the annals of mining.

The reader interested in pursuing the subject of the porphyry copper deposits will find good summaries in Titley and Hicks (1966), Ander-

son (1969), and Lowell and Guilbert (1970), which will also provide further references to an extensive literature.

Table 4 lists most of the important porphyry copper deposits of Arizona. A few relatively small deposits and some undeveloped large prospects are not included.

TABLE 4
**Porphyry Copper Mines of Arizona**

| Name of Mine | County |
|---|---|
| Bagdad | Yavapai |
| Castle Dome | Gila |
| Christmas | Gila |
| Copper Cities (Miami) | Gila |
| Dos Pobres | Graham |
| Esperanza | Pima |
| Inspiration | Gila |
| Lakeshore | Pima |
| Lavender Pit (Bisbee) | Cochise |
| Mineral Park (Ithaca Peak) | Mohave |
| Mission | Pima |
| Morenci | Greenlee |
| New Cornelia | Pima |
| Pima | Pima |
| Ray | Pinal |
| Sacaton | Maricopa |
| San Juan | Graham |
| San Manuel | Pinal |
| San Xavier North | Pima |
| Sierrita | Pima |
| Silver Bell | Pima |
| Twin Buttes | Pima |

# 2. Porphyry Copper-Related Deposits

Nearly all of the large, low-grade, porphyry copper ore bodies have associated with them peripheral underground mines that exploited ores of higher grade than those of the porphyry bodies themselves.

Rather typically, these mines had been known and worked for years before technological advances and changing economic conditions made exploitation of the lean porphyry copper ores profitable. Not uncommonly the presence of the older mines helped delineate target areas in the search for the large, low-grade, ore bodies. The metal production of the porphyry copper segment of a given area almost always is substantially greater than the contribution from the associated underground mines which work smaller ore bodies of higher grade. This is the case, for example, at Morenci, San Manuel, and Ajo. Bisbee, however, is a notable exception. The total metal production of the vast complex of underground workings, mined since the late 1870s, substantially exceeded the copper production of the relatively short-lived open pit copper mines of the district. The scope of its underground workings, a long and fascinating mining history, and the great variety and beauty of its minerals make Bisbee rather special in mining annals. The cessation of most mining activity in the Bisbee area in 1975 brought to an end a century of continuous activity in a region that has few peers for the amount of nonferrous metal mined and for the quality and quantity of mineral specimens recovered and preserved.

Although, like Bisbee, the Mammoth–St. Anthony mine at Tiger was discovered in the late 1870s, the huge nearby porphyry copper ore body at San Manuel was not brought into production until 1956. The geological relationships between the San Manuel deposit and the peripheral Mammoth–St. Anthony mine are not as obvious as those between the underground mines and the Sacramento Hill and Lavender porphyry ore bodies at Bisbee, but the unusual variety and exceptional quality of the suite of colorful and beautifully crystallized secondary minerals from the mine at Tiger earned it an international reputation as one of the world's exceptional mineral localities.

## THE BISBEE DISTRICT

The Bisbee district has long been a premier mineral-producing camp. Actually, Warren is the formal name of the mining district in which the celebrated mining area of Bisbee is situated, and the name Warren has had priority for years. The name Bisbee is used in this discussion because few persons who are not in the mining profession would be aware that the famous Bisbee mineral locality lies in the Warren mining district. In Part Three, the catalog section, however, the name Warren is retained.

Ores were found here in about 1876, and for three years thereafter mining was carried out on the Hendricks claim. Bisbee has become particularly famous for its copper minerals, although interestingly enough the first mining was for lead from cerussite ores in limestone. In 1880 the Copper Queen mine began operations on a large body of copper ore in limestones outcropping on Queen Hill. Ever since that time, into the 1970s, the district has been a steady producer of copper. The early production was from underground mines in oxidized and sulfide ores in limestone; more recently this production has been overshadowed by mining operations in the Lavender pit.

The copper ores at Bisbee were introduced by an intrusive porphyritic igneous rock approximately 180 million years ago. Much of the copper and iron released by this intrusive replaced the enclosing limestones, forming irregular masses of ore which often were totally isolated within barren, seemingly unaffected, limestone. Scattered sulfide grains were also retained in the intrusive, and these lower grade ores were those later mined by open-pit methods.

Large masses of some of these sulfides have been encountered in the mines. Pyrite is common everywhere, often as granular masses of great size; but good crystals, especially large ones, are uncommon. Sizable masses of other sulfides have been discovered in the underground workings. Veins of alabandite are occasionally found; bornite-chalcocite intergrowths are common, as is chalcopyrite. Very nice pseudohexagonal chalcocite crystals have been found in the Cole shaft. The rarer sulfides generally are seen only under the microscope in polished samples.

The original primary ores were sulfides, and the variety of minerals is relatively limited. In most cases these minerals are not of specimen quality, and even the richest ores may be quite drab in appearance. The following primary minerals, mostly sulfides, have been found at Bisbee:

| | | |
|---|---|---|
| aikinite | djurleite | pyrite |
| alabandite | enargite | rickardite |
| bornite | famatinite | sphalerite |
| chalcocite | galena | stromeyerite |
| chalcopyrite | gold | tennantite |
| cinnabar | greenockite | tetrahedrite |
| covellite | molybdenite | uraninite |
| digenite | powellite(?) | wittichenite |

The rock mass that introduced ores at Bisbee is a quartz monzonite porphyry. In most places the original silicate minerals in this rock (quartz, orthoclase, plagioclase, biotite) have been destroyed, and the rock is recrystallized beyond recognition. Usually the altered rock consists of granular quartz enclosing pyrophyllite (locally sericite) scales loosely bunched together so as to show the outlines of earlier plagioclase and biotite phenocrysts. Rutile usually accompanies the altered biotite. Of the sulfide grains scattered throughout the altered porphyry, pyrite is by far the most common.

Quartzites invaded by the porphyry are altered strongly and often closely resemble the intrusive, for they too consist largely of quartz, sericite, and scattered pyrite grains. Shaly rocks in the walls have behaved similarly but contain considerably more sericite than the quartzites. Perhaps the most pronounced effects are in calcareous rocks near the intrusive. These have been converted to calc-silicate tactites which may be rich in minerals such as epidote, garnet (usually near grossular-andradite), diopside, and tremolite. Calcite occurs in these tactites, and, if the rock is leached in weak acids, one may see good microcrystals of the calc-silicate minerals, but just as often they are tightly intergrown grains showing poor development of crystal faces.

Sulfide ores in these metamorphosed wallrocks usually occur in irregular bodies, as widely scattered grains, or as heavily disseminated grains which may replace one particular bed or lamina in the original sediment. However, sulfides also occur in well-defined veins, and the gangue minerals in these include quartz and calcite (common), or minerals such as barite, chlorite, and magnetite. Sulfides, together with minerals such as tremolite, epidote, or even diopside, may also occur in veins that formed during the peak of metamorphism.

Sulfides occur in calcareous rocks at such great distance from the porphyry that effects of metamorphism may not be seen. Some of these sulfide "plums" are mystifying, for they may be encased in massive limestone showing neither the effects of recrystallization nor even trace quantities of the metals occurring in such abundance a few inches away. Ore bodies of this nature can be found only by luck and persistence, or by geophysical methods.

Rock-forming, gangue, and alteration minerals found at Bisbee include:

| | | |
|---|---|---|
| allanite | dolomite | orthoclase |
| allophane | enstatite | penninite |
| alunite | epidote | plagioclase |
| apatite | fluorite | pumpellyite |
| aragonite | graphite | pyrophyllite |
| augite | grossular | rhodochrosite |
| barite | halloysite | rutile |
| biotite | hematite | sanidine |
| calcite | hornblende | sericite |
| celadonite | hydrobiotite | siderite |
| chromite | kaolinite | sphene |
| clinochlore | laumontite | stevensite |
| clinochrysotile | magnetite | thomsonite |
| delessite | metahalloysite | tourmaline |
| diaspore | microcline | tremolite |
| dickite | olivine | zircon |
| diopside | | |

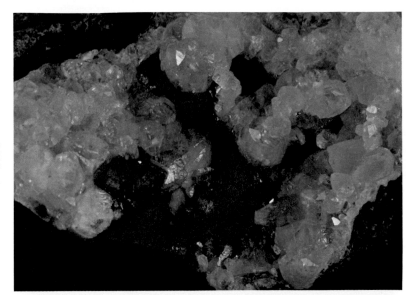

**Cuprite (chalcotrichite) in calcite.**
Ajo, Pima County. National
Museum of Natural History
collection. Jeff Kurtzeman.

**Malachite in calcite.**
Bisbee, Cochise County. University
of Arizona collection.
Jeff Kurtzeman.

The fame of Bisbee as a mineral locality is largely due to the effects of weathering in relatively recent geological times. As the abundant pyrite dissolved, immense quantities of acid were provided which furthered dissolution not only of other sulfides but of encasing rocks, particularly the limestones.

Where ore pods or veins occurred in limestone, the acids (carrying iron and copper) were eventually neutralized, but in the process some leaching of limestone occurred and caverns developed. The copper and iron remaining in these "spent" acids were deposited along the walls of these cavities, usually as thick layers of iron oxides (goethite, hematite) and then as crusts of copper carbonates or oxides. Perhaps the most spectacular specimens formed in this way are those magnificent crystalline specimens of azurite and malachite like those found in the Copper Queen mine. The Phelps Dodge Corporation (which controls the Bisbee mines) has in its geological offices specimens of these minerals several feet across; azurite crystals up to one-half inch or more in size occur in crusts up to an inch or more thick and show varying stages of alteration to crystalline or fibrous malachite.

In pods of supergene clays in shaly lime-

**Malachite replacing an azurite crystal.**
Bisbee, Cochise County. Richard Graeme collection.
Jeff Kurtzeman.

stones one often finds large masses or nuggets of crystalline cuprite and native copper. Cavities in the cuprite provide good hunting for small but spectacular crystals of rarer minerals such as connellite, brochantite, spangolite, atacamite, and chlorargyrite. These masses are often thickly rimmed with tenorite, chrysocolla, and malachite. It was doubtless in similar material that paramelaconite was found. This rare mineral was described in 1891 by Koenig, and no specimens have been found since at Bisbee. The original crystals were splendent black prisms an inch long. Spangolite, described

**Cuprite.**
Bisbee, Cochise County. Julius Weber.

by Penfield in 1890, doubtless came from Bisbee originally (Palache and Merwin, 1909), although Tombstone has been cited as the locality. (In the early days the Tombstone mining district included Bisbee; it became a separate district later.) The largest spangolite crystal on Penfield's specimen was 8 mm in diameter and 5.5 mm high.

Azurite also occurs in clayey seams, often as balls or nodules of exceptionally large (to 4½ inches) curved crystals embedded in clay. When cleaned these are handsome specimens indeed!

In many places the acid copper-bearing waters were not completely neutralized, and sulfates such as brochantite formed. Spectacular sprays of long, slender, brochantite prisms have been found, especially in the early days; they sometimes were accompanied by small but brilliant cerussite crystals.

Often the sulfate-rich waters were not thoroughly neutralized because they reached clay-rich seams in the limestones; a variety of copper-aluminum sulfates was formed as a result. Usually these minerals are not coarsely crystalline, but they provide superb micromount material. Mining activities at Bisbee in the early 1970s encroached upon old Copper Queen workings and provided anew a suite of well-crystallized minerals such as chalcoalumite, cyanotrichite, antlerite, and brochantite. These minerals occur with basaluminate, some unidentified aluminum sulfates, and gibbsite.

At the Shattuck mine some copper silicates (in addition to chrysocolla) were formed. A mineral named for the mine, shattuckite, was found and described by Schaller (1915). It was noted as a replacement product of malachite. He also described bisbeeite from this mine, but in the mid 1970s bisbeeite was still not fully understood and may be invalid.

In areas where pyrite was abundant and acids from its dissolution were not neutralized, one may find a variety of iron-bearing and other sulfates. Some of these minerals are forming in old mine workings; others are older and more stable sulfates. Well-crystallized voltaite, roemerite, rhomboclase, and coquimbite have been described in the past and continue to turn up. These minerals also form stalactites or stalagmites and may be associated with chalcanthite. Recently, spectacular masses of epsomite crystals, in part altered to hexahydrite, have been found.

In the porphyry responsible for all the copper at Bisbee, there is, oddly enough, a remarkable paucity of interesting or well-crystallized oxidation products. By far the most common mineral is jarosite, which occurs in veinlets or as warty crusts lining cavities left by leaching of the pyrite. Some turquoise has been recovered from the porphyry and nearby rocks.

The mineralogy of Bisbee cannot be given full justice here. Many other species occur in microcrystals, but not cabinet specimens, of high quality. A list of oxide zone minerals found at Bisbee is given below:

| | |
|---|---|
| alunite | hetaerolite |
| anglesite | hexahydrite |
| anthonyite | hisingerite |
| antlerite | hydrobasaluminite |
| atacamite | hydrohetaerolite |
| aurichalcite | jarosite |
| azurite | kornelite |
| basaluminite | langite |
| bayleyite | leadhillite |
| bindheimite | lepidocrocite |
| bisbeeite | malachite |
| botryogen | melanterite |
| braunite | metavoltine |
| brochantite | mimetite |
| bromargyrite | mottramite |
| carbonate- | murdochite |
| cyanotrichite | paramelaconite |
| cerussite | paratacamite |
| chalcanthite | pharmacosiderite |
| chalcoalumite | plattnerite |
| chalcophanite | psilomelane |
| chalcophyllite | pyrolusite |
| chalcosiderite | pyromorphite |
| chlorargyrite | ransomite |
| chrysocolla | rhomboclase |
| conichalcite | roemerite |
| connellite | rosasite |
| copiapite | scngierite |
| copper | shattuckite |
| coquimbite | silver |
| cuprite | smithsonite |
| cyanotrichite | spangolite |
| delafossite | stibiconite |
| descloizite | sulfur |
| devilline | szomolnokite |
| dioptase | tenorite |
| embolite | tilasite |
| epsomite | turquoise |
| gibbsite | tyuyamunite |
| goethite | uraninite |
| graemite | variscite |
| groutite | voltaite |
| gypsum | willemite |
| hausmannite | wulfenite |
| hemimorphite | |

## THE MAMMOTH–ST. ANTHONY MINE, TIGER

This complex base and precious metal deposit was staked by Frank Shultz in 1879–1882 as a "mammoth lode gold vein," hence the Mammoth mine. The town of Shultz, Arizona, was established at the mine site in 1896, and gold alone was produced from quartz through 1912. During World War I, 1916–1919, the mines were reopened primarily for molybdenum production, although gold also was recovered. The Mammoth, Collins, and Mohawk-New Years mines were consolidated in 1934 by the St. Anthony Mining and Development Co., Ltd. The town of Shultz was re-established as Tiger in 1939. Through the time of closing of the mine in 1954, substantial amounts of gold, silver, lead, zinc, copper, molybdenum, and vanadium were mined.

The area is underlain by Oracle granodiorite, which contributed material for younger conglomerate and arkose. These younger rocks are interbedded with basalt and tuff and are collectively known as the Cloudburst Formation. Following intrusion of rhyolite and andesite in Tertiary time, this entire sequence was faulted to form the main vein structure. This fault was later filled with quartz, adularia, barite, and fluorite, and became mineralized with pyrite, chalcopyrite, galena, and sphalerite.

These primary sulfides later were subjected to supergene weathering in the upper portions of the vein; the sulfide minerals followed a course of alteration common in many Southwestern deposits. Pyrite oxidized to goethite and hematite; secondary copper sulfides, oxides, carbonates, and silicates were developed; galena altered to anglesite and cerussite; and sphalerite formed smithsonite, hemimorphite, and willemite. The vein was then cross-faulted. The former upper, highly-oxidized part became designated as the Mammoth vein, and the lower oxide-sulfide portion became the Collins vein.

Locally, further development of additional mineral species took a remarkable direction. Suites of complex minerals were formed containing lead, copper, and silver combined with carbonate, sulfate, chloride, and water. Various hypotheses of their formation involve retention or reintroduction of some of these components from hydrothermal solutions; derivation of some of the components from supergene alteration; or possibly their supply from groundwater. Deposition of quartz sheathing around these zones probably provided a closed chemical sys-

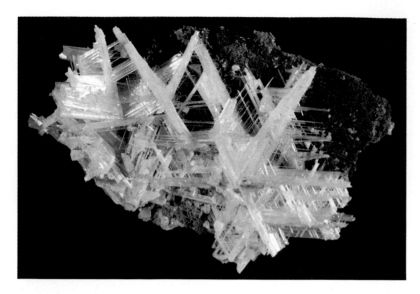

**Cerussite.**
Mammoth–St. Anthony mine,
Pinal County. Bideaux collection.
Jeff Kurtzeman.

tem for their formation and in part assured their survival. Metastability of some of the minerals of this suite is indicated by evidence of extensive corrosion and pseudomorphism.

This anomalous suite includes the minerals which are primarily responsible for the mine's fame among mineralogists and collectors. While the occurrence of these minerals in such relative abundance and often large crystal sizes is unique, their existence is not. Other deposits such as at the Rowley, Apache, and Grand Reef mines in Arizona; at Leadhills, Scotland; and in several small districts around Anarak, Iran, display comparable parageneses.

Finally, solutions containing molybdenum, vanadium, chromium, arsenic, and phosphorus reacted with lead and copper in the vein to form secondary molybdate, vanadate, and other minerals, as in many other Arizona deposits. Upper portions of the San Manuel copper-molybdenum porphyry deposit, which is earlier and somewhat more than a mile away, are depleted in at least molybdenum. Other of these elements probably were supplied by dissolution of trace minerals in the surrounding host rocks.

Following is a listing of known mineral species from Tiger, categorized into somewhat arbitrary divisions:

### Gangue Minerals

| | | |
|---|---|---|
| adularia | calcite | magnetite |
| amesite | chlorite | plagioclase |
| antigorite | clays | quartz |
| barite | epidote | stilbite |
| biotite | fluorite | tourmaline |
| | heulandite | |

### Primary Sulfides

| | | | |
|---|---|---|---|
| chalcopyrite | galena | pyrite | sphalerite |

### Secondary Sulfides

| | | |
|---|---|---|
| acanthite | chalcocite | djurleite |
| bornite | covellite | wurtzite |

### Oxidized Minerals: "Normal Sequence"

| | | |
|---|---|---|
| anglesite | goethite | plancheite |
| aurichalcite | gold | ramsdellite |
| azurite | hematite | rosasite |
| cerussite | hemimorphite | silver |
| chrysocolla | hisingerite | smithsonite |
| creaseyite | hollandite | sulfur |
| cuprite | malachite | tenorite |
| devilline | minium | willemite |
| dioptase | murdochite | |

### Oxidized Minerals: "Anomalous Sequence"

| | | |
|---|---|---|
| alamosite | connellite | melanotekite |
| atacamite | diaboleite | paralaurionite |
| beaverite | embolite | paratacamite |
| bideauxite | hydrocerussite | phosgenite |
| boleite | iodargyrite | plumbonacrite |
| brochantite | leadhillite | pseudoboleite |
| caledonite | linarite | wherryite |
| | matlockite | yedlinite |

### Oxidized Minerals: Late Stage

| | | |
|---|---|---|
| descloizite | mottramite | vanadinite |
| fornacite | pyromorphite | wulfenite |
| mimetite | tsumebite | |

Anyone who examines an extensive collection of Tiger minerals gains the impression that combinations of these minerals are endless, espe-

**Cerussite on dioptase.**
Mammoth–St. Anthony mine, Pinal
County. A. L. Flagg Foundation
collection. Jeff Kurtzeman.

cially on specimens with minute crystals. Outstanding cabinet specimens of white botryoidal smithsonite, azurite crystals with cerussite, or azurite altered to malachite were produced from the mine. Associations of dioptase with orange-to-red wulfenite, willemite, and cerussite; complexly twinned or reticulated cerussite, sometimes water-clear or of large size; blue leadhillite; cleavable masses and microcrystals of matlockite and phosgenite; and large vanadinite crystals with incrustations of descloizite-mottramite are among the best specimens of these minerals produced from any locality in the United States.

Rich blue specimens of crystalline caledonite, diaboleite, and linarite with white hydrocerussite and yellowish paralaurionite are the finest examples of these species yet found in the world, as are the greenish black fornacite crystals. The species wherryite, bideauxite, yedlinite, and creaseyite are unique, or nearly so, to Tiger. Beyond doubt, yet other new species await description among the Tiger suites distributed throughout many collections.

# 3. Uranium and Vanadium Deposits

The variety and sometimes colorful appearance of the uranium and vanadium minerals make them one of the most interesting assemblages of Arizona. Although these minerals do not characteristically form coarse crystals, as do some of the oxidized copper minerals, they have, nevertheless, considerable appeal to the collector. As is true of the minerals of the base metals, the ultimate reason for the discovery and exploitation of this group of minerals is its importance as a source of mineral wealth of the state. The great number of species represented, together with the unusual nature of the deposits in which they occur, warrants a brief discussion of the geological provenance of this group of minerals.

In contrast to the porphyry copper deposits of Arizona, which are largely confined to the Basin and Range physiographic province of southern Arizona, the principal uranium and vanadium deposits are in northern Arizona, although many other lesser deposits and prospects are widely scattered throughout other parts of the state. Northeastern Arizona is the southwest portion of the large, topographically high, Colorado Plateau, a remarkable uranium-vanadium province extending over parts of Arizona, New Mexico, Colorado, and Utah. Here the earth's surface is underlain by thick sequences of relatively undisturbed, flat-lying, sedimentary rocks which have been strongly dissected into spectacular steep-walled canyons by the river systems draining the region. Major uranium- and vanadium-rich areas occur in Monument Valley of extreme northeastern Arizona; others are centered around Cameron, where they extend in a belt about 10 miles wide

and 75 miles long along the Little Colorado River. A third important producer of uranium is the isolated Orphan mine, perched on the south rim of the Grand Canyon. This mine is of the breccia pipe type and is geologically quite different from the majority of the Plateau-type deposits.

## MONUMENT VALLEY

The famous Monument No. 2 mine, situated in Monument Valley, has been studied more intensely than any other Arizona uranium-vanadium deposit; it is representative of the so-called Plateau-type deposits. The description which follows is based primarily on the geology of this mine and is largely drawn from the work of Witkind and Thaden (1963).

The sedimentary rocks of most concern with respect to uranium and vanadium occurrence in Monument Valley are in the Shinarump Member of the Triassic Chinle Formation which maintains a rather uniform thickness of about 50 to 75 feet over the area. The rocks of this unit are of continental origin and are typically light gray, medium-grained to coarse-grained, sandstones with subordinate yellowish-brown conglomerates and some mudstones. Shinarump sandstones are commonly cross-stratified and contain an abundance of scour channels. The distribution of rock types within the Shinarump is often irregular, and lateral gradations of one rock type into another within distances of a few hundred feet are common. Mixed with the pebbles and sands of the conglomerate are large quantities of fossil plant remains, among which are large logs. The surface of the underlying

mudstones and siltstones of the Moenkopi Formation were scoured and channeled by ancient
stream action before the Shinarump rocks were
deposited. The long axes of these channels have
an average northwesterly trend. In places the
scouring reached down through the underlying
Triassic-age Moenkopi rocks to incise the
DeChelly sandstones of the Permian Cutler
Formation. Sands, pebbles, and fossil plant remains of the Shinarump filled these ancient
troughs.

There appears to be a higher concentration
of fossil woody material in the channels at the
base of the Shinarump than elsewhere in the
unit. The fossil wood has been replaced by a
variety of substances: silica replacement, which
formed the well-known "petrified wood," is
most common, but malachite, azurite, copper
sulfides, uranium minerals (especially uraninite,) and vanadium minerals (montroseite) have
also been locally active in this respect. Some of
the wood was converted to black coaly or carbonaceous substances, and concentrations of uranium and vanadium minerals are frequently
associated with this material. Clay minerals are
present in the channels, some from included
mudstones derived from the underlying Moenkopi Formation, and others from the devitrification of the volcanic ash which abounds in the
Chinle Formation. The clay minerals appear
not to be related to the solutions which transported uranium and vanadium into the rocks.

The ore deposits are largely confined to these
channels and take several forms. One type is
termed the "rod" deposit because the ore concentrations are roughly cylindrical, log-shaped
masses. A second type, the "tabular" ore bodies,
usually occurs at or near the bases of the channels; these bodies are somewhat elongate parallel to the length of the channels. A third type
consists of irregular masses of rock impregnated
and replaced by the dark vanadium minerals of
the corvusite type. The fourth ore occurrence is
the so-called "roll" deposit, named for the characteristic curved bands of channel fill impregnated by yellow uranium minerals which cut
the bedding planes of the sandstone.

High concentrations of uranium minerals are
characteristic of the rod deposits which form
the richest ore bodies in the Monument No. 2
mine. The rods are annular in cross section and
typically contain an inner core of light gray
sandstone surrounded by a sheaf of tyuyamunite impregnation, outside of which is a limo-
nite-rich zone. The transition from the high
uranium mineral concentration typical of the

rods to the barren or weakly mineralized sandstone is usually abrupt.

The tabular ore bodies are irregular in out-
line and contain large quantities of uranium
minerals impregnating conglomeratic sandstone
and fossil plant matter. Rich concentrations of
ore minerals may occur in depressions in the
channel floors. The bodies are of moderate size,
about 60 feet long, 20 to 40 feet wide, and up
to 6 feet thick.

Ore bodies of the corvusite type may be
rather large (up to 600 feet long, 100 feet wide,
and 40 feet in thickness) and consist of irregular
impregnations of sandstone by dark corvusite
and associated minerals. The ore mineral distribution within these bodies is erratic, but a
rude zoning has been noted: blue-black central
areas, typical of high concentrations of corvusite, gradually merge into surrounding areas of
high limonite and low vanadium mineral content. While this type of ore concentration is
most abundant near the bases of the channels,
some occur at and near the tops. Where these
have been exposed to weathering at the present
surface, oxidation of the dark-colored corvusite
to hewettite has pervasively colored the rocks
a deep red.

Although the roll deposits are of great importance elsewhere in the uranium-vanadium province of the Colorado Plateau, they are relatively
minor in the Monument Valley region. These
curved mineralized bands consist of imprega-
tions of sandstone ranging in thickness from a
few inches to several feet and contain either
yellow uranium or dark vanadium minerals.
The colored bands alternate with barren or
weakly mineralized zones of sandstone. Mineralization in the rolls may have been localized
at the intersections of fractures with cross-strati-
fications in the sandstone.

Reasons for the concentration of the uranium
and vanadium minerals in these kinds of ore
bodies are often obscure. It is thought that two
factors may have been most effective in bring-
ing about precipitation of the elements from
solution: (1) chemical reaction of the trans-
porting solutions with the carbonaceous plant
remains so abundant in the channel fill and (2)
changes in the porosity and permeability of the
host sandstone and conglomerate. There is evidence to suggest that each may have been in-
strumental in causing precipitation locally. Some
geologists who have studied this problem believe the rod deposits were localized at the sites
of carbonaceous fossil logs. Subsequent oxidation altered primary uraninite to secondary

uranium minerals such as becquerelite and uranophane; the oxidation of pyrite may have provided the limonite found in the outer parts of the rod deposits. In the vanadium-rich zones, the primary mineral was probably montroseite which subsequently oxidized to corvusite; further oxidation converted corvusite to hewettite and other secondary minerals.

Many of the base metal deposits are so closely related in time and space to known centers of igneous activity that these are regarded with a considerable degree of confidence as being genetically involved with the generation of the ore fluids. Igneous intrusions to which one can point as likely sources of the ore-bearing fluids are characteristically absent near deposits of the Plateau type, and the ultimate sources of these metals are conjectural. Several possible origins have been proposed, among which are hypothetical, deep-seated, igneous bodies and the devitrification of volcanic ash present in the over-lying Chinle Formation. It has also been suggested that the high concentrations of uranium and vanadium required to produce the deposits were attained by a reiterative process involving exposure by erosion, weathering, leaching and downward percolation, precipitation, and reconcentration of an initially low metal concentration originally derived from weathered igneous materials. Whatever the ultimate source of the elements, they were subsequently transported in aqueous solutions through the sedimentary rocks to their present sites.

The following composite list of minerals found in the Monument No. 2 and No. 1 mines, and the Mitten No. 2 mine, is from the data of D. H. Johnson and A. G. King (Witkind and Thaden, 1963), with additions from other sources:

*Uranium and Vanadium Minerals*

| | |
|---|---|
| autunite | navajoite |
| becquerelite | pascoite |
| carnotite | rauvite |
| coconinoite | roscoelite |
| corvusite | schoepite |
| doloresite | steigerite |
| fernandinite | torbernite |
| fourmarierite | tyuyamunite |
| hewettite | uraninite |
| metahewettite | uranophane |
| metatorbernite | vandendriesschite |
| metatyuyamunite | vesignieite |
| metazeunerite | volborthite |
| montroseite | zippeite |

*Associated Minerals*

| | | |
|---|---|---|
| alunite | chrysocolla | montmorillonite |
| apatite | galena | opal |
| azurite | gypsum | pyrite |
| bornite | hydromica | quartz |
| calcite | ilsemannite(?) | sphalerite |
| chalcanthite | jarosite | sulfur |
| chalcedony | kaolinite | wad (lithium- |
| chalcocite | limonite | bearing) |
| chlorite | malachite | |

Uraninite and montroseite were the first primary uranium and vanadium minerals to form. At the Monument No. 2 mine they appear to have been deposited independently of one another, for they are seldom found in close association. Their formation was accompanied by the deposition of pyrite and small quantities of other sulfides including bornite, galena, and sphalerite, and at about this time the host rocks were modified by the addition of silica and calcium carbonate. The valence of uranium in uraninite is +4, the lowest state known in minerals; that of vanadium is +3, the lowest state of known vanadium minerals. Both of these minerals are readily susceptible to oxidation. Presumably it was not until the later periods of uplift and erosion of the Colorado Plateau area that these primary minerals were sufficiently close to the surface to be subjected to oxidizing conditions.

The great variety of secondary uranium and vanadium minerals is ultimately the product of oxidation of primary uraninite and montroseite. (In some deposits of the Plateau type the silicate coffinite is also an abundant primary uranium source.) While the oxidation of these minerals was a complex process, a general progressive sequence has been recognized. Montroseite probably oxidized first to form corvusite in which the vanadium is in the +4 state. Corvusite is the most abundant vanadium mineral in the Monument No. 2 mine, and its blue-black or dark greenish to brownish ores are accompanied by smaller amounts of rauvite, navajoite, and hewettite, together with subsidiary becquerelite, tyuyamunite, pyrite, and residual uraninite. Uraninite appears to have undergone oxidation at a somewhat slower rate than the vanadium minerals, and relicts of it are fairly common scattered among these oxidation products. Uraninite typically oxidizes to becquerelite and uranophane in which the valence state of uranium is +6.

Continued oxidation of the ores progressively yielded more and more minerals in which the valence states of uranium and vanadium were high, and minerals in which both elements are present began to appear. In the later stages of oxidation the stable mineral tyuyamunite locally became very abundant. Rich tyuyamunite occurrences are often accompanied by lesser amounts of relict uraninite and many other, oxidized, minerals. During the middle and late stages of oxidation, secondary solutions bearing uranium mixed with vanadium solutions to produce minerals such as carnotite, tyuyamunite, and rauvite, which contain both these elements. The most thorough oxidation produced ores which consist of tyuyamunite, limonite, and a little hewettite. This simple mineral assemblage is typical of the ores in the upper parts of the Monument No. 2 mine.

## THE CAMERON AREA

The economic importance of the uranium deposits centered in the vicinity of Cameron is substantially less than that of deposits in Monument Valley. The Cameron occurrences are noted here because the low vanadium and copper content ores have fostered a somewhat different type of mineralogy. The deposits are relatively small and widely scattered through an area about 10 miles wide by 75 miles long, extending northwest and southeast from Cameron along the Little Colorado River. The most productive part of the area lies within a few miles of Cameron. The data on which this description is based are largely drawn from the work of S. R. Austin (1964).

Although some uranium mineralization is found in the lower Triassic Moenkopi Formation and the Shinarump Member, the richest and most abundant concentrations are in the overlying Petrified Forest Member of the upper Triassic Chinle Formation. The most productive horizon consists of soft shales and mudstones with which are interbedded coarse-grained sandstones localized in scour channels. The clay content of the sands is fairly high, and it characteristically occurs in the form of pellets about the size of the quartz grains of the sandstone or a little larger.

The Cameron deposits are composed largely of secondary uranium minerals, of which the most abundant are boltwoodite, uranocircite, meta-uranocircite, and the vanadium-bearing metatyuyamunite. Ore mineralization is most commonly associated with carbonaceous fossil logs which were flattened by the weight of overlying sediments, and with sandstones adjacent to them. The most abundant primary minerals are marcasite, pyrite, and uraninite, which replaced the fossil wood and filled the interstices between grains in the sandy lenses nearby. (Coffinite has been recognized locally in small amounts but apparently is not of much importance in the Cameron area.) Other primary minerals include galena, greenockite, calcite, covellite, and smaltite. Microscopic examination of the mineralized wood reveals that its cellular structures were preserved in the finest detail, even though completely replaced by the primary minerals. Secondary barite and gypsum fill shrinkage cracks in the fossil logs.

Removal of overlying rock strata by erosion eventually brought the mineralized logs and sandstones near the surface where oxygen in the atmosphere and in underground waters oxidized and broke down the susceptible uraninite and sulfide minerals. The paucity of vanadium in the primary ores gave rise to predominantly vanadium-free, secondary uranium minerals with which were deposited a variety of secondary iron sulfates.

Chemical considerations suggest that the oxidation products of primary uranium minerals should be sufficiently soluble to migrate readily away from the source. Vanadium-bearing compounds, on the other hand, tend to precipitate quickly because of their low solubilities. While there is evidence for some local redistribution of uranium, the secondary minerals in these deposits did not, as a rule, migrate far from foci of primary mineralization. Reasons for this fixation of the secondary uranium minerals in the absence of vanadium are thought to include (1) lack of water due to local aridity, (2) restricted movement of groundwaters, and (3) geologically recent exposure of the deposits to oxidation which may not have allowed sufficient time for migration to have taken place. (Several lines of evidence suggest that the formation of minerals during the limited redistribution of uranium took place as recently as 4,500 to 9,000 years ago.)

The advanced degree of oxidation of the Cameron deposits makes the sequence of events during oxidation of the primary minerals difficult to establish. A partial paragenetic sequence has been worked out, however, for secondary minerals which migrated into geologically young gravels overlying some of the eroded deposits. Limestone pebbles in these gravels were progressively coated (outward) by layers consisting of cobaltian wad; gypsum and alunite; schoepite; betazippeite; a mixture of uranophane, beta-uranophane, and metatorbernite.

The secondary minerals found in these deposits are given in the following list (Austin, 1964):

### Uranium and Vanadium Minerals

beta-uranophane
"betazippeite"
boltwoodite
carnotite
coconinoite
coffinite
gummite
meta-autunite
metatorbernite
metatyuyamunite
meta-uranocircite

phosphuranylite
sabugalite
schoepite
schroeckingerite
torbernite
tyuyamunite
umohoite
uraninite
uranocircite
uranophane

several unidentified species

### Associated Minerals

alunite
atacamite
barite (strontian)
bieberite
calcite
chalcedony

copiapite
covellite
dolomite
ferrimolybdite(?)
galena
greenockite

gypsum
halotrichite
hematite
ilsemannite
jarosite
limonite
malachite
marcasite

metasideronatrite
opal (uraniferous)
pyrite
smaltite(?)
sphaerocobaltite
sulfur
wad (cobaltian)

Locally, the generally grayish rocks of the Petrified Forest Member have been converted to yellowish to buff colors, a phenomenon that has served as a guide to ore in the district. This so-called "bleaching" results in part from the impregnation of the rock by products formed from the oxidation of primary iron sulfides, including the sulfates jarosite, copiapite, and halotrichite, and the oxides hematite and limonite. Acidic sulfate solutions released during oxidation also attacked calcareous and aluminous rocks to form gypsum and alunite, whose white colors contribute to the bleaching effect, as do illites formed by the alteration of montmorillonite clays.

# 4. "Mine Fire" Minerals of the United Verde Mine, Jerome

A mine fire that started in the massive sulfide ore body in the United Verde mine at Jerome in 1894 burned for several decades despite concerted efforts to extinguish it. The fire was thought to have been caused by spontaneous combustion of unstable sulfide minerals on exposure to air. Surface stripping operations later exposed rocks above the fire area to reveal a suite of newly-formed hydrated sulfate minerals. Among the eleven mineral species formed directly as a consequence of the mine fire, seven were shown to be previously unknown species. Of these seven, all but one are unique products of the local conditions in the United Verde mine.

The geology and mineralogy of the Jerome area have been described in the papers of Reber (1922), Anderson (1927), Lausen (1928), Palache (1934), Anderson and Creasey (1958), Hutton (1959), and Anderson and Nash (1972). The main United Verde ore body in which the fire burned is a northerly-trending, steeply-plunging, pipe-like massive sulfide deposit composed of fine-grained pyrite and lesser amounts of chalcopyrite in a matrix of quartz and/or carbonate gangue. Minor amounts of arsenopyrite, sphalerite, galena, bornite, and tennantite are also present. The country rocks enveloping the ore body are metamorphosed bedded volcanic tuffs into which a gabbro mass was intruded. Unlike the majority of metallic mineral deposits of Arizona, both country rocks and sulfide ore body of the United Verde mine are of Precambrian age. Reinterpretation of the geology of the mine by Anderson and Nash (1972) led them to suggest that the massive sulfides were deposited contemporaneously with the volcanic tuffs enclosing them. The source of the sulfides is thought to have been hydrothermal emanations fed into a submarine basin

floored by volcanic tuffs; the sulfides were soon buried beneath accumulating volcanic materials. The ores and enclosing rocks were later structurally deformed by folding, metamorphosed, and modified by hydrothermal activity. These post-depositional events were thought to have mobilized some of the copper sulfides, which migrated to form veins, and also created the rich chlorite mineral assemblages of the local "black schist."

The burning stopes were sealed off by bulkheads, and some success was achieved in containing the fire by the use of air introduced under pressure. Later, when bulkheads were removed, the exposed rocks were red hot (Talley, 1917). Unsuccessful methods employed in attempts to contain the fire included the injection of water, carbon dioxide, and later, steam under pressure into the affected areas. The water thus introduced was probably gasified under the high temperatures which prevailed and reacted with the abundant iron and copper sulfide ores. The gases were thought by Lausen (1928) to have contained, in addition to water vapor and sulfur dioxide, small amounts of ferrous and ferric iron, copper, sodium, potassium, and aluminum. Their upward migration through fractures in the overlying rocks led to the precipitation of the unusual suite of secondary minerals.

Lausen reported that copious quantities of gas and smoke were issuing from natural vents in the floor of the mine pit at the time he collected samples. Study of the chemistry of these gases convinced him that the observed minerals could not have formed from them. He concluded that the new minerals were directly precipitated from gas phases of quite different composition which were generated immediately after the introduction of water into the mine. Lausen's paragenetic observation also demonstrated that the sulfate minerals could not have been co-precipitated but must have been formed during several distinct periods of sulfate precipitation.

The secondary sulfates were most abundant on the north and west sides of the massive sulfide pipe and were deposited in fractures in areas of abundant fine-grained, iron-oxide-stained quartz (jasper) as well as in cracks in pyrite.

Lausen recognized nine sulfate compounds in material he collected from the pit floor, five of which he proposed as new species. One of these, louderbachite, was later shown by Pearl (1950) to be identical with roemerite. The new minerals were named butlerite, guildite, ransomite, and rogersite. Rogersite was renamed lausenite because of the prior use of the name rogersite (Butler, 1928).

To our knowledge only two of the minerals have been recognized elsewhere. Butlerite has been reported from the Santa Elena mine, La Alcaparrosa, San Juan, Argentina, and from Chuquicamata, Chile. Ransomite occurs at Bisbee.

Lausen also described an amorphous dark-reddish material composed largely of arsenic, sulfur, and selenium, which formed as thin films and globules on the interior of large iron hoods placed about active gas vents in the pit to protect the miners. He gave the name jeromite to the substance, whose standing as a mineral species is debatable.

Other sulfates found by Lausen were alunogen, copiapite, coquimbite, and voltaite. (The voltaite occurrence had been previously described by C. A. Anderson in 1927 as the first U.S. and the seventh world occurrence of the species.)

In 1934 Charles Palache added to the list of mine fire minerals the rare species claudetite ($As_2S_3$), whose dimorph, arsenolite, was later identified at the United Verde mine, and he reported the first natural occurrence of elemental selenium. C. O. Hutton (1959) later described the new species yavapaiite ($KFe[SO_4]_2$) from material collected in the mine fire area.

Thus a total of six (seven, if jeromite is included) new mineral species were formed under these unusual circumstances.

# 5. Arizona Meteorites

Data on known Arizona meteorites were obtained from the files of the Center for Meteorite Studies, Arizona State University; Catalogue of Meteorites (Hey, 1966); and the literature cited. Arizona meteorites in the collections of Arizona State University (ASU) and the University of Arizona (UA) are noted.

*Apache County: Ganado* (iron, hexahedrite). Recognized as a meteorite in 1938. About 40 g. ASU collection (Nininger, 1940). *Houck* (iron, coarse octahedrite). Found in 1927. 66.7 kg. *Navajo* (iron, nickel-poor ataxite). In 1921 a 3,306-pound mass was found buried in talus 13 miles from Navajo; a second part was discovered four years later buried in soil near by (Merrill, 1922).

*Coconino County: Canyon Diablo* (iron, coarse octahedrite). Found in 1891. A total weight of meteoritic material in excess of 30 tons has been recovered from the vicinity of Meteor Crater, about 10 miles southeast of Canyon Diablo. UA 7639 (300+ pounds); 8336 (581 g); 323 (281 g); 8333–8335, 8337 (5 pieces totaling 439 g); ASU collection (Foote, 1891; Nininger, 1949). *Coon Butte* (stone, brecciated grey olivine-hypersthene chondrite). Found in 1905, about one mile west of Coon Butte (Meteor Crater). 636 g. ASU collection (Mallet, 1906). *Monument Rock* (iron, coarse octahedrite). Small fragments weighing in aggregate 90 g were found in 1948 near the crest of the east rim of Meteor Crater; ΛSU collection. *San Francisco Mountains* (iron, finest octahedrite). Found in 1920. About 1700 g. ASU collection (Perry, 1934). *Seligman* (iron, coarse octahedrite). Found in 1947. UA 8375 (16 g). ASU collection. *Winona* (stony-iron, mesosiderite). Found in 1928 in the ruins of the Eldon Indian pueblo at Winona. UA 818 (11 fragments totaling

885.8 g). (Brady, 1929; Heineman and Brady, 1929; Buddhue, 1940).

*Gila County: Clover Springs* (stony-iron, mesosiderite). Discovered by Mr. Jim Wingfield in the late 1940s or early 1950s, about 13 miles southwest of Clover Springs in the Mogollon Rim country. 7700 g. ASU collection. *Gun Creek* (iron, medium octahedrite). Found in 1909, about 70 miles northeast of Globe in the Sierra Ancha Mountains. About 50 pounds. (Palache, 1926).

*Maricopa County: Weaver Mountains* (iron, nickel-rich ataxite). Found near Wickenburg in 1898. UA 5628 (main mass, 28.195 kg). ASU collection (2.587 kg) (Henderson and Perry, 1951). *Wickenburg* (iron, coarse octahedrite). Found in 1940. 250 g. Possibly a transported fragment of Canyon Diablo. ASU collection (Nininger, 1940). *Wickenburg* (stone, black olivine-hypersthene chondrite). Recognized in 1940. 9.2 kg. ASU collection (Nininger, 1940).

*Mohave County: Bagdad* (iron, medium octahedrite). Found by Mr. Donald Stout on Burro Creek in 1959. 2.2 kg. ASU collection (Moore and Tackett, 1963). *Wallapai* (iron, fine octahedrite). Two pieces weighing 306 and 124 kg were found in 1927. UA 7638 (250+ pounds); 8351 (11 g); ASU collection (Merrill, 1927). *Wikieup* (stone, olivine-bronzite chondrite). Found in 1965 by Mr. A. B. Walker on the eastern slope of McCracken Peak, about 20 miles south-southwest of Wikieup. 372 g. ASU collection.

*Navajo County: Holbrook* (stone, olivine-hypersthene chondrite). Observed to fall July 9, 1912. An estimated 14,000 stones in the fall were inferred to have a composite mass of 481 pounds. The largest stone weighed 14.5 pounds. UA 817 (730.0 g); 824 (61.0 g); 819 (3 pieces — 6.31, 1.64, 1.22 g); 8338 (3 pieces — 41.6, 25.2, 3.3 g) (Foote, 1912; Gibson, 1970).

*Pima County: Pima County* (iron, nickel-poor ataxite). Acquired in 1947 by S. H. Perry from Eldred D. Wilson, Arizona Bureau of Mines. History unknown, but supposedly found near Tucson. 210 g (Henderson and Perry, 1949). *Silver Bell* (iron, coarsest octahedrite). Known before 1947. UA 877 (3.438 kg). *Tucson* (iron, nickel-rich ataxite). Known before 1850. Two large irons: one, the celebrated Ring meteorite (Irwin-Ainsa), weighs 688 kg; the other, the Carleton iron, weighs 287 kg. Both are on display in the Smithsonian Institution (Fletcher, 1890).

*Yavapai County: Ash Fork* (iron, coarse octahedrite). A 60 pound mass was found in 1901 by Charles Quitzow in Cedar Glade, about 25 miles south-southwest of Ash Fork (Reeds, 1937). *Bloody Basin* (iron, coarse octahedrite). Found in 1964 by Mr. Arthur Morrison while on a hunting trip. 5074 g. It has been suggested that possibly this iron may be a fragment of Canyon Diablo transported by Indians (Buseck and Moore, 1966). ASU collection. *Camp Verde* (iron, coarse octahedrite). Found in 1915, but recognized as a meteorite in 1939. 61.5 kg. ASU collection (59.79 kg)

**Cyanotrichite.**
Grandview mine, Grand Canyon, Coconino County.
Stanley B. Keith.

(Nininger, 1940). *Cottonwood* (stone, olivine-bronzite chondrite). Found in 1955. About 800 g. ASU collection. *Fair Oaks* (iron, coarse octahedrite). Found in 1937 by John Coates while hunting. 787 g. ASU collection.

*Yuma County: Ehrenberg* (iron, coarse octahedrite). A ten pound mass was discovered in 1862 near the Colorado River by Herman Ehrenberg, a German engineer employed in surveying a townsite (Palache, 1926). *Kofa* (iron, coarsest octahedrite). Found in 1893. 490 g.

PART THREE

# CATALOG OF
# ARIZONA MINERAL OCCURRENCES

## THE MEANINGS OF HEADING STYLES USED IN THIS SECTION

ACANTHITE — Names in capital letters indicate mineral species widely accepted as valid by most authorities.

ACMITE (AEGERINE) — Names in capital letters followed by a second capitalized name in parentheses indicate substantial question as to which name is preferred; both names are reputable.

Chrysoprase (see QUARTZ); Adularia (a variety of ORTHOCLASE) — Names in lower case indicate the mineral (or substance) is not a valid species; an explanation follows in parentheses.

ANTIGORITE (see Serpentine); CHRYSOTILE (see Serpentine) — In a few instances the description of a species may be given in a common place in the catalog.

ARIZONITE (?); Cornuite (?) — Names whose validity is seriously questioned by some mineralogists but not by all are followed by a query in parentheses.

*ALAMOSITE — An asterisk precedes those minerals which have been added to the Arizona Mineral List (p. 8) since publication of the 1959 edition of *Mineralogy of Arizona* by Galbraith and Brennan.

# Arizona Mineral Occurrences

## ACANTHITE

Silver sulfide, monoclinic $Ag_2S$. Dimorphous with the isometric argentite. Argentite, which may form in hydrothermal veins at elevated temperature, inverts upon cooling to acanthite so that most, if not all, naturally occurring silver sulfide is the monoclinic dimorph, acanthite. An important silver ore mineral; commonly associated with other silver minerals, galena, tetrahedrite, and nickel-cobalt ores. Also as a secondary mineral in the zone of sulfide enrichment, with chalcocite, native silver, and silver halogen minerals.

*Cochise County:* Tombstone district, in oxidized ores formed from alteration of argentiferous tetrahedrite (Butler et al., 1938; Romslo and Ravitz, 1947). Pearce district, with chlorargyrite, bromargyrite, embolite, and iodargyrite (in quartz veins) at the Commonwealth mine (Endlich, 1897).

*Gila County:* Richmond Basin; the chief primary silver mineral, in masses up to several pounds in weight.

*Graham County:* Aravaipa district, in veins of the Grand Reef system (Ross, 1925a); Copper Creek district, Blue Bird mine (Simons, 1964).

*Mohave County:* Cerbat Range, Keystone, Golden Star, and Queen Bee mines, Mineral Park district (Bastin, 1925); Prince George mine and veins of the Banner group, Stockton Hill district; also found at various properties in the Cerbat district (Schrader, 1909); Rawhide mine, McConnico district, altering to chlorargyrite (UA 1091).

*Pima County:* Santa Rita Mountains, Blue Jay mine, Helvetia-Rosemont district (Schrader and Hill, 1915). Sierrita Mountains, Papago district. Quijotoa Mountains, Morgan mine.

*Pinal County:* Pioneer district (Romslo and Ravitz, 1947), Silver King mine, in large quantity on the upper levels; Belmont property, as small blebs in galena. Dripping Spring Mountains, Little Treasure mine, Saddle Mountain district. Mammoth district, Mammoth-St. Anthony mine, rarely as minute monoclinic crystals on leadhillite and native silver.

*Santa Cruz County:* Santa Rita Mountains, the Alto, Eureka, Ivanhoe, Montezuma, and Empress of India mines, Tyndall district; Augusta, Happy Jack, and Anaconda mines, Wrightson district. Patagonia Mountains, La Plata and Meadow Valley mines, Red Rock district; January, Blue Eagle, Flux, and American mines, Harshaw district (Schrader and Hill, 1915; Schrader, 1917).

*Yavapai County:* Bradshaw Mountains, Dos Oris mine, Hassayampa district, with native silver and chlorargyrite. Wickenburg Mountains, Monte Carlo mine, in primary ores with native silver, skutterudite, and proustite (Bastin, 1922); Stonewall Jackson mine (Guild, 1917). Big Bug district, Hillside mine, with arsenopyrite, chalcopyrite, galena, sphalerite, pyrite, and tetrahedrite (Axelrod et al., 1951); Arizona National mine, in galena with freibergite and in cavities with wire silver (Lindgren, 1926).

*Yuma County:* Silver district, said to have been mined in important quantities in the Princess and other veins (Wilson, 1933).

## *ACMITE (AEGIRINE)

Sodium iron silicate, $NaFeSi_2O_6$. A rock-forming mineral of the pyroxene group which occurs primarily as a product of late crystallization of alkaline magmas.

*Apache County:* Monument Valley, Garnet Ridge, with diopsidic jadeite and pyrope-almandine garnet in eclogite inclusions from kimberlite pipes (Watson and Morton, 1969).

*Pima County:* Gunsight Mountain, northern end of the Sierrita Mountains, with sphene and scapolite in a pulaskite dike (Bideaux et al., 1960).

## ACTINOLITE

Calcium magnesium iron silicate hydroxide, $Ca_2(Mg,Fe)_5Si_8O_{22}(OH)_2$. A member of the amphibole group of rock-forming minerals. Forms a continuous substitutional series with the iron-free tremolite. Occurs in both thermally metamorphosed gneisses, schists, and marbles, and in contact metamorphic rocks, particularly limestones.

*Cochise County:* Little Dragoon Mountains, in metamorphosed limestones, Johnson district.

*Graham County:* Aravaipa and Stanley districts, as gangue of contact-metamorphosed ores (Ross, 1925a). Galiuro Mountains, Ash Peak.

*Greenlee County:* Clifton-Morenci district, in the northern part of the district where Paleozoic limestones are in contact with the main intrusion, with garnet, diopside, epidote, tremolite, and other calc-silicate minerals (Moolick and Durek, 1966).

*Mohave County:* Black Mountains, Oatman district, as thin sheets between layers of quartz in the Big Jim vein.

*Pima County:* Sierrita Mountains, abundant in contact rocks, Twin Buttes district; present in skarns and tactites at the Twin Buttes mine (Stanley B. Keith, pers. comm., 1973). Helvetia-Rosemont district, in contact metamorphosed sedimentary rocks at a number of mines (Creasey and Quick, 1955).

*Pinal County:* Globe district, Old Dominion mine, along bedding planes in Mescal Limestone.

*Santa Cruz County:* Patagonia Mountains, Westinghouse property.

*Yavapai County:* Bradshaw Mountains, in country rock, Iron Queen mine; as the fibrous variety with bournonite at the Boggs mine, Big Bug district (Lindgren, 1926).

*Yuma County:* Harcuvar Mountains, Yuma Copper and Cabrolla properties, as a replacement of limestone beds. Cemetery Ridge area, east of Deadman Tank, as bladed green crystals in amphibolite dikes with an asbestiform variety, developed in schist (Wilson, 1933).

## *ADAMITE

Zinc arsenate hydroxide, orthorhombic $Zn_2(AsO_4)(OH)$. Dimorphous with paradamite. A rare secondary mineral found in the oxidized portion of some base metal deposits.

*Coconino County:* Grandview (Last Chance) mine, Horseshoe Mesa, Grand Canyon National Park, in very small amount associated with zeunerite, scorodite, and olivenite (Leicht, 1971).

*Yavapai County:* Found at a small prospect in the Mayer area (Sec. 17, T12N, R2W) as pale green crystalline crusts in vein quartz.

## Adularia
## (a variety of ORTHOCLASE)

## *AEGIRINE-AUGITE

Sodium calcium iron magnesium aluminum silicate, $(Na,Ca)(Fe,Mg,Al)(Si_2O_6)$. A rock-forming silicate mineral that occurs primarily as a late product of the crystallization of alkaline magmas.

*Pima County:* Gunsight Mountain, north end of the Sierrita Mountains, with acmite, scapolite, and sphene in a pulaskite dike (Bideaux et al., 1960).

## Agate (see QUARTZ)

## AIKINITE

Lead copper bismuth sulfide, $PbCuBiS_3$. A rare mineral found at few localities in the world; may be associated with native gold and galena.

*Cochise County:* At the Black Prince claims, as blebs and veinlets between garnet crystals in limestone. From an unnamed locality about 15 miles northeast of Tombstone (UA 3572). Also reported from the Warren district.

*Gila County:* Globe-Miami district, Miami mine, in veinlets cutting chalcopyrite, with tennantite and enargite (Legge, 1939).

*Pima County:* Roskruge Range, in small quantities at the Roadside mine (S 12049).

## AJOITE

Copper aluminum silicate hydrate, $Cu_6Al_2Si_{10}O_{29} \cdot 5H_2O$. A rare mineral intimately associated with shattuckite in the oxidized zone of copper deposits. The occurrence at Ajo is the type locality.

*Maricopa County:* At several localities south of Wickenburg, including the Moon Anchor mine and Potter-Cramer property, associated with wickenburgite, mimetite, willemite, and phoenicochroite (Williams, 1968).

*Pima County:* Ajo district, New Cornelia open-pit mine, as pale aquamarine tufts and crystals filling interstices between radiating spherulitic, dark blue crystalline shattuckite. (Schaller and Vlisidis, 1958; Hutton and Vli-

**Ajoite.**
New Cornelia mine, Ajo,
Pima County. Julius Weber.

sidis, 1960; Sun, 1961; Newberg, 1964). The chemical analysis of the type specimen by A. C. Vlisidis gave the following result:

| | | | | | |
|---|---|---|---|---|---|
| $SiO_2$ | 45.90 | CaO | 0.99 | $As_2O_5$ | 1.33 |
| CuO | 33.90 | BaO | 0.50 | $SO_3$ | 0.23 |
| FeO | 0.78 | $Al_2O_3$ | 7.30 | $H_2O(-)$ | 2.55 |
| MnO | 0.10 | $TiO_2$ | 0.16 | $H_2O(+)$ | 5.03 |
| MgO | 0.66 | | | TOTAL | 99.43% |

After deducting conichalcite and barite as impurities:

| | | | | | |
|---|---|---|---|---|---|
| $SiO_2$ | 45.90 | MgO | 0.66 | $TiO_2$ | 0.16 |
| CuO | 32.98 | CaO | 0.34 | $H_2O(-)$ | 2.55 |
| FeO | 0.78 | BaO | — | $H_2O(+)$ | 4.93 |
| MnO | 0.10 | $Al_2O_3$ | 7.30 | TOTAL | 95.70% |

Specific gravity: 2.96.

## *AKAGANEITE

Iron oxide hydroxide chloride, $\beta$-FeO(OH,-Cl). A rare secondary mineral which is thought to have formed by the alteration of pyrrhotite at the type locality, the Akagané mine, Iwate Prefecture, Japan.

*Mohave County:* Mineral Park district, reported as possibly present at the Ithaca Peak mine (Eidel et al., 1968).

*Santa Cruz County:* As thin warty reddish brown films on fractured vein quartz containing partly leached pyrite and molybdenite from a small prospect north of the Santo Niño mine.

## ALABANDITE

Manganese sulfide, MnS. An uncommon primary mineral found in vein deposits, commonly associated with sphalerite, galena, pyrite, rhodochrosite, rhodonite, quartz, and calcite.

**Alamosite in diaboleite.**
Mammoth–St. Anthony mine,
Pinal County. R. A. Bideaux.

*Cochise County:* Tombstone district, Lucky Cuss mine (Moses and Luquer, 1892; Blake, 1903; Hewett and Rove, 1930; Butler et al., 1938); Oregon-Prompter mine (Hewett and Fleischer, 1960). Warren district, Higgins mine, Copper Queen mine, in veinlets in limestone (UA 954). Chiricahua Mountains, Humboldt mine (Hewett and Rove, 1930); Hilltop mine (UA 5417).

*Santa Cruz County:* Northern Patagonia Mountains, Harshaw district, Trench mine, with sphalerite, galena, and rhodochrosite (Hewett and Rove, 1930); Patagonia district, World's Fair mine, with rhodonite (UA 9924).

## *ALAMOSITE

Lead silicate, $PbSiO_3$. A rare secondary mineral, found with wulfenite and leadhillite at the type locality near Alamos, Sonora, Mexico. The occurrence noted here is believed to be the third in the world.

*Pinal County:* Mammoth district, Mammoth-St. Anthony mine, on a single specimen, as white crystalline sprays and balls to 5 mm; associated with diaboleite and willemite.

## ALLANITE

Cerium calcium yttrium aluminum iron silicate hydroxide, $(Ce,Ca,Y)_2(Al,Fe)_3Si_3O_{12}(OH)$. A member of the epidote group. Occurs commonly as an accessory mineral in granites; also found in pegmatites, schists, and as a detrital mineral.

*Cochise County:* Tombstone district, as microscopic crystals in granodiorite (Butler et al., 1938).

*Maricopa County:* Estrellas Mountains, southwest of Phoenix, in very coarse granite as phenocrysts up to 1 x ¼ inch.

*Mohave County:* Aquarius Range, Rare Metals mine and Columbite prospect, in pegmatite with gadolinite (Heinrich, 1960); McConnico district, Consolidated Spar mine, in granite pegmatite (Frondel, 1964). Cerbat Mountains, Kingman Feldspar quarry. Greenwood Mountains, near Signal (Robert O'Haire, pers. comm.).

*Pima County:* Near Willow Springs Ranch, Oracle Junction, in pegmatite with schorlite (Robert O'Haire, pers. comm., 1972). Eleven miles southwest of San Xavier Mission on the Cottonwood Ranch, sparingly distributed in black sands over an area several miles across (Adams and Staatz, 1969). As spectacular masses in a vein with tourmaline, actinolite, and calcite southeast of Covered Wells (Williams, 1960).

*Yavapai County:* Eureka district, in pegmatite knots on the 7U7 ranch near Bagdad, with triplite and bermanite (Hurlbut, 1936; Leavens, 1967). White Picacho district, as a rare accessory mineral in crystals up to 4 inches long (Jahns, 1952). Reported from near Yarnell. At a locality three miles west of Congress Junction (UA 6638).

## *ALLOPHANE

Amorphous hydrous aluminum silicate, perhaps near $Al_2SiO_5 \cdot nH_2O$. A widespread constituent of clays; usually amorphous to x-rays because of small particle size or disordered structure. Allophane clays are certainly much more abundant in Arizona than the few documented localities would suggest.

*Cochise County:* Warren district, Sacramento Hill, in a hydrothermally altered granite porphyry stock with sericite, hydromuscovite, kaolinite, and alunite (Schwartz, 1947).

*Gila County:* Globe-Miami district, Inspiration mine, as a hydrothermal alteration product in granite porphyry (Schwartz, 1947, 1956); Castle Dome mine, a product of hydrothermal alteration of quartz monzonite (Peterson et al., 1951). Van Dyke claim (AM 30585).

*Greenlee County:* Copper Mountain district (Morenci), as pseudomorphs after plagioclase, with halloysite, in altered porphyry copper ore and in granite porphyry (Schwartz, 1947,

1958), especially around the periphery of the ore body (Moolick and Durek, 1966).

*Mohave County:* Hualpai Mountains, in gangue at the Antler mine (Romslo, 1948).

*Pinal County:* Mammoth district, San Manuel mine, in hydrothermally altered monzonite and quartz monzonite porphyries (Lovering et al., 1950; Schwartz, 1953). Ray district, in hydrothermally altered porphyry stock as well as in sericitized veins (Schwartz, 1947, 1952).

## ALMANDINE

Iron aluminum silicate, $Fe_3Al_2(SiO_4)_3$. A member of the garnet group. Occurs typically in regionally metamorphosed argillaceous sediments, but also in the contact metamorphic environment, and in some igneous rocks.

*Coconino County:* Grand Canyon, in Archean rocks of the Inner Gorge; as crystals over an inch in diameter on Phantom Creek.

*Mohave County:* Aquarius Range, as crystals in light-colored volcanic rock at the southern end of the range. As pink material in migmatite in the Cerbat Range (Thomas, 1953).

*Pima County:* Santa Catalina Mountains, in the Front Range and crestal portions of the main range, in gneiss and granite complexes and in pegmatite; the garnets contain varying amounts of the end-member components almandine, spessartine, and pyrope (Pilkington, 1961). Helvetia-Rosemont district, Peach-Elgin copper deposit, with grossular and diopside in bedded replacement deposits (Heyman, 1958).

## ALUNITE

Potassium aluminum sulfate hydroxide, $KAl_3(SO_4)_2(OH)_6$. A mineral of widespread occurrence formed in the wall rocks of sulfide ore bodies by processes related to hydrothermal activity.

*Apache and Navajo Counties:* Monument Valley, with uranium-vanadium ores in channels at the base of the Shinarump Conglomerate and below channels in the Moenkopi Formation and DeChelly Member of the Cutler Formation (Mitcham and Evensen, 1955).

*Cochise County:* Warren district, Sacramento Hill, in the hydrothermally altered granite porphyry stock, as an alteration product of feldspar (Schwartz, 1947), and as large masses from the Cole mine resembling variscite.

*Coconino County:* Cameron area, in uranium ores as white powdery crusts and masses; as small prismatic crystals in sandstone near the Black Peak breccia pipe near Cameron (Barrington and Kerr, 1961); at the Black Point-Murphy mine where it is closely associated with gypsum and secondary uranium minerals in Pleistocene gravels (Austin, 1964).

*Gila County:* Globe district, as veins in diabase at the Old Dominion mine (Lausen, 1923). Dripping Spring Mountains, Apex mine, Banner district.

*Graham County:* Lone Star district, Gila Mountains, intimately associated with jarosite and turquoise in the oxidized zone of the Safford porphyry copper deposit (Robinson and Cook, 1966).

*Greenlee County:* Clifton-Morenci district, Ryerson mine, as grains, irregular masses, and fibrous aggregates in altered porphyry (Lindgren, 1905; Reber, 1916).

*Maricopa County:* Eastern part of the Vulture district, about 3 miles west of Morristown on the west side of Hassayampa River, where it constitutes a major phase within hydrothermally altered rhyolite in an area of about one half square mile, associated with kaolinite (Sheridan and Royse, 1970).

*Mohave County:* Mineral Park district, Ithaca Peak mine, as nodules in a clay-turquoise-sulfide vein traversing an igneous host rock (Field, 1966; Eidel et al., 1968).

*Pima County:* Associated with jarosite in veins in the Oxide pit, Silver Bell mine (Kerr, 1951). In innumerable narrow veins cutting the Concentrator Volcanics at Ajo (Gilluly, 1937; Hutton and Vlisidis, 1960). Pima district, Esperanza mine, in veinlets with or without turquoise, in the oxidized capping over the predominantly chalcopyrite-chalcocite ore body (Loghry, 1972).

*Pinal County:* Mammoth district, San Manuel mine, abundant in hydrothermally altered monzonite and quartz monzonite porphyries with kaolinite and quartz (Schwartz, 1947, 1953, 1958, 1966). Of uncommon occurrence in the Ray area, associated with the oxidation of sulfides (Ransome, 1919).

*Santa Cruz County:* Patagonia Mountains, Palmetto district, Evening Star prospect and Three-R mine, disseminated in an altered granite porphyry (Schrader, 1913, 1914, 1917; Wilson, 1944). Red Mountain copper prospect, where it is thought to be of both primary and secondary origin (Loghry, 1972).

*Yuma County:* Sugarloaf Butte, about 5 miles west of Quartzsite and 1 mile south of U.S. Highway 60, where it occurs as branching, irregular veins which constitute a substantial deposit within dacite; unusual in containing up to 4.3 percent $Na_2O$ (Heineman, 1935; Thoenen, 1941; Wilson, 1944; Omori and Kerr, 1963) (H 108800). A partial analysis by F. S. Wartman gave:

| | | |
|---|---|---|
| $Al_2O_3$ 36.5 | $K_2O$ 4.85 | $Na_2O$ 4.3 |
| $SO_3$ 38.1 | | $SiO_2$ 1.2 |

## ALUNOGEN

Aluminum sulfate hydrate, $Al_2(SO_4)_3 \cdot 18H_2O$. A water-soluble secondary mineral commonly formed by the decomposition of pyrite or under fumerolic conditions. May be associated with a variety of other secondary sulfates. Probably more widespread in the state than suggested by the single locality noted below.

*Yavapai County:* Formed in the United Verde mine as a result of the burning of pyritic ore (Lausen, 1928).

## AMBLYGONITE

Lithium sodium aluminum phosphate fluoride hydroxide, $(Li,Na)Al(PO_4)(F,OH)$. Occurs in granite pegmatites of the lithium- and phosphate-rich types, commonly associated with spodumene, lithiophylite-triphylite, lepidolite, and tourmaline.

*Maricopa and Yavapai Counties:* In the pegmatites of the White Picacho district, associated with spodumene and zinnwaldite; Midnight Owl mine (UA 2675) (Jahns, 1952).

*Maricopa County:* Mitchells Wash, northeast of Morristown (UA 5623). San Domingo district, near San Domingo Wash, northeast of Wickenburg, in several pegmatite bodies as rough crystals up to 4 or 5 feet in diameter (Jahns, 1953).

## *AMESITE

Magnesium iron aluminum silicate hydroxide, $(Mg,Fe)_4Al_4Si_2O_{10}(OH)_8$. A member of the septachlorite group, related chemically to the chlorites and structurally to the kaolinite group. Occurs under conditions similar to those under which the chlorites form.

*Pinal County:* Mammoth district, Mammoth-St. Anthony mine, as a white powdery matrix on which wulfenite occurs.

## Amethyst (see QUARTZ)

## AMPHIBOLE

A group of rock-forming silicates which are abundant in certain igneous and metamorphic rocks. Refer to the alphabetical listing of the individual species.

## ANALCIME

Sodium aluminum silicate hydrate, $NaAlSi_2O_6 \cdot H_2O$. A fairly common mineral which is much like the members of the zeolite group in its chemistry and in some of its occurrences. Found in some igneous rocks of intermediate and mafic composition; also formed as a result of hydrothermal processes.

*Apache County:* At a locality near Nutrioso, associated with clinoptilolite, as cement in Tertiary sandstone (Wrucke, 1961). Widespread in the mafic volcanic rocks of the Hopi Buttes volcanic field (Williams, 1936).

*Cochise County:* Along the San Simon River, associated with chabazite, clinoptilolite, erionite, and herschelite in tuffs of late Cenozoic age (Sand and Regis, 1966; Regis and Sand, 1967). Willcox Playa, in mudstone of Pleistocene age; of authigenic origin (Pipkin, 1967).

*Coconino County:* Found in the baked border zone of the Tuba monchiquite dike, near Cameron; associated with chlorite and illite (Barrington and Kerr, 1962).

*Mohave County:* Near Wikieup, where it comprises the bulk of a friable green "sandstone" formerly thought to be composed almost entirely of glauconite. The analcime grains are coated with glauconite (Wilson, 1944; Robert O'Haire, pers. comm., 1972). East of Big Sandy Wash in E. ½, T.16N, R.13W, in Pliocene tuff, associated with chabazite, clinoptilolite, erionite, and phillipsite (Ross, 1928, 1941). Maggie Canyon, in Sec. 30, T.12N, R.13W, as cement in sandstone of the Chapin Wash Formation (Lasky and Webber, 1949).

*Pima County:* Santa Rita Mountains, in cavities in amygdaloidal basalts, Rosemont area.

*Pinal County:* Near Eloy, where it was noted in a diamond drill hole in silty claystone in Sec. 25, T.7S, R.8E (Sheppard, 1969).

*Yavapai County:* An analysis of material from the Aquarius Cliffs by J. G. Fairchild (Wells, 1937) gave the following result:

| | | | | | |
|---|---|---|---|---|---|
| $SiO_2$ | 60.61 | MgO | 0.05 | $K_2O$ | 1.02 |
| $Al_2O_3$ | 18.03 | CaO | 0.04 | $H_2O(-)$ | 0.34 |
| $Fe_2O_3$ | 1.01 | $Na_2O$ | 10.98 | $H_2O(+)$ | 8.36 |

TOTAL 100.44%

## ANATASE

Titanium oxide, $TiO_2$. Trimorphous with rutile and brookite. An uncommon secondary mineral found in veins and cavities in schists and gneisses;

**Analcime with hematite inclusions.**
Horseshoe Dam, Maricopa County. Julius Weber.

formed from titanium derived through leaching of country rock by hydrothermal solution.

*Gila County:* Reported in the suite of authigenic heavy minerals in Precambrian quartzites in the Diamond Butte quadrangle (Gastil, 1958).

*Graham County:* Stanley district, Friend mine.

*Pima County:* Ajo district, as tiny, well-formed, pyramidal crystals present in minor amounts associated with papagoite in altered rock (Hutton and Vlisidis, 1960).

*Santa Cruz County:* Patagonia Mountains, with brookite, chalcopyrite, wulfenite, and molybdenite, all as micro-crystals in interstices between adularia crystals, in the vicinity of the Santo Niño mine, near Duquesne (collected by William Hunt).

## ANDALUSITE

Aluminum silicate, orthorhombic $Al_2SiO_5$. Trimorphous with kyanite and sillimanite. Commonly associated with kyanite or sillimanite in regionally metamorphosed rocks such as slates, schists, and gneisses; also typically found in thermally altered rocks where it may be associated with cordierite. Rarely found in granites.

*Cochise County:* Cochise district, as square, prismatic porphyroblasts up to 2 inches in length in schist near the Texas Canyon stock; locally as the variety chiastolite (Cooper and Silver, 1964). Apache Pass, Dos Cabazas Mountains, as metacrysts in metamorphosed mudstones of Mesozoic age. Chiricahua Mountains, variety chiastolite (UA 8166).

*Gila and Pinal Counties:* Locally abundant in the Pinal Schist near post-Cambrian granitic rocks (Ransome, 1919).

*Gila County:* Banner district, Christmas mine, in small amounts in metamorphosed siltstones of the Naco Formation, associated with muscovite, biotite, quartz, and orthoclase (Perry, 1969).

*Mohave County:* Red Lake district, Grand Wash Cliffs, in pegmatite. Hualpai Mountains, Cedar district, 11 miles east of Yucca, in quartz veins in schist.

*Pima County:* Ajo district, in small quantities in the Cardigan Gneiss (Gilluly, 1937).

*Pinal County:* Gila River Indian Reservation, Sacaton Mountains, about 10 miles north of Casa Grande, in exometamorphosed sediments encased in granodiorite, as the variety titanandalusite, associated with sillimanite, corundum, and cordierite (Bideaux et al., 1960).

*Yavapai County:* Bradshaw Mountains, as scattered lenses and disseminations in schist; large pinkish crystals have been found near Middleton, on the Crown King road; also reported near Cleator. From near Granite Mountain in extensive veins. Santa Maria Mountains, near Camp Wood, as flakes and nodules in schist. Also found at Bagdad.

*Yuma County:* About 3 miles southeast of Quartzsite, with kyanite, sillimanite, and dumortierite, as prismatic crystals in schist (Wilson, 1929, 1933; Duke, 1960). Near La Paz, at Sec. 34, T.4N, R.21W, as small crystalline masses in granite (David Shannon, pers. comm., 1971).

## ANDERSONITE

Sodium calcium uranyl carbonate hydrate, $Na_2Ca(UO_2)(CO_3)_3 \cdot 6H_2O$. A very rare water-soluble secondary mineral which occurs as efflorescences with gypsum and other secondary oxidized minerals. The Hillside mine occurrence is the type locality.

*Coconino County:* Cameron area (Bollin and Kerr, 1958).

*Yavapai County:* Eureka district, Hillside mine, with gypsum, schroeckingerite, bayleyite, swartzite, johannite, and uraninite; as an efflorescence on the walls of mine workings in crusts about one-eighth inch thick on gypsum (Axelrod et al., 1951). A chemical analysis by F. S. Grimaldi on 3.8 mg gave the following results:

| | | | | | |
|------|-----|--------|------|---------|------|
| MgO | 0.5 | $UO_3$ | 43.4 | $SO_3$ | 1.6 |
| CaO | 8.9 | $CO_2$ | 19.6 | $(H_2O)$ | 16.7 |
| $Na_2O$ | 9.3 | | | TOTAL | 100.0% |

Specific gravity: 2.8.

## ANDRADITE

Calcium iron silicate, $Ca_3Fe_2(SiO_4)_3$. A commonly occurring member of the garnet group; typically found in contact metamorphic deposits and skarns formed in impure limestones.

*Cochise County:* Dragoon Mountains, where it is common in the wall rocks of pyritic ores in Abrigo Limestone, Turquoise district (Perry, 1964).

*Gila County:* Dripping Spring Mountains, Banner district, in large massive beds at the Christmas mine; 79 mine, as an abundant silicate in contact metamorphosed limestones (Keith, 1972). Intergrown with epidote, calcite, and chalcopyrite at Harrington's claims, near the East Verde River (Ross, 1925a).

*Graham County:* Aravaipa and Stanley districts. Stanley Butte, as massive material and as crystals up to 2 inches in diameter; crystals are generally brown, but some display greenish hues because of platy inclusions distributed in layers just beneath their surfaces (Sinkankas, 1964, 1966).

*Greenlee County:* Clifton-Morenci district, in layered limestone, forming masses from 50 to 100 feet thick (Lindgren, 1905; Guild, 1910; Reber, 1916; Moolick and Durek, 1966).

*Pima County:* Empire Mountains, as zones of massive material at contacts between Paleozoic limestones and quartz monzonite intrusives, Helvetia-Rosemont district, King mine, in contact metamorphosed limestones (Michel, 1959). Santa Rita Mountains, as the most common silicate mineral in limestones of the Rosemont district (Schrader and Hill, 1915; Schrader, 1917). Sierrita Mountains, as zones up to 200 feet in width in limestone at the Twin Buttes district; found in the skarns of the Twin Buttes mine (Stanley Keith, pers. comm., 1973); Pima district, where it is the primary constituent of tactite formed in Paleozoic limestones at the Mission mine (Kinnison, 1966).

*Santa Cruz County:* Patagonia Mountains, as crystals up to 2 inches in diameter in metamorphosed limestone at the Westinghouse property (Schrader and Hill, 1915; Schrader, 1917).

## ANGLESITE

Lead sulfate, $PbSO_4$. Abundant in oxidized lead deposits, most commonly as masses surrounding galena and altering, in turn, to cerussite. Only a few of the many occurrences in Arizona are listed here.

*Cochise County:* Tombstone district, Tombstone Extension mine (Butler et al., 1938). Warren district, Shattuck mine, and in the 1,800-foot level of the Campbell mine as glassy spearhead-shaped crystals with leadhillite (S 114586). Turquoise district, Silver Bill mine. Huachuca Queen mine (UA 7777). Gunnison Hills, Texas Arizona mine (Cooper and Silver, 1964).

*Gila County:* Globe district, Lost Gulch and Defiance mines (Wilson et al., 1950; Peterson, 1962); Castle Dome mine (Peterson et al., 1951). Apache mine (Wilson et al., 1950).

*Graham County:* Stanley district (Ross, 1925a); Aravaipa district (Simons, 1964), Grand Reef mine, where a few well-formed translucent crystals to an inch in size have been found with linarite in quartz-lined vugs (Richard L. Jones, pers. comm., 1969).

*Maricopa County:* As nodules 1 to 3 inches in diameter in the dumps of the Montezuma and Prodigal mines, west of Morristown. Painted Rock Mountain, at the Rowley mine, near Theba (Wilson and Miller, 1974). Tonopah Belmont mine (Robert O'Haire, pers. comm.).

*Mohave County:* Cerbat Range, at the Tennessee-Schuylkill mine, Wallapai district (Thomas, 1949). In several properties of the Mineral Park district.

*Pima County:* Empire Mountains, at several of the mines of the Hilton group. At the King mine, Santa Rita Mountains (UA 7776). Sierrita Mountains, abundant at the Paymaster mine, Olive Camp. Quijotoa Mountains, at the Morgan mine (Ransome, 1922). Cababi district, South Comobabi Mountains, Mildren and Steppe claims (Williams, 1963).

*Pinal County:* Galiuro Mountains, at the Saddle Mountain group. Mammoth district, Mammoth-St. Anthony mine as spear-shaped crystals up to about ¼ inch (Fahey et al., 1950) (H 98082).

*Santa Cruz County:* Patagonia Mountains, Westinghouse and Mowry mines (Schrader and Hill, 1915). Tyndall district, Santa Rita Mountains, Cottonwood Canyon, associated with cerussite and wulfenite in oxidized lead ore at the Glove mine (Bideaux et al., 1960; Olson, 1966).

*Yavapai County:* Bradshaw Mountains, in the Copperopolis mine, Castle Creek district (Lindgren, 1926). Eureka district, Hillside mine (Axelrod et al., 1951). Big Bug district, Iron King mine (Anderson and Creasey, 1958). Jerome (UA 7711).

*Yuma County:* Castle Dome Mountains, Castle Dome district (Foshag, 1919; Wilson, 1933); Brush (1873) chemically analyzed a compact variety of anglesite from the district.

## ANHYDRITE

Calcium sulfate, $CaSO_4$. Formed in extensive sedimentary deposits with halite, gypsum, and other salts from evaporation of oceanic waters of inland seas; not uncommon as a gangue mineral in some sulfide mineral deposits.

*Apache and Navajo Counties:* In the subsurface of the southern portions of the counties in the Supai Salt Basin which embraces about 2,300 square miles; with halite, dolomite and clastic red beds (Peirce and Gerrard, 1966; Peirce, 1969).

*Gila County:* Dripping Spring Mountains, Christmas mine, as a fairly abundant constituent of diamond drill cores between the 1,000 and 1,200 foot levels (Peterson and Swanson, 1956), and in ore bodies replacing dolomite, in veinlets and interbanded with layers of magnetite (Perry, 1969).

*Pima County:* Ajo district, New Cornelia mine, sparingly as minute crystals of hypogene origin in the ore body (Gilluly, 1937), and as large lilac-colored pieces up to 1½ inches across (Sinkankas, 1964) (H 107494). Pima district, as a hypothermal mineral at the Twin Buttes mine (Stanley B. Keith, pers. comm., 1973).

*Pinal County:* Mammoth district, San Manuel and Kalamazoo ore bodies, in the inner alteration zone of mineralized monzonite and quartz monzonite porphyry, associated with quartz, sericite, and sulfides (Lowell, 1968). A drill hole sunk by the Humble Oil and Refining Company in 1972 penetrated 80 feet of halite and about 6,000 feet of anhydrite, just west of the Picacho Mountains, Sec. 5, T.8S, R.8E. The lateral extent of this enormous body of salts is not known at present (H. Wesley Peirce, pers. comm., 1973).

## ANKERITE

Calcium iron magnesium manganese carbonate, $Ca(Fe,Mg,Mn)(CO_3)_2$. A member of the dolomite group of minerals in which there is extensive substitution among iron, magnesium, and manganese. Probably rather widely distributed in metamorphosed limestones and metallic veins in the state, but few localities have been noted.

*Gila County:* Globe-Miami district, as a manganian variety in the Ramboz deposit, with rhodochrosite (Peterson, 1962). Payson district, with tetrahedrite at the Silver Butte mine (Lausen and Wilson, 1927). At the Salt River, between Cibecue Creek and Salt River Draw, as drusy coatings on walls of the central fracture in the Tomato Juice and Rock Canyon deposits (Granger and Raup, 1969); also in the Sorrel Horse and Horse Shoe deposits.

*Pima County:* Esperanza mine, Pima district (UA 9371). Ajo district, New Cornelia ore body, in veins or as druses coated by calcite (Gilluly, 1937).

*Yavapai County:* In pyritic ore of the United Verde mine. Bradshaw Mountains, Big Bug district, at the Arizona National mine; Iron King mine, where it was formed during the alteration of a massive sulfide deposit, as disseminated grains and in veinlets (Creasey, 1952). Black Canyon district, Howard Copper and Kay Copper properties. M and M veins, Tiger district. Tillie Starbuck mine, Hassayampa district, as small rhombs associated with dolomite in cavities.

*Yuma County:* Northern Mohawk Mountains, as thin brownish-gray veins cutting schist and gneiss (Wilson, 1933).

## *ANTHONYITE

Copper hydroxide chlorite hydrate, $Cu(OH,Cl)_2 \cdot 3H_2O$. A rare, water-soluble, secondary species, of which this is only the second reported occurrence.

*Cochise County:* Warren district, 1,300 level of the Cole mine, where it occurs as large (5 mm or more) corroded crystals of a vivid violet color. These encrust crumbly pyritic ores and are associated with an unknown copper hydroxide. Chemical analysis shows this anthonyite to be virtually halogen-free in contrast to the type locality mineral. The occurrence may, in some way, be related to burning of sulfide ores in a nearby stope.

After the first few specimens were found, the occurrence was washed down prior to a collecting trip; all the remaining anthonyite was destroyed.

## ANTHOPHYLLITE

Magnesium iron silicate hydroxide, $(Mg,Fe)_7Si_8O_{22}(OH)_2$. An orthorhombic member of the amphibole group. A metamorphic mineral common in schists and gneisses; less common in contact metamorphosed rocks.

*Cochise County:* Cochise district, Johnson Camp, asbestiform, in a narrow vein cutting Horquilla Limestone (Cooper and Silver, 1964).

*Mohave County:* Hualpai Mountains, Antler

mine (Romslo, 1948). Kingman area, with phlogopite (UA 1193).

*Pima County:* Santa Rita Mountains, Blue Jay mine (UA 5233). Santa Catalina Mountains, Kielberg's Iron Mountain claim (UA 5288).

*Yavapai County:* Reported from the Eureka district.

## ANTIGORITE (see Serpentine)

## ANTLERITE

Copper sulfate hydroxide, $Cu_3(SO_4)(OH)_4$. A rare secondary mineral formed in the oxidized zone of copper deposits; easily mistaken for brochantite, which it resembles. The mineral takes its name from the Antler mine which is the type locality.

*Apache County:* Monument Valley, Monument No. 2 mine (UA 2279).

*Cochise County:* Warren district, as small crystals of excellent quality from an unspecified mine (probably the Copper Queen) emplanted on brochantite (Palache, 1939a) (S 95724); also from the west side of the Holbrook pit.

*Gila County:* Banner district, 79 mine, as pale green crusts in the 31 stope area (Thomas Trebisky, pers. comm., 1975).

*Graham County:* Lone Star district, in small amounts in veins in metamorphosed latites and andesites (Hutton, 1959b).

*Greenlee County:* Morenci district, reported from the Clay ore body (UA 6259).

*Mohave County:* Hualpai Mountains, at the Antler mine on material from which it was originally described as a new species by Hillebrand (1889a; Romslo, 1948); occurs as soft green lumps (S 6075, part of the type material). The original analyses by Hillebrand gave:

(1)

| CuO | 68.19 | CaO | 0.05 | SO₃ | 20.46 |
|-----|-------|-----|------|-----|-------|
| ZnO | 0.29  |     |      | H₂O | 11.11 |

TOTAL 100.10%

(2)

| CuO | 67.64 | CaO | 0.04 | SO₃ | 21.49 |
|-----|-------|-----|------|-----|-------|
| ZnO | 0.04  |     |      | H₂O | 10.76 |

TOTAL 99.97%

Specific gravity: 3.93.

*Yavapai County:* Jerome, as highly perfect crystals up to 3 mm long on fracture surfaces in chlorite schist which contains chalcopyrite; abundant in the lower part of the oxidized zone, with cyanotrichite and brochantite.

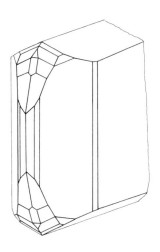

**Antlerite.**
Bisbee (Palache, 1939).

## APATITE

A group name, the most common members of which can be expressed by the general composition: calcium phosphate fluoride hydroxide chloride, $Ca_5(PO_4)_3(F,OH,Cl)$. Members of the group are the most common of the phosphorus-bearing minerals. Formed under a variety of conditions but particularly common as accessory minerals in igneous rocks, as tiny crystals. The name apatite is frequently used synonymously with fluorapatite, the most common member of the group. The particular species has apparently seldom been determined for the apatite minerals of the state.

*Apache County:* Garnet Ridge, an accessory mineral in ejection boulders of garnet gneiss, associated with zircon and muscovite (Gavasci and Kerr, 1968). Monument Valley, Monument No. 2 mine, as carbonate apatite — $Ca_{10}(PO_4)_6$-$(CO_3) \cdot H_2O$, probably related to bone material (Witkind and Thaden, 1963).

*Apache and Navajo Counties:* Monument Valley, as one of the most common heavy minerals in uranium-bearing Shinarump Conglomerate, associated with barite, leucoxene, tourmaline, and zircon (Young, 1964).

*Gila County:* Globe-Miami district, as a common accessory mineral in the igneous rocks of the district; also as veinlets of hydrothermal origin cutting the Scanlan Conglomerate and arkoses of the Pioneer Formation in the Castle Dome area (Peterson, 1962). Dripping Spring Mountains, Banner district, Christmas mine, as remnant euhedral grains in metamorphosed diorite (Perry, 1969).

*Graham County:* Aravaipa district, in micropegmatite at the Fisher prospect, Turnbull Mountain. Lone Star district, near Safford, as a minor constituent in metasomatized volcanic rocks (Hutton, 1959b).

*Greenlee County:* Clifton-Morenci district, as a primary mineral in green mica-bearing rock, as small rods embedded in green mica, with pyrite and magnetite (Reber, 1916).

*Mohave County:* Hualpai Mountains, as white fluorapatite crystals which exhibit golden yellow fluorescence, associated with muscovite. Aquarius Range, in pegmatite in a granite dike, with sphene, chevkinite, monazite, and cronstedtite (Kauffman and Jaffe, 1946).

*Pima County:* Ajo district, in minor amounts in altered rock (Hutton and Vlisidis, 1960). An accessory mineral in rocks of the Silver Bell area (Kerr, 1951).

*Pinal County:* Bunker Hill district, Copper Creek, Childs-Aldwinkle mine, as a gangue mineral in a breccia pipe; one crystal on the 820 foot level was 5 inches long (Kuhn, 1941) (UA 535); Old Reliable mine, with rutile (UA 9606). Mammoth district, in the potassium silicate phase of hydrothermal alteration of monzonite and quartz monzonite porphyries; also in the argillic phase of alteration (Creasey, 1959).

*Santa Cruz County:* Patagonia Mountains, as large crystals at the Four Metals mine (UA 9298); Palmetto district, as an accessory mineral in granite porphyry (Schrader, 1913).

*Yavapai County:* As large crystals in granodiorite, Springfield group, Pine Grove district. Eureka district, as an accessory mineral in titaniferous magnetite bodies near Bagdad (Ball and Broderick, 1919; Schwartz, 1947; Anderson, 1950). In the pegmatites of the White Picacho district (Jahns, 1952). Copper Basin district, as an accessory mineral in several of the igneous rocks (Johnston and Lowell, 1961). Big Bug district, Iron King mine, introduced during hydrothermal alteration of the massive sulfide deposit; as disseminated grains and needle-like crystals (Creasey, 1952).

## *APOPHYLLITE

Potassium calcium silicate fluoride hydroxide hydrate, $KCa_4Si_8O_{20}(F,OH) \cdot 8H_2O$. Apophyllite occurs primarily as a secondary mineral in amygdules in basalts, commonly associated with zeolites; it is also found in contact metamorphic limestones bordering intrusives.

*Cochise County:* Tombstone district, as subhedral grains with idocrase and wollastonite in tactite (Bideaux et al., 1960).

*Gila County:* Banner district, coating kinoite at the Christmas open pit mine.

*Pima County:* Santa Rita Mountains, near Helvetia, where it was observed in a drill core which penetrated skarn developed in paleozoic limestones and dolomites; as vuggy, crystalline masses in which are embedded copper and kinoite (Anthony and Laughon, 1970).

## ARAGONITE

Calcium carbonate, orthorhombic $CaCO_3$. Trimorphous with calcite and vaterite. Metastable under standard conditions, tending to revert to the stable calcite. Formed in spring deposits and from sulfate-bearing saline solutions; in beds with gypsum, in cavities in lavas, and in limestone caverns.

*Cochise County:* Warren district, as magnificent coralloid groups of flos ferri, in limestone caverns. Dragoon Mountains, as stalactites and stalagmites lining solution cavities in silver-lead deposits, Turquoise district.

*Mohave County:* 35 miles southeast of Hackberry, in granite pegmatite, with gadolinite (Palmer, 1909).

*Pima County:* Santa Rita Mountains, as flos ferri in Onyx Cave. In the Pima district as fine crystals from the San Xavier West mine (Arnold, 1964) (UA 5782), from Mineral Hill (UA 7630, H 89687), and copper-stained in fissures in the Sierrita Mountains. Reported from the Silver Bell mine, Silver Bell Mountains (UA 6466).

*Yavapai County:* United Verde Extension mine, Jerome (UA 7626). As pseudohexagonal crystals from Castle Hot Springs (UA 220). From basalts northeast of Camp Verde (E. R. Brenizer, pers. comm., 1974) (UA 10185).

*Yuma County:* Castle Dome district, in channels and vugs with smithsonite, hydrozincite, wulfenite, vanadinite, and mimetite.

## *ARIZONITE (?)

A mineral species of questioned validity. First described by Palmer (1909) on material collected by F. L. Hess of the U.S. Geological Survey from a pegmatite about 25 miles southeast of Hackberry in Mohave County, where it occurred with gadolinite (H 86617). Palmer found the material to be an iron titanium oxide and ascribed to it the formula $Fe_2Ti_3O_9$. Subsequently Overholt and co-workers (1950) and Ernst (1943) reported that the material was a mixture of hematite, anatase, ilmenite, and rutile, and concluded that it was a weathering product of ilmenite, findings with which Bolfa et al. (1961) were in general agreement.

However, the experimental work of Karkhanavala (1959) led him to state that arizonite is "a specific and unique, though rare and unstable chemical compound." Flinter (1959) also believes that the material has validity as a distinct mineral species. Type material is represented by S 86973.

## ARGENTITE (see ACANTHITE)

## ARSENIC

Arsenic, As. Found in hydrothermal veins, most commonly associated with cobalt, nickel, and silver ores but also in other sulfide deposits.

*Santa Cruz County:* Patagonia district, Washington Camp, Double Standard mine, as reniform masses in metamorphosed dolomitic limestone; masses of 50 pounds or more were observed in the mine (Warren, 1903; Struthers, 1904; Guild, 1910; Schrader and Hill, 1915).

## *ARSENOLITE

Arsenic oxide, $As_2O_3$. The dimorph of claudetite. A secondary mineral formed by alteration of primary arsenides or arsenic-bearing sulfides; also as a product of mine fires.

*Yavapai County:* United Verde mine, Jerome, as octahedral crystals on "burned ore" matrix; formed during the mine fires (UA 6708).

## ARSENOPYRITE

Iron arsenic sulfide, FeAsS. The most common of the arsenic-bearing minerals. Formed under a wide variety of conditions: in high temperature gold-quartz veins, in contact metamorphic sulfide deposits; less commonly in pegmatites and low temperature veins.

*Maricopa County:* Sparingly in the pegmatites of the White Picacho district (Jahns, 1952). As the cobaltian variety, danaite, from an unspecified locality near Tempe (UA 6193).

*Mohave County:* Cerbat Range, in some of the mines of the Chloride and Mineral Park districts, notably the Minnesota-Conner, Windy Point, and Queen Bee properties (Schrader, 1909).

*Pima and Santa Cruz Counties:* Santa Rita and Patagonia Mountains, especially the Duquesne and Washington Camp areas, Patagonia district, in a number of contact metamorphic deposits; Mowry mine (Schrader, 1917).

*Yavapai County:* Sparingly at the United Verde mine (Lausen, 1928) and Shea property, Verde district. Bradshaw Mountains, as crystals in the Boggs mine. Big Bug district, Iron King mine, abundant in the massive sulfide ores of *en echelon* vein deposits, as subhedral grains up to 1.5 mm showing diamond-shaped sections (Creasey, 1952). Eureka district, in a vein with bismuthinite, near the Hillside mine (Axelrod et al., 1951). Sparingly in the pegmatites of the White Picacho district (Jahns, 1952). Near Prescott, as the cobaltian variety, danaite (UA 6193). In vein quartz with pyrite at the Old Dick mine as the cobaltian danaite. These danaite-bearing quartz veins may cut earlier massive arsenopyrite.

Asbestos (see Serpentine)

Attapulgite (see PALYGORSKITE)

## ATACAMITE

Copper chloride hydroxide, $Cu_2Cl(OH)_3$. A secondary mineral formed from the oxidation of other secondary copper minerals, especially under arid, saline conditions. Commonly associated with malachite, cuprite, chrysocolla, brochantite, gypsum, and limonite. Probably more widely represented in Arizona than would be indicated by the limited number of occurrences reported.

*Gila County:* Globe-Miami district, Castle Dome mine (H 108296); Inspiration mines (Olmstead and Johnson, 1966).

*Maricopa County:* Painted Rock Mountains, at the Rowley mine, near Theba, in very small amounts, associated with caledonite, and thought possibly to have been derived by alteration from caledonite or linarite (Wilson and Miller, 1974).

*Pima County:* Cerro Colorado Mountains, Cerro Colorado mine. Southern side of Saginaw Hill, about 7 miles southwest of Tucson, associated with other oxidized minerals, including brochantite, pseudomalachite, malachite, libethenite, cornetite, and chrysocolla (Khin, 1970). Cababi district, South Comobabi Mountains, Mildren and Steppe claims (Williams, 1963).

*Pinal County:* Galiuro Mountains, as small green crystals with olivenite on the main level of the Old Reliable mine, Copper Creek district. Mammoth district, Mammoth–St. Anthony mine, as coarse, granular aggregates of deep green color on specimens from the 400 foot level, Collins vein; San Manuel mine, as laths in chrysocolla and cornuite (a variety of chrysocolla),

**Atacamite and anglesite.**
Mammoth–St. Anthony mine, Pinal County.
R. A. Bideaux.

and along fractures (Bideaux et al., 1960; Thomas, 1966) (UA 3180 and others).

*Yavapai County:* Black Hills, in small quantities at the United Verde Extension mine (Guild, 1910).

## AUGITE

Calcium magnesium iron titanium aluminum silicate, $(Ca,Mg,Fe^{2+},Fe^{3+},Ti,Al)_2$-$(Si,Al)_2O_6$. A common rock-forming mineral of the pyroxene group which occurs in a wide variety of mafic igneous rocks, such as gabbro, diabase, and basalt. Only a few localities can be mentioned here.

*Apache County:* Monument Valley, Garnet Ridge, in a breccia dike associated with loose garnets and as fragments in sand and soil (Gavasci and Kerr, 1968).

*Cochise County:* Tombstone district, in diorite porphyry dikes and basaltic rocks (Butler et al., 1938).

*Gila and Pinal Counties:* Abundant in diabase sills which intrude rocks of the Apache Group over large areas in central Arizona. In the Dripping Spring Mountains, with olivine and iddingsite in Tertiary basalts.

*Navajo County:* North of Bidahochi Butte, as loose crystals in an unnamed diatreme (Bideaux et al., 1960).

*Santa Cruz County:* Patagonia Mountains, Duquesne and Washington Camp, in metamorphosed limestones (Schrader and Hill, 1915).

*Yavapai County:* Copper Basin district, as an accessory mineral in igneous rocks of the area (Johnston and Lowell, 1961). Big Bug district, Iron King mine, as phenocrysts in porphyritic basalt flows, with olivine (Creasey, 1952).

*Yuma County:* Reported to be abundant in metamorphosed limestones of the northern part of the county; this mineral may well be misidentified diopside.

## AURICHALCITE

Zinc copper carbonate hydroxide, $(Zn,Cu)_5$-$(CO_3)_2(OH)_6$. An uncommon secondary mineral found in the oxidized zones of lead and copper deposits with other oxidized minerals.

*Cochise County:* Warren district, in the upper portions of the Copper Queen ore body (Ransome, 1904; Guild, 1910), "in beautiful tubes lining cavities" (Kunz, 1885); with hemimorphite and calcite, 1,200–1,300 foot levels of the Cole shaft. Tombstone district, as plumose aggregates of pale blue crystals on the west side of the Quarry "roll" (Butler et al., 1938). Turquoise district, as incrustations and drusy linings of cavities in oxidized lead-silver deposits; Mystery mine (UA 8942). Little Dragoon Mountains, Cochise district, Johnson Camp, with chrysocolla, malachite, tenorite, native copper, and hemimorphite, in the oxidized portions of pyrometasomatic sulfide deposits (Cooper and Huff, 1951; Cooper and Silver, 1964).

*Coconino County:* Grand Canyon National Park, Horseshoe Mesa, Grandview (Last Chance) mine, on or near Red Wall Limestone, as greenish-blue, lath-like crystals forming tufted incrustations (Leicht, 1971).

*Gila County:* The beautiful aurichalcite specimens from the 79 mine, Banner district, are among the finest in the world, rivaling those from Mapimi, Mexico. The most notable occurrence is on the fourth level of the mine where the mineral occurs as delicate sprays of acicular and lath-like crystals which range in color from pale sky-blue to deep sea blue-green. The sprays typically occur on hemimorphite, smithsonite, and wulfenite, and, in turn, may have perched on them cerussite, calcite, plattnerite, or murdochite; formed contemporaneously with rosasite (UA 1201) (Keith, 1972).

*Pima County:* Empire Mountains, as small radiating fibrous masses and seams with smithsonite and hemimorphite in the Lone Mountain mine. Pima district, in fissures in garnetized limestone in the Sierrita Mountains, at the Queen mine at Twin Buttes (UA 5565), and at the San Xavier West mine (Arnold, 1964). Waterman Mountains, Silver Hill mine, with plattnerite and murdochite (Bideaux et al., 1960). Cababi district, South Comobabi Mountains, from the east Silver-Lead claim, as small (up to 1 mm) crystals on quartz (Williams, 1962).

*Pinal County:* Vekol Mountains, Reward

**Aurichalcite.**
79 mine, Gila County. Julius Weber.

mine. Mammoth district, Mammoth–St. Anthony mine, found on a few specimens as radiating sprays of crystals.

*Santa Cruz County:* Patagonia district, Flux mine, as light blue clots of rather thick acicular crystals on limonite, with hemimorphite (Thomas Trebisky, pers. comm., 1972).

*Yuma County:* Plomosa Mountains, Black Mesa mine, Sec. 16, T.3N, R.17W, as well-formed small crystals with crystalline malachite in limonite gangue with small brilliant hexagonal willemite crystals (David Shannon, pers. comm., 1972).

## *AUSTINITE

Calcium zinc arsenate hydroxide, $CaZn(AsO_4)(OH)$. A rare secondary mineral found in the oxidized zones of some base metal deposits.

*Pinal County:* Galiuro Mountains, Table Mountain mine, as intergrowths with conichalcite (Bideaux et al., 1960).

## AUTUNITE

Calcium uranyl phosphate hydrate, $Ca(UO_2)_2(PO_4)_2 \cdot 10–12H_2O$ (water content variable between about 10 and 12 molecules). The most common of the secondary uranium minerals, of wide distribution; occurs not uncommonly in the oxidized zones of hydrothermal mineral deposits.

*Apache County:* Monument Valley, Monument No. 2 mine, where it is associated with a wide variety of oxidized uranium and vanadium minerals (Witkind and Thaden, 1963).

*Gila County:* Red Bluff prospect, with uranophane and uraninite (R. Robinson, pers. comm., 1954).

*Maricopa County:* Lucky Find group, in a mafic dike cutting Precambrian granite (Robinson, 1956).

*Navajo County:* Monument Valley, Monument No. 1 and Mitten No. 2 mines (Witkind, 1961; Witkind and Thaden, 1963).

*Santa Cruz County:* With uranophane in a vein with lead ore, near Alamo Spring. Santa Rita Mountains, Duranium claims, in arkosic sandstone with kasolite and uranophane (Robinson, 1954).

## *AXINITE
## (see also MANGANAXINITE and TINZENITE)

Calcium manganese iron aluminum borosilicate hydroxide, $(Ca,Mn,Fe^{2+})_3Al_2BSi_4O_{15}(OH)$. Most common occurrence is in contact metamorphic aureoles where intrusive rocks have invaded sediments, especially limestones; commonly associated with other calcium silicate minerals.

*Graham County:* Landsman's Camp, Aravaipa Canyon, as crystals up to 5 mm across on johannsenite (UA 9480).

## AZURITE

Copper carbonate hydroxide, $Cu_3(CO_3)_2$-$(OH)_2$. A widely distributed secondary mineral in the oxidized zones of copper deposits; frequently associated with malachite, to which it alters, cuprite, native copper, and limonite. Widespread but usually sparse in the copper deposits of the State.

*Cochise County:* Warren district, as magnificent crystallized specimens of extraordinary size (to 4½ inches) at the Copper Queen mine, one of the world's premier localities (Kunz, 1885; Douglas, 1899; Ransome, 1903, 1904)

(UA 79, 177, 736, 1153, etc.; BM 81130, 1968, 624). Excellent crystallized specimens were noted from the deeper workings of the mine "implanted as a secondary growth in parallel position upon well-formed pseudomorphs of malachite after azurite of longer dimensions (length parallel to *b* axis, 7 cm)" (Palache and Lewis, 1927); some crystals exhibit a blue exterior but have green cores (Schwartz and Park, 1932); Junction mine (Mitchell, 1920). Tombstone district, Lucky Cuss and Toughnut mines (Butler et al., 1938). Turquoise district, as large crystallized masses at the Maid of Sunshine mine. Cochise district, Little Dragoon Mountains, in the Johnson Camp area (Kellogg, 1906).

*Coconino County:* Kaibab Plateau, as extensive impregnations in chert beds; Apex copper property (Tainter, 1947a). Grand Canyon National Park, Horseshoe Mesa, Grand View (Last Chance) mine, as short prismatic crystals on clay and in vugs in sandstone; some crystals of azurite altering to malachite are up to 3 by 1 cm in size (Leicht, 1971). Cameron area, with malachite, anhydrite, and pyrite in mineralized silica plugs (Barrington and Kerr, 1963).

*Gila County:* Globe-Miami district, a not uncommon secondary mineral in many deposits of the area (Woodbridge, 1906; Schwartz, 1921, 1934); Castle Dome mine (Peterson, 1947); Copper Cities deposit, where it is associated with malachite and turquoise (Peterson, 1954); Sleeping Beauty mine (UA 9707); "Blue Ball" mine, as nodular concretions of considerable size (Sinkankas, 1964), often with a central cavity containing malachite crystals. Banner district, 79 mine (Kiersch, 1949; Keith, 1972). Payson district, as crystallized masses at the Silver Butte, Golden Wonder, and Bishops Knoll mines.

*Greenlee County:* Clifton-Morenci district, as large bodies in the Longfellow, Detroit, Man-

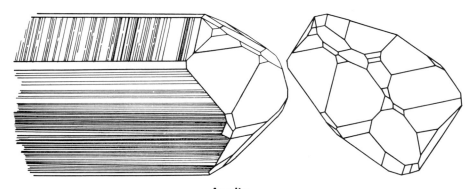

**Azurite.**
Bisbee (Palache, 1927).

ganese Blue, and Shannon mines; sheaf-like and spherical masses up to 40 pounds in weight were found in kaolinized shale (Kunz, 1885; Farrington, 1891; Lindgren, 1903, 1904, 1905; Reber, 1916; Schwartz, 1934).

*Maricopa and Gila Counties:* Mazatzal Mountains, at Pine Mountain, near Mount Ord and Saddle Mountain, on Alder and Slate Creeks, found with malachite and chalcopyrite in mercury deposits in a schist belt (Lausen, 1926).

*Mohave County:* Cerbat Mountains, a sparse mineral, widely distributed throughout the Wallapai district where it is associated with sulfide vein deposits and disseminations (Thomas, 1949). Bentley district, at the Grand Gulch, Bronze L and Copper King mines (Hill, 1914).

*Pima County:* Santa Rita Mountains, widely distributed at mining properties throughout the range including, primarily, the Old Baldy district (Schrader, 1917); as fine crystalline specimens from the Copper Mountain prospect of the Anaconda group (Schrader and Hill, 1915). Helvetia, as massive material, crystals, and as pseudomorphs of malachite after azurite to about one-half inch. Sierrita Mountains, common in the Pima district; as fine rosettes from the Banner mine (UA 6759). Santa Catalina Mountains, Cañada del Oro (UA 4126). Cababi district, South Comobabi Mountains, Mildren and Steppe claims (Williams, 1963). Ajo district, as granular aggregates having concentric structures, with malachite and quartz, also as pseudo-cubic crystals (Schwartz, 1934). Waterman Mountains, Silver Hill mine (UA 1646). Silver Bell Mountains, Silver Bell mine, with cuprite and malachite in ore bodies in limestone (*Eng. Min. Jour.*, 1904).

*Pinal County:* Pioneer district, as small but beautifully crystallized groups in the open cut of the Silver King mine. Galiuro Mountains, Copper Creek district, Childs-Aldwinkle mine (Kuhn, 1941). Mineral Creek district, northeast of Ray Hill and near the Ray deposit, one of several oxidized minerals including malachite, cuprite, tenorite, jarosite, and goethite in Holocene gravels, locally replacing a fossil log; crystals up to 3 mm long (Phillips et al., 1971). Mammoth district, San Manuel mine, present in small amounts in the oxidized portion of the ore body, with chrysocolla, malachite and cuprite (Schwartz, 1949); Mammoth–St. Anthony mine, as deep blue crystals to two inches, associated with reticulated cerussite, and as pseudomorphs of malachite after azurite (UA 9239; 8549).

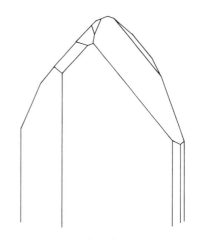

**Azurite.**
Mildren and Steppe area (Williams, 1962).

*Santa Cruz County:* Patagonia district, at Duquesne and Washington Camp (Schrader, 1917). Tyndall district (Schrader, 1917).

*Yavapai County:* Black Hills, as fine quality specimens from the Yeager mine, Black Hills district (Anderson and Creasey, 1958). Copper Basin district, found in the oxidized zones of a number of sulfide ore deposits in the area (Johnston and Lowell, 1961); botryoidal specimens have been found at Sec. 20 and 21, T.13N, R.3W (David Shannon, pers. comm., 1971). Good crystallized material has been found lining vugs and as small, radiating, spherical aggregates irregularly distributed with malachite in limonitic clay in the replacement ore body of the United Verde Extension mine at Jerome (Schwartz, 1938).

*Yuma County:* Muggins Mountains, Red Knob claims, where it is associated with weeksite, opal, vanadinite, carnotite, gypsum, and calcite (Outerbridge et al., 1960). Gila Mountains, Blue Butte vein (Wilson, 1933).

# B

## *BABINGTONITE

Calcium iron manganese silicate hydroxide, $Ca_2(Fe^{2+},Mn)Fe^{3+}Si_5O_{14}(OH)$. An uncommon mineral known from relatively few localities about the world. Found in skarns, granite pegmatites, and hydrothermal veins; generally believed to be of low temperature origin.

*Graham County:* Santa Teresa Mountains, Aravaipa district, Landsman claims, as dark brown, bladed, euhedral crystals up to 7 mm long. Found in a vein with galena and sphalerite in a contact metamorphic zone developed in limestone and shale: also associated with euhedral diopside-hedenbergite and yellow garnet (material collected by Raymond Rhodes and Richard L. Schick, 1972).

## BARITE

Barium sulfate, $BaSO_4$. A widespread, relatively common mineral which forms under low to moderate temperature conditions in hydrothermal veins; also in sedimentary rocks as replacements, veins, and cavity fillings formed by hypogene or meteoric solutions. Barite occurrences of the state are summarized by Stewart and Pfister (1960).

*Apache and Navajo Counties:* Monument Valley, with the uranium-vanadium ores in Shinarump conglomerate-filled channels at the base of the Chinle Formation (Mitcham and Evensen, 1955; Young, 1964).

*Cochise County:* Tombstone district, Ground Hog mine as a vein, and near the Lucky Cuss mine as white crystals (Butler et al., 1938; Stewart and Pfister, 1960). Western slope of the Dragoon Mountains, at the Johnnie Boy No. 1 claim as veins and replacements in limestone (Stewart and Pfister, 1960). Warren district, occasionally occurs with manganese ores (Palache and Shannon, 1920; Taber and Schaller, 1930). Hopeful claim, located north of the Mule Mountains in NW ¼, Sec. 4, T.22S, R.23E, as strong veins of nearly pure barite cutting Cretaceous sediments (Tenney, 1936; Funnell and Wolfe, 1964); Ramirez group of claims located on the north side of Gadwell Canyon in the foothills of the northern end of the Mule Mountains, in veins of nearly pure barite as much as three feet wide cutting Cretaceous sedimentary rocks (Stewart and Pfister, 1960).

*Coconino County:* Grand Canyon National Park, Horseshoe Mesa, Grandview (Last Chance) mine, in a fault zone in Red Wall Limestone, associated with aurichalcite (Leicht, 1971); Orphan mine (UA 7098).

*Gila County:* Richmond Basin, where it is abundant in veins. In small amounts with fluorite at the Castle Dome mine (Peterson et al., 1946, 1951). Reported in a vein about 500 feet long near Coolidge Dam, associated with galena. Superior district, Magma mine, as brilliant, tabular, brown to golden-yellow transparent crystals on chalcocite (Bideaux et al., 1960). South

of Payson "barite occurs sporadically in more or less parallel fractures for about 15 miles along an east-west zone in granitic rocks"; occurrences include the Top Hat, Gilmore Spring, Gisela, Baronite, Grey Fox, Zulu, Green Valley, and Lone Pine claims and groups of claims (Stewart and Pfister, 1960).

*Graham County:* Stanley district, abundant in veins near Stanley Butte (Ross, C. P., 1925a). Also reported in veins southeast of Klondyke on the Marcotte group of claims (Stewart and Pfister, 1960). At the Barium King group of claims, located in Sec. 19, T.4S, R.20E, and Sec. 24, T.4S, R.19E, about 5 miles east of Turnbull Mountain (Funnell and Wolfe, 1964).

*Greenlee County:* Ash Peak district, Luckie mine, associated with fluorite and psilomelane in a vein in andesite porphyry (Hewett, 1964).

*Maricopa County:* Goldfield Mountains, 14 miles north of Mesa in Secs. 4 and 5, T.2N, R.7E, as veins at the Granite Reef (Arizona Barite) mine (Wilson, 1944; Stewart and Pfister, 1960) (UA 5955). Painted Rock Mountains, Rowley mine, near Theba, as colorless crystals up to 2 cm long and 2 to 4 mm in thickness, often lining cavities (Wilson and Miller, 1974). Aguila district, Valley View mine, where it forms layers of crystals with fluorite which alternate with black calcite in a hypogene vein cutting an andesite flow (Hewett, 1964).

*Mohave County:* Reported as veins in the Aquarius Range. Cerbat Mountains, Wallapai district (UA 4925). Near Alamo Crossing, as veins. Artillery Mountains, with fluorite in veins with manganese oxides cutting the Artillery Formation; in the Barbee vein (Sec. 1, T.11N, R.14W), associated with chalcedony and mammilary layers of psilomelane(?) in basalt (Hewett and Fleischer, 1960). McCracken Mountains, McCracken mine, about 8 miles west of Signal, where it is reported to be present in large quantities in veins with quartz, carbonates, and galena (Bancroft, 1911; Funnell and Wolfe, 1964). On the Rucker group of claims located about 30 miles east of Kingman in Sec. 2, T.20N, R.12W, as sporadic pods, segregations, and stringers, and as veins up to 8 feet wide, in granitic rocks (Stewart and Pfister, 1960).

*Navajo County:* Monument Valley, Starlight No. 3 mine, as druses of minute brown crystals on fracture surfaces in sandstone (Atomic Energy Comm., Grand Junction office, mineral collection).

*Pima County:* Ajo district, sparingly in veins cutting Concentrator Volcanics and quartz monzonite (Gilluly, 1937). South Comobabi Mountains, Cababi district, at the Mildren and Steppe claims (Williams, 1963). Picacho de Calera Hills. One half mile east of Colossal Cave, at the Heavy Boy mine, as cleavable white masses (Stewart and Pfister, 1960). Silver Bell Mountains, found in House Canyon as large, bladed crystals (Kerr, 1951); coating fractures in the Silver Bell mine. Santa Rita Mountains, as an abundant gangue mineral in some sulfide deposits (Schrader and Hill, 1915; Schrader, 1917). Quijotoa Mountains, White Prince claims (Stewart and Pfister, 1960), the Morgan mine (UA 7036) and at the Quijotoa mine (Sec. 33, T.15S, R.2E) (Stewart and Pfister, 1960; Funnell and Wolfe, 1964), where bladed crystals form the matrix for galena crystals. As large, quartz-encrusted bladed crystals from the northwest ridge of the Santa Catalina Mountains (UA 9601); Santa Catalina Mountain foothills, north of Campbell Avenue, Tucson. Coyote Mountains (UA 1145).

*Pinal County:* Mammoth district, in a vein with calcite and psilomelane on the west side of Tucson Wash (Schwartz, 1953); Mammoth–St. Anthony mine, as groups of large, tabular crystals (Peterson, 1938). Galiuro Mountains, as tabular crystals and crystal groups from the Old Reliable mine, Copper Creek district. Pioneer district, Magma mine, as crystals on chalcocite (UA 2015); Gonzales Pass deposit (Secs. 16 and 17, T.2S, R.11E) in a vein following a fault in Pinal Schist (Stewart and Pfister, 1960).

*Santa Cruz County:* Santa Rita and Patagonia Mountains, one of the principal gangue minerals in base and precious metal deposits, commonly associated with quartz, fluorite, rhodochrosite, and other carbonates, notably in the Tyndall, Old Baldy, and Patagonia districts (Schrader, 1917).

*Yavapai County:* Bradshaw Mountains, as a gangue mineral in a number of properties of the districts; French Creek deposit (Sec. 29, T.9N, R.1W) (Stewart and Pfister, 1960). Eureka district, Bagdad mine (Anderson, 1950). Bullard district, Hatton mine (Hewett and Fleischer, 1960; Hewett, 1964). At the Red Chief prospect, near Bouse, with fluorite (UA 2716). With native silver, chalcopyrite, and nickel arsenides at the Monte Cristo mine, near Wickenburg (Bastin, 1922); on the MGM claims, a few miles northeast of Wickenburg, as filling and replacement in a brecciated zone in volcanic breccia and granite (Stewart and Pfister, 1960).

*Yuma County:* Trigo Mountains, Mendevil

claims, in veins. Castle Dome Mountains, in many veins, and as large, clear crystals with wulfenite and fluorite, Castle Dome mines (Foshag, 1919). Mohawk Mountains, at the Barite mine as white to pink, radiating crystal aggregates in calcite veins. At Cottonwood Pass, near Salome (Wilson, 1944). Bouse district, east slope of the Plomosa Mountains at a number of deposits, in veins which form an arc around veins of manganese oxides, in layered volcanic host rock (Stewart and Pfister, 1960; Hewett, 1964).

## *BASALUMINITE

Aluminum sulfate hydroxide hydrate, $Al_4(SO_4)(OH)_{10} \cdot 5H_2O$. A rare secondary mineral associated with other sulfates.

*Cochise County:* Warren district, as white earthy masses with hydrobasaluminite and other aluminum sulfates encrusting silicified limestones in the Holbrook pit.

## *BASSANITE

Calcium sulfate hydrate, $2CaSO_4 \cdot H_2O$. An uncommon mineral which occurs as pseudomorphs after gypsum in cavities in volcanic rocks and as a result of fumerolic activity. Also observed among alteration products in porphyry copper deposits.

*Pima County:* Sierrita Mountains, Sierrita open pit mine, as tiny, subhedral, fibrous crystals in veinlets filled with quartz, calcite, and chlorite in hydrothermally altered diorite and quartz diorite porphyry (Roger Lainé, written comm., 1972).

*Yuma County:* In veins in andesite with chrysocolla and tenorite in several small prospects near Bouse.

## BASSETITE

Iron uranyl phosphate hydrate, $Fe(UO_2)_2(PO_4)_2 \cdot 8H_2O$. A secondary mineral commonly associated with uraninite.

*Gila County:* A fairly common mineral in the uranium deposits in Dripping Spring Quartzite of northwest Gila County; at the Sue mine, associated with saleeite, and, locally, metanovacekite (Granger and Raup, 1969); Sierra Ancha Mountains, Red Bluff mine (Granger and Raup, 1962).

## *BAYLDONITE

Lead copper arsenate hydroxide, $PbCu_3(AsO_4)_2(OH)_2$. A rare oxide zone mineral found only in a few localities.

*Pima County:* Found at the Frijole prospect in the Santa Rita Mountains (T.18S, R.15E, sec. 24). Fairly common in coarse milky vein quartz and derived from sulfosalts. Associated with bindheimite, cerussite, wulfenite, and mimetite.

## BAYLEYITE

Magnesium uranyl carbonate hydrate, $Mg_2(UO_2)(CO_3)_3 \cdot 18H_2O$. A rare, water-soluble, secondary mineral found with other secondary uranium minerals, and gypsum as an efflorescence in a tunnel at the Hillside mine at Bagdad, which is the type locality.

*Cochise County:* Warren district, Cole mine (Richard Graeme, pers. comm., 1973) (UA 10078).

*Yavapai County:* As an eighth-inch-thick efflorescence on the walls of the Hillside mine, associated with schoeckingerite, andersonite, swartzite, gypsum, johannite, and uraninite; occurs locally as well-formed crystals (Axelrod et al., 1951). Chemical analyses by F. S. Grimaldi gave the following results:

(1)

| MgO | 9.03 | $UO_3$ | 30.80 | $H_2O$ | 35.19 |
|-----|------|--------|-------|--------|-------|
| CaO | 3.42 | $CO_2$ | 14.60 | Acid | |
| $Na_2O$ | — | $SO_3$ | 4.43 | Insol. | 2.27 |
| $K_2O$ | — | | | Ignited | |

TOTAL 99.74%

(2)

| MgO | 8.97 | $UO_3$ | 32.42 | $H_2O$ | 36.60 |
|-----|------|--------|-------|--------|-------|
| CaO | 2.75 | $CO_2$ | 15.36 | Acid | |
| $Na_2O$ | 0.19 | $SO_3$ | 3.95 | Insol. | 0.45 |
| $K_2O$ | 0.09 | | | Ignited | |

TOTAL 100.78%

Specific gravity: 2.05.

## BEAVERITE

Lead copper iron aluminum sulfate hydroxide, $Pb(Cu,Fe,Al)_3(SO_4)_2(OH)_6$. An uncommon secondary mineral formed in the oxidized portion of lead-copper deposits in arid regions; commonly occurs with plumbojarosite.

*Cochise County:* Tombstone district, in small quantities with cerussite in the Empire and Toughnut mines (Butler et al., 1938).

*Pinal County:* Mammoth district, Mammoth–St. Anthony mine, rarely as shining golden-yellow scales around the bases of linarite crystals.

## BECQUERELITE

Calcium uranium oxide hydrate, $CaU_6O_{19} \cdot 11H_2O$. A secondary mineral usually closely associated with uraninite from which it is commonly derived; associated with other secondary uranium minerals such as schoepite, fourmarierite, and curite.

*Apache County:* Monument Valley, Monument No. 2 and Cato Sells mines, where it is an important ore mineral; associated with uraninite (Frondel, 1956; Finnell, 1957; Witkind and Thaden, 1963).

## BEIDELLITE

Sodium calcium aluminum silicate hydroxide hydrate, $(Na,Ca/2)_{0.33}Al_2(Al,Si)_4O_{10}(OH)_2 \cdot nH_2O$. There is concern as to the propriety of retention of the name beidellite by some authorities (see, for example, Grim, 1968) because of uncertainties as to the purity of the material on which the species was originally described. The name is retained by Fleischer (1975), and it is a member of the montmorillonite (smectite) group of clay minerals. Beidellite occurs as a component of bentonitic clays and is common in hydrothermally altered areas associated with mineral deposits.

*Gila County:* Globe-Miami district, Castle Dome mine, where it occurs in veinlets, fine aggregates, and as scattered flakes in the altered quartz monzonite (Peterson et al., 1946; 1951).

*Greenlee County:* Morenci district, in less intensely altered zones in the vicinity of intense hydrothermal alteration of porphyry copper ores, associated with kaolinite, montmorillonite, and allophane (Schwartz, 1947, 1958).

*Pima County:* Ajo district, New Cornelia mine, commonly occurs as pseudomorphic replacement of plagioclase phenocrysts in the outlying parts of the mineralized area (Gilluly, 1937).

*Pinal County:* Mammoth district, San Manuel mine, a minor constituent of hydrothermally altered quartz monzonite and monzonite porphyries, associated with kaolinite, hydromuscovite, leucoxene, and rutile in a quartz-sericite-pyrite-chalcopyrite aggregate (Schwartz, 1947).

## *BEMENTITE

Manganese silicate hydroxide, $Mn_8Si_6O_{15}$-$(OH)_{10}$. An uncommon mineral known to occur in contact metamorphosed manganiferous limestones.

*Pinal County:* Mineral Creek district, Ray mines, near the Emperor tunnel, as minute clear platy crystals with hematite in oxidized capping in the Granite Mountain Porphyry (Loghry, 1972).

## *BERLINITE

Aluminum phosphate, $AlPO_4$. A very rare mineral reported from the Westana iron mine, Kristianstad, Sweden, where it occurs with other phosphate minerals.

*Gila County:* Globe-Miami district, Inspiration mine, as a single light brown, hollow pod less than 2 mm in size, completely enclosed by hematite in oxidized and leached capping (Loghry, 1972).

## BERMANITE

Manganese phosphate hydroxide hydrate, $Mn^{2+}Mn_2^{3+}(PO_4)_2(OH)_2 \cdot 4H_2O$. A rare mineral formed in granite pegmatites. The 7U7 ranch area is the type locality.

*Yavapai County:* On the 7U7 ranch property near the Bagdad Copper mine, in narrow veins cutting a spherical mass of triplite which occurs as a segregation in pegmatite lenses in granite; as single crystals and subparallel aggregates with fan-shaped or rosette-like appearance (Hurlbut, 1936). Associated minerals include phosphosidcrite (metastrengite), leucophosphite, hureaulite, and other phosphate minerals (Leavens, 1967; Hurlbut and Aristarian, 1968). A more recent analysis of the type bermanite by Jun Ito (Hurlbut and Aristarian, 1968) gave the following result:

| | | | | | |
|---|---|---|---|---|---|
| $Al_2O_3$ | 0.19 | MnO | 12.8 | $Na_2O$ | — |
| $Fe_2O_3$ | 3.2 | MgO | 1.05 | $H_2O$ | 20.2 |
| $Mn_2O_3$ | 30.6 | CaO | 0.75 | $P_2O_5$ | 31.4 |

TOTAL 100.19%

## BERTRANDITE

Beryllium silicate hydroxide, $Be_4Si_2O_7(OH)_2$. Occurs in granite pegmatites, associated with beryl.

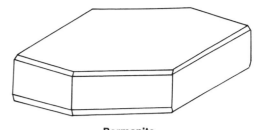

**Bermanite.**
7U7 Ranch area (Hurlbut, 1936).

*Maricopa County:* Reported from the Independence claim, White Picacho district.

## BERYL

Beryllium aluminum silicate, $Be_3Al_2Si_6O_{18}$. An important ore mineral of beryllium, most commonly found in cavities in granite and pegmatite; also associated with tin ores and mica schists.

*Cochise County:* Swisshelm Mountains, east of Elfrida, as euhedral crystals in vugs with fluorite and as beryl-muscovite masses in veins (Diery, 1964; Staatz et al., 1965). Dragoon Mountains, as small, colorless crystals with fluorite at the Boericke tungsten property; also at the Gordon and Abril mines in garnet tactite (Balla, 1962; Shawe, 1966). Beryl Hill claims; Tungsten Blue Bird mines, in scheelite-bearing quartz veins; in the Texas Canyon stock, near Dragoon. Little Lulu and Silver Drip claims, the latter as aquamarine in aplite dikes in granite; Thompson Beryl claims (Meeves, 1966).

*Graham County:* At Goodwin Wash (Shawe, 1966).

*Maricopa County:* Near Aguila (UA 6664).

*Mohave County:* Aquarius Range, as crystals in pegmatite, Rare Metals mine and the Columbite prospect (Heinrich, 1960). Beryl Wash, near Kingman, as greenish-blue crystals up to 12 inches long and 4 inches in diameter. G and M pegmatite, 15 miles southwest of Wikieup, as bluish-green crystals up to 9 feet long and 18 inches in diameter at the Boriana mine (Hobbs, 1944). 15 miles south of Peach Springs, 4 miles east-southeast of Wright Creek ranch, in several irregular pegmatite dikes in schist and gneissic granite, with microcline, albite, quartz, fluorite, schorl, sphene, and muscovite; locally, as well-formed prismatic crystals (Schaller et al., 1962). This beryl is quite unusual in its chemical composition and has been assigned the formula

$(Na,Cs)Be_3Al(Fe^{2+},Mg)Si_6O_{18}$. The average of several analyses follows.

| | | | | | |
|---|---|---|---|---|---|
| $SiO_2$ | 59.52 | FeO | 2.24 | $K_2O$ | 0.16 |
| BeO | 12.49 | MnO | 0.29 | $Cs_2O$ | 6.68 |
| $Al_2O_3$ | 10.63 | MgO | 2.16 | CaO | 0.11 |
| $Fe_2O_3$ | 2.08 | $TiO_2$ | 0.05 | BaO | — |
| $Cr_2O_3$ | 0.09 | $Li_2O$ | 0.23 | $H_2O$ | 1.62 |
| $Sc_2O_3$ | 0.10 | $Na_2O$ | 1.16 | $P_2O_5$ | 0.27 |

TOTAL 99.88%

This is the lowest silica content recorded for beryl.

Cerbat Range, in pegmatite (UA 6309); Chloride district, with gadolinite in pegmatite (Thomas, 1953). Virgin Mountains, Hummingbird claims, where it locally constitutes up to 3 percent of quartz-muscovite pegmatite; in crystals up to 15 inches long (Olson and Hinrichs, 1960). Hualpai Mountains, Kingman district (Meeves et al., 1966).

*Pima County:* Sierrita Mountains, as massive material and as beautiful blue-green crystals on the Bella Donna claim, near the Frielinger Feldspar property, in quartz veins in granite; gem quality aquamarine crystals affording stones to 40 carats have been found in some quantity from pegmatites in the vicinity of the Sierrita Mountains. Reported from the Baboquivari Mountains (Shawe, 1966). Agua Verde, in a vein in granite (Staatz et al., 1965). Santa Catalina Mountains, Apache Peak, in quartz veins (Robert O'Haire, pers. comm., 1972).

*Yavapai County:* Eureka district, Black Pearl mine, in veins which traverse biotite granite (Brownell, 1959; Dale, 1961; Staatz et al., 1965); Bagdad vicinity. Bradshaw Mountains, 4 miles southeast of Wagoner, in pegmatite veins. Peck mining district, in a pegmatite dike, 3 miles east of the Crown King post office. White Picacho district, as crystals up to 11 inches long in pegmatites; associated with lithium minerals (Jahns, 1952; Meeves et al., 1966). Lawler Peak area, in tabular masses of quartz.

*Yuma County:* Gila Mountains, about one and one-half miles east of the Fortuna mine, as small, lavender, and rose-colored crystals in a matrix of yellowish quartz. Associated with this beryl are small masses of an unidentified black mineral containing columbium.

## *BETA-URANOPHANE

Calcium uranyl silicate hydrate, $Ca(UO_2)_2$-$Si_2O_7 \cdot 6H_2O$. A secondary mineral, dimorphous with uranophane.

*Coconino County:* Near Cameron (Bollin and Kerr, 1958).

*Gila County:* Found in Dripping Spring Quartzite in the Red Bluff, Lucky Stop, and Hope deposits, in northwest Gila County (Granger and Raup, 1969).

## BEUDANTITE

Lead iron arsenate sulfate hydroxide, $PbFe_3$-$(AsO_4)(SO_4)(OH)_6$. A rare secondary mineral found in the oxidized portions of mineralized veins.

*Yavapai County:* Pine Grove district, Crown King mine, as greenish crusts on vein matter, with jarosite and iron oxides.

## BEYERITE

Calcium lead bismuth carbonate oxide, $(Ca,Pb)Bi_2(CO_3)_2O_2$. A secondary mineral formed by the alteration of bismutite or other primary bismuth minerals, usually found in pegmatites.

*Maricopa and Yavapai Counties:* White Picacho district, as grayish green films on bismutite and bismuth(?), and as dense masses and pearly white flakes in small cavities (Jahns, 1952).

## *BIDEAUXITE

Lead silver chloride fluoride hydroxide, $Pb_2$-$AgCl_3(F,OH)_2$. An extremely rare secondary mineral known only from the Mammoth–St. Anthony mine where it occurs in the oxide zone with other secondary lead minerals.

*Pinal County:* Mammoth district, Mammoth–St. Anthony mine, as colorless crystals, 2 to 7 mm in maximum dimension, found enveloping and replacing boleite in the oxide zone of the Collins vein(?); associated also with cerussite and galena, from which it was ultimately derived, and with leadhillite, matlockite, and anglesite. The type specimen is S 114583 (Williams, 1970). Chemical analysis gave:

| | | | | | |
|---|---|---|---|---|---|
| Pb | 62.66 | Cl | 14.74 | F | 3.26 |
| Ag | 15.74 | | | OH | (2.78) |

TOTAL 99.18%

Specific gravity: $6.274 \pm 0.008$.

## *BIEBERITE

Cobalt sulfate hydrate, $CoSO_4 \cdot 7H_2O$. A moderately soluble, uncommon mineral formed through oxidation of cobalt-bearing sulfide and arsenide ores.

*Coconino County:* Cameron area, as pink- to rose-colored efflorescences in most of the mines in the Petrified Forest Member of the Chinle Formation (Austin, 1964). Huskon No. 1 mine, where the powdery mineral stains cross-vein fibrous gypsum or halotrichite (H, unnumbered specimen).

*Pima County:* Ajo district, found as a pink efflorescence in a ditch draining the approach to the New Cornelia open pit mine (Gilluly, 1937).

## BINDHEIMITE

Lead antimony oxide hydroxide, $Pb_2Sb_2O_6$-(O,OH). An uncommon secondary mineral generally in small quantities in oxidized antimonial lead ores; often replaces tetrahedrite.

*Cochise County:* Tombstone district, as yellowish gray spots in siliceous ores (Butler et al., 1938).

*Santa Cruz County:* Patagonia Mountains, in small amounts at the Mowry mine (Schrader and Hill, 1915; Schrader, 1917).

## BIOTITE

Potassium magnesium iron aluminum silicate hydroxide fluoride, $K(Mg,Fe)_3(Al,Fe)Si_3O_{10}$-(OH,F)$_2$. An important and abundant rock-forming mineral; the most common of the mica minerals. Formed under a wide variety of conditions and found in rocks of nearly all types. Also a product of hydrothermal alteration in mineral deposits. Widespread in the rocks of the state.

## *BIRNESSITE

Sodium calcium manganese oxide hydrate, $(Na,Ca)Mn_7O_{14} \cdot 3H_2O$. An uncommon mineral apparently of secondary origin formed by the breakdown of primary manganese minerals; it may be associated with other manganese oxide minerals, rhodonite, and rhodochrosite.

*Cochise County:* Turquoise district, as a constituent of black, scummy crusts on altered quartz monzonite from several prospects immediately north of Courtland. Todorokite is the other major component of these crusts.

## BISBEEITE (?)

Copper silicate hydrate, $CuSiO_3 \cdot H_2O$(?). A rare, ill-defined, secondary mineral. Defined by Schaller in 1915 from the Shattuck mine at Bisbee, the species has been reviewed by Laurent and Pierrot (1962) who studied type material and compared it with African occurrences at Kambove, Katanga, and Rénéville, Congo. Their studies confirmed bisbeeite as a distinct species and identical to material from the African locali-

ties to which they assigned the formula (Cu,Mg)-$SiO_3 \cdot nH_2O$. After study of material from the Azurite mine (see *Pinal County,* below) and a review of the literature, Oosterwyck-Gastuche (1967), however, concluded bisbeeite to be of doubtful validity.

*Cochise County:* Warren district, Shattuck mine, as pseudomorphs after shattuckite which is, in turn, pseudomorphous after malachite (Schaller, 1915) (H 87653 includes part of the type material). Chemical analyses by W. T. Schaller (Wells, 1937) gave the following results:

(1)

| | | | | | |
|---|---|---|---|---|---|
| SiO$_2$ | 36.71 | CaO | 0.48 | H$_2$O(+) | 8.32 |
| Fe$_2$O$_3$ | 1.31 | H$_2$O(−) | 4.37 | CuO | 49.45 |

TOTAL 100.64%

(2)

| | | | | | |
|---|---|---|---|---|---|
| SiO$_2$ | 36.93 | CaO | 0.39 | H$_2$O(+) | 7.46 |
| Fe$_2$O$_3$ | — | H$_2$O(−) | 8.56 | CuO | 46.66 |

TOTAL 100.00%

Specific gravity: (1) 3.05; (2) 2.88.

*Pima County:* Ajo district, in veins at the New Cornelia mine.

*Pinal County:* Tortolita Mountains, Azurite mine, associated with plancheite, shattuckite, malachite, and chrysocolla in quartz (Bideaux et al., 1960) (UA 6342).

## BISMITE

Bismuth oxide, $Bi_2O_3$. A secondary mineral formed by the oxidation of other bismuth minerals. The validity of the species at some of the occurrences listed may be open to question.

*Maricopa and Yavapai Counties:* As massive material in the pegmatites of the White Picacho district (UA 8970).

*Yavapai County:* Bradshaw Mountains, as an alteration product of bismuthinite at the Swallow mine, Castle Creek district (Lindgren, 1926). Eureka district, Bagdad mine. Bumblebee area (UA 6113).

*Yuma County:* Reported from a locality north of Vicksburg.

## BISMUTH

Bismuth, Bi. An uncommon primary mineral found in hydrothermal veins, pegmatites, and topaz-bearing quartz veins.

*Maricopa County:* Vulture district, Cleopatra mine. At a locality southeast of Granite Reef dam.

*Mohave County:* Aquarius range, 30 miles south of Hackberry, with gadolinite in pegmatite.

*Pima County:* Sierrita Mountains, Esmeralda mine (UA 7163).

*Yavapai County:* Bradshaw Mountains, in the Humbug Creek placers and on Minnehaha Flats. Reported from Buckhorn Wash, east of Brooks Hill. A rare constituent of the pegmatites of the White Picacho district; as thin flakes and irregular masses, one of which weighed 2½ pounds (Jahns, 1952).

## BISMUTHINITE

Bismuth sulfide, $Bi_2S_3$. A comparatively rare mineral formed under moderately high temperature conditions in hydrothermal veins and in tactites, often with chalcopyrite.

*Cochise County:* As bladed crystals up to 1 inch long in chalcopyrite, with pyrophyllite, from an unspecified locality (S 269).

*Mohave County:* Aquarius range, in small quantities with native bismuth and gadolinite in pegmatite.

*Yavapai County:* Bradshaw Mountains, Swallow mine, Castle Creek district, altering to bismite (Lindgren, 1926). Eureka district, 45 miles west of Prescott, in pegmatite (Dale, 1961). Midnight Owl mine, White Picacho district, in pegmatite (Jahns, 1952).

## BISMUTITE

Bismuth carbonate oxide, $Bi_2(CO_3)O_2$. A secondary mineral formed by the alteration of bismuthinite, native bismuth, or other bismuth minerals.

*Maricopa County:* In the Outpost pegmatite, White Picacho district (Jahns, 1952).

*Mohave County:* Hualpai Mountains, east of Yucca.

*Yavapai County:* Eureka district, in prospects at the Granites, on the 7U7 Ranch as large (3 inch) pseudomorphs with bismuthinite (Frondel, 1943) (H 100738). An uncommon constituent of the pegmatites of the White Picacho district (Jahns, 1952).

## *BIXBYITE

Manganese iron oxide, $(Mn,Fe)_2O_3$. Commonly occurs in cavities in rhyolite with garnet, topaz, beryl, and hematite; also in metamorphosed manganese ores as at Langban, Sweden.

*Pinal County:* Encrusting and replacing spessartine, near Saddle Mountain, Winkelman area (UA 8967, 8968); also as loose crystals in stream beds (Richard L. Jones, pers. comm.).

## BOLEITE

Lead copper silver chloride hydroxide hydrate, $Pb_9Cu_8Ag_3Cl_{21}(OH)_{16} \cdot H_2O$. A rare secondary mineral formed in small amounts in oxidized lead-copper deposits.

*Gila County:* Globe-Miami district, Apache mine, with cerussite, brochantite, and matlockite (Bideaux et al., 1960).

*Maricopa County:* Painted Rock Mountains, Rowley mine near Theba, as minute spheres (0.01 to 0.03 mm in diameter) on cerussite; also associated with linarite, leadhillite, atacamite, diaboleite, and other oxidized zone minerals (Wilson and Miller, 1974).

*Pinal County:* Mammoth district, Mammoth–St. Anthony mine, as dark blue cubes often on diaboleite in the Collins vein; in a leadhillite vug in a small block faulted from the Mammoth vein, associated with cerussite, phosgenite, paralaurionite, quartz, hydrocerussite, diaboleite, matlockite, wherryite, and chrysocolla (Palache, 1941b; Fahey et al., 1950).

## *BOLTWOODITE

Potassium uranyl silicate hydroxide hydrate, $K_2(UO_2)_2(SiO_3)_2(OH)_2 \cdot 5H_2O$. A moderately common secondary mineral formed by oxidation of black primary uranium ores.

*Coconino County:* One of the more common uranium minerals in the Cameron area, as yellow areas in a blackish sandstone, Huskon Nos. 17 and 20 mines (Daphne Ross, cited *in* Honea, 1961); also at the Ramco Nos. 20 and 22 mines, Jack Daniels No. 1 and Yazzie No. 102 mines (Austin, 1964).

## BORNITE

Copper iron sulfide, $Cu_5FeS_4$. An important ore mineral of copper in many mines of the state. Commonly intimately associated with either chalcocite or chalcopyrite, usually with both. Principally of primary origin, but small amounts of secondary bornite are common in secondarily enriched ores; also in contact metamorphic deposits.

*Apache and Navajo Counties:* Monument Valley, with chalcopyrite, chalcocite, and native copper in sandstone, associated with uranium-vanadium ores at several properties (Evensen and Gray, 1958).

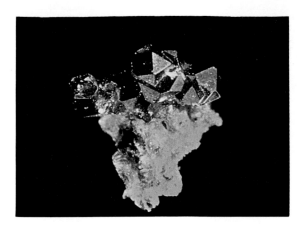

**Boleite.**
Mammoth–St. Anthony mine, Pinal County.
Julius Weber.

*Cochise County:* Warren district, where it is one of the important ore minerals in hydrothermal deposits in limestone, at the Campbell mine and other ore bodies (Trischka et al., 1929; Schwartz and Park, 1932, 1939; Yagoda, 1945; Bain, 1952; Bryant, 1968); as quarter-inch crystals, 1,300–1,400 foot levels in the Cole shaft. Cochise district, Little Dragoon Mountains (Romslo, 1949), Johnson Camp, in the Black Prince-Peabody mines area (Kellogg, 1906; Cooper and Huff, 1951; Cooper, 1957; Baker, 1960; Cooper and Silver, 1964). Turquoise district, at the Leadville, Great Western, Copper Belle, and Tejon mines. In small quantities in quartz veins in the Whetstone Mountains. Huachuca Mountains (UA 1080).

*Coconino County:* Reported from several localities in ore bodies with chalcocite and cuprite.

*Gila County:* Globe-Miami district, common as a primary ore mineral at the Old Dominion mine; also of secondary origin, forming a distinct blanket beneath the chalcocite zone (Schwartz, 1947, 1958; Peterson, 1962). Banner district, Christmas mine, where it is a common ore mineral (Peterson and Swanson, 1956; Knoerr and Eigo, 1963; Perry, 1969).

*Graham County:* Lone Star district, Gila Mountains, with pyrite and chalcopyrite, but much less abundant than the latter; in veins and disseminations in the Safford porphyry copper deposit (Robinson and Cook, 1966).

*Mohave County:* Chloride district, Atlanta and Pinkham mines (Schrader, 1909). Grand Wash Cliffs, Bronze L mine.

*Pima County:* Ajo district, where it is concentrated around pegmatite bodies in the New Cornelia Quartz Monzonite (Joralemon, 1914; Gilluly, 1937, 1942; Schwartz, 1947, 1958). Tucson Mountains, disseminated in a porphyry at the Arizona Tucson property. Pima district, a minor constituent of the primary sulfide mineralization at the Twin Buttes mine (Stanley B. Keith, pers. comm., 1973), and as a minor constituent at the Pima mine (Himes, 1972). Silver Bell Mountains, in the oxidized ores of the Silver Bell district (*Engr. Mining Jour.*, 1904; Kerr, 1951). Santa Catalina Mountains (Dale et al., 1960), in the contact ores at the Stratton-Daily camp. Near Helvetia and Rosemont, in drill core with djurleite and chalcopyrite, also associated with younger kinoite and apophyllite in skarn formed in Paleozoic limestone and dolomite (Anthony and Laughon, 1970). Comobabi Mountains, Cababi district, with silver and chalcocite (UA 1079).

*Pinal County:* Pioneer district, as exceedingly rich ore found to the deepest levels of the Magma mine (Ransome, 1914; Harcourt, 1937, 1942; Short et al., 1943; Brett and Yund, 1964; Morimoto and Gyobu, 1971). An analysis of bornite from the Magma mine by E. T. Allen (1916) gave:

| Cu 62.99 | S 25.58 | Pb 0.10 |
|----------|---------|---------|
| Fe 11.23 |         | Ag none |

TOTAL 99.90%

Galiuro Mountains, Copper Creek district, Childs-Aldwinkle mine (Kuhn, 1941; Denton, 1947a). Dripping Spring Mountains, Adjust mine, Saddle Mountain district. Mammoth district, San Manuel mine, sparingly present in the sulfide ores (Schwartz, 1947, 1953; Chapman, 1947; Lovering, 1948; Lowell, 1968). Globe-Miami district, Silver King mine (Guild, 1917).

*Santa Cruz County:* At several mining properties in the Santa Rita and Patagonia Mountains (Schrader and Hill, 1915; Schrader, 1917; Marshall and Joensuu, 1961).

*Yavapai County:* Common in the copper-bearing veins of the Bradshaw Mountains; often carries free gold. Black Hills, Yeager mine, as a shoot of high-grade ore (Lindgren, 1926; Anderson and Creasey, 1958). Eureka district, United Verde mine, present in small amounts with chalcopyrite, tennantite, and pyrite as both a primary and a secondary mineral (Lausen, 1928; Schwartz, 1938).

*Yuma County:* Buckskin Mountains, Planet mine. Harquahala Mountains, with chalcopyrite (UA 9344).

## BOTRYOGEN

Magnesium iron sulfate hydroxide hydrate, $MgFe(SO_4)_2(OH) \cdot 7H_2O$. Found typically in arid climates; formed as a result of oxidation of pyritic ores.

*Cochise County:* Warren district, on the 2,200 level of the Campbell mine as porous botryoidal crusts of orange crystals associated with copiapite (Fabien Cesbron, pers. comm., 1975).

## BOURNONITE

Lead copper antimony sulfide, $PbCuSbS_3$. One of the most common sulfosalts, formed in hydrothermal veins at moderate temperatures and commonly associated with galena, tetrahedrite, sphalerite, chalcopyrite, and other sulfides and sulfosalts.

*Cochise County:* Tombstone district, sparingly, with other copper-antimony minerals (Butler et al., 1938).

*Pima County:* Santa Rita Mountains, Busterville mine, Helvetia-Rosemont district.

*Santa Cruz County:* With tetrahedrite at the Hosey and Augusta mines, Wrightson district.

*Yavapai County:* Bradshaw Mountains, Big Bug district, as masses in quartz and as crystals with pyrite, chalcopyrite, siderite, and actinolite at the Boggs mine (Blake, 1890; Guild, 1910; Lindgren, 1926). (W. P. Blake's observation of this mineral probably constituted its first recognition in the United States.)

## BRACKEBUSCHITE

Lead manganese iron vanadate hydrate, $Pb_4MnFe(VO_4)_4 \cdot 2H_2O$. An extremely rare secondary mineral found with vanadinite, descloizite, and cerussite; only one well established locality (in Argentina) is known.

*Cochise County:* Reported to occur in minute amounts at the Gallagher Vanadium property near Charleston. The original identifier is unknown; the mineral may have been confounded with black descloizite or plattnerite crystals known to occur in this locality, and the occurrence must be regarded as suspect.

## *BRANNERITE

Uranium calcium cerium titanium iron oxide, $(U,Ca,Ce)(Ti,Fe)_2O_6$. A rare mineral found in placer gravels, granitic rocks, and quartz veins.

*Cochise County:* Swisshelm Mountains, with powellite in granite (AEC).

## BRAUNITE

Manganese oxide silicate, $Mn^{2+}Mn_6^{3+}SiO_{12}$. Occurs as veins and lenses as a product of the metamorphism of other manganese minerals, and with pyrolusite, psilomelane, and wad as a secondary mineral formed by weathering.

*Cochise County:* Warren district, Higgins mine, as radiating masses and compact needles (Palache and Shannon, 1920; Taber and Schaller, 1930; Hewett and Rove, 1930; Hewett and Fleischer, 1960); as coarse (up to one inch in length) cleavable crystals in irregular replacement pods in limestone around the shaft of the White Tail Deer mine.

*Gila County:* Globe-Miami district, present in supergene oxides, probably derived from alabandite (Hewett and Fleischer, 1960).

*Mohave County:* Rawhide Mountains, associated with a variety of manganese oxide minerals at the Artillery Peak deposit (Head, 1941).

*Pima County:* Arivaca district, COD mine, forming a matrix cut by veinlets of hausmannite (Hewett, 1972). Tucson Mountains, north end of the Juan Santa Cruz picnic grounds, in some of the piemontite in sandstone (Guild, 1935). With psilomelane and pyrolusite in fractures in andesite, Coyote and Baboquivari Mountains (Havens et al., 1954).

*Santa Cruz County:* Patagonia Mountains, Patagonia district, present in supergene oxides which may have been derived from alabandite (Hewett and Fleischer, 1960); Harshaw district (Havens et al., 1954).

## *BRAVOITE

Nickel iron sulfide, $(Ni,Fe)S_2$. A member of the pyrite group in which it may form a solid solution series.

*Coconino County:* Cameron district, Huskon mine, associated with "pitchblende," greenockite, pyrite, marcasite, calcite, and siliceous gangue replacing a Triassic conifer in the Chinle Formation (Maucher and Rehwald, 1961).

## *BREZINAITE

Chromium sulfide, $Cr_3S_4$. A very rare mineral described from the Tucson iron meteorite.

*Pima County:* Near Tucson (?), as tiny (5–80 microns) anhedral grains in the metal matrix and contiguous to silicate inclusions.

A chemical analysis (Bunch and Fuchs, 1969) is as follows:

| Cu | 48.3 | Ti | 0.96 | Ni | 0.08 |
|----|------|----|------|-----|------|
| Fe | 3.9 | Mn | 0.86 | S | 45.0 |
| V | 1.61 | | | TOTAL | 100.71% |

## BROCHANTITE

Copper sulfate hydroxide, $Cu_4SO_4(OH)_6$. A fairly common secondary mineral found in oxidized copper deposits especially in arid regions. Associated with a variety of common oxidized copper minerals including malachite with which it may be confused upon cursory examination.

*Cochise County:* Warren district, widely distributed as an intergrowth with malachite; as magnificent, coarse, crystalline masses at the Shattuck mine; with cuprite at the Copper Queen mine, and in the Calumet and Arizona mine (Ransome, 1904; Holden, 1922; Palache, 1939b; Omori and Kerr, 1963). Tombstone district, Toughnut mine, as needle-like crystals lining vugs in cuprite with connellite and malachite (Butler et al., 1938).

**Brochantite.**
Apache mine, Gila County. Julius Weber.

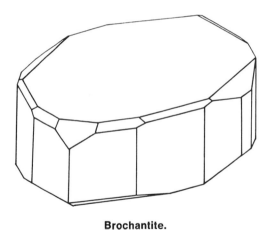

**Brochantite.**
Bisbee (Palache, 1939).

*Coconino County:* Grand Canyon National Park, Horseshoe Mesa, Grandview (Last Chance) mine, associated with cyanotrichite and other sulfate minerals, as radiating groups of acicular crystals and as shorter prismatic crystals (Leicht, 1971).

*Gila County:* Banner district, 79 mine, where it is intimately associated with oxidizing sulfide minerals (Kiersch, 1949; Keith, 1972). Apache mine, near Globe, with cerussite, matlockite, and boleite (Bideaux et al., 1960).

*Graham County:* Lone Star district, Gila Mountains, near Safford, in metasomatized

latites and andesites with pseudomalachite, malachite, antlerite, carbonate-apatite, chrysocolla, jarosite, and lepidocrocite (Hutton, 1959b). Also at the Safford porphyry copper deposit where it is an important constituent of the oxide mineralization, although less so than chrysocolla (Robinson and Cook, 1966).

*Greenlee County:* Clifton-Morenci district, abundant as an intergrowth with malachite, less common as crystals (Lindgren and Hillebrand, 1904; Lindgren, 1904, 1905; Guild, 1910; Moolick and Durek, 1966).

*Mohave County:* Grand Wash Cliffs, Grand Gulch mine, Bentley district (Hill, 1914). Noted by Genth (1868) as minute crystals associated with cuprite and chrysocolla, at Bill Williams Fork.

*Pima County:* Pima district, at the Mission mine as well-formed crystals (William E. Rhodes, pers. comm.), the San Xavier West mine at Twin Buttes (Arnold, 1964), with chrysocolla and cuprite at the Twin Buttes open pit mine (UA 9608), and at the Banner mine (UA 9354). South Comobabi Mountains, Cababi district, at the Mildren and Steppe claims (Williams, 1963). Uncommon as crusts in the weathered zone of the New Cornelia ore body at Ajo (Gilluly, 1937). Waterman Mountains, Silver Hill mine (UA 6799). On the south side of Saginaw Hill, about 7 miles southwest of Tucson, associated with other oxidized zone minerals including malachite, pseudomalachite, libethenite, atacamite, and chrysocolla (Khin, 1970).

*Pinal County:* Mammoth district, Mammoth–St. Anthony mine where it is relatively abundant and often associated with other sulfate-bearing

**Brochantite with azurite.**
Silver Hill mine, Pima County. Julius Weber.

minerals such as linarite, caledonite, and lead-hillite (UA 6114). Galiuro Mountains, Copper Creek district. Sacaton Hill, in drill core (Robert O'Haire, pers. comm., 1972).

*Yavapai County:* Black Hills, United Verde mine, in the lower part of the oxidized zone in chlorite schist which contains chalcopyrite, with antlerite and cyanotrichite (Phelps Dodge Corp., pers. comm., 1972).

*Yuma County:* Buckskin Mountains, Mineral Hill property. Associated with gypsum at the Venegas prospect, Thule Mountains (Wilson, 1933). Apache mine, near Salome (UA 7826).

## BROMARGYRITE (bromyrite)

Silver bromide, AgBr. A complete substitutional series extends from bromargyrite to chlorargyrite (AgCl). Formed in the oxidized zones of silver deposits from primary silver minerals.

*Cochise County:* Tombstone district, fourth level, Skip shaft, Empire mine, as individual gray-green crystals and irregular scales throughout limonitic gangue (Butler et al., 1938) (Tombstone, UA 7517). An analysis by R. Carrillo gave:

| | | |
|---|---|---|
| Cl  0.6 | I  2.6 | Ag 56.7 |
| Br  38.9 | | TOTAL 98.8% |

Pearce district, with chlorargyrite, embolite, iodargyrite, and acanthite in quartz veins at the Commonwealth mine (Endlich, 1897). Warren district, Shattuck mine, on the 200 level as greenish-yellow crusts and ill-formed crystals replacing native silver.

Bronzite (ferroan ENSTATITE)

## BROOKITE

Titanium dioxide, $TiO_2$. Trimorphous with rutile and anatase. An accessory mineral in igneous and metamorphic rocks; also in hydrothermal veins and as a detrital mineral.

*Gila County:* Reported as occurring in concentrates derived from the Globe area, with biotite, quartz, and ilmenite.
*Santa Cruz County:* Patagonia Mountains, with anatase, chalcopyrite, wulfenite, and molybdenite, all as microcrystals in interstices between adularia crystals. In the vicinity of the Santo Niño mine, near Duquesne (collected by William Hunt).

## BRUCITE

Magnesium hydroxide, $Mg(OH)_2$. Forms as an alteration product of periclase in contact metamorphosed limestones and dolomites; also a low-temperature hydrothermal vein mineral.

*Mohave County:* Oatman district, 3 miles northwest of Oatman, in veins with magnesite and serpentine cutting volcanic rocks (Wilson and Roseveare, 1949; Funnell and Wolfe, 1964).

## *BRUNSVIGITE (a member of the CHLORITE group)

## BUETSCHLIITE

Potassium calcium carbonate hydrate, $K_6Ca_2$-$(CO_3)_5 \cdot 6H_2O$. Found with calcite as a product of the hydration of fairchildite which formed in clinkers of fused wood ash in partly burned trees. Since buetschliite and fairchildite were characterized on material from Arizona as well as Idaho, the occurrence noted here constitutes a co-type locality.

*Coconino County:* Grand Canyon National Park. Discovered by ranger William J. Kennedy in an unspecified, partly burned tree "at a fire on the north side of Kanabownits Canyon one-quarter mile from the Point Sublime road and one-half mile from the North Entrance road" (Milton, 1944; Milton and Axelrod, 1947; Mrose et al., 1966).

## BUTLERITE

Iron sulfate hydroxide hydrate, $Fe(SO_4)(OH) \cdot 2H_2O$. Monoclinic, dimorphous with parabutler-

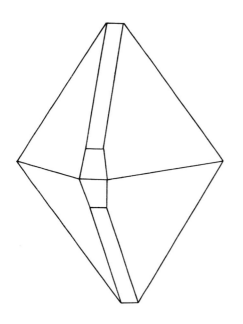

**Butlerite.**
United Verde mine (Lausen, 1928).

ite. Formed at the United Verde mine (the type locality) as a result of a mine fire.

*Yavapai County:* Black Hills, United Verde mine, as a thin crystalline coating formed as a result of the burning of pyritic ores (Lausen, 1928; Cesbron, 1964; Fanfani et al., 1971) (UA 52, 7867; H 90539; S 95953). A chemical analysis by T. F. Buehrer (Lausen, 1928) gave:

| | | |
|---|---|---|
| $H_2O$ 22.83 | $Fe_2O_3$ 36.31 | FeO 0.41 |
| $SO_3$ 38.63 | $Al_2O_3$ 0.55 | $Na_2O$ 2.73 |

TOTAL 101.46%

Specific gravity: 2.55.

## *BUTTGENBACHITE

Copper nitrate chloride hydroxide hydrate, $Cu_{37}(NO_3)_4Cl_8(OH)_{62} \cdot 8H_2O$ (McLean and Anthony [1972], by analogy to the revised formula of connellite with which the mineral is isostructural). A very rare secondary hydroxide mineral known from Likasi, Katanga, Zaire, Africa. This is the second reported world locality.

*Pima County:* South Comobabi Mountains, Cababi district, Mildren and Steppe claims, as a product of the oxidation of sulfide ores in quartz veins cutting andesite. Identification based on optical properties and qualitative chemical analysis on very limited material (Williams, 1962, 1963).

# C

## *CACOXENITE

Basic iron phosphate hydrate, $Fe_4(PO_4)_3$-$(OH)_3 \cdot 12H_2O$. An uncommon mineral of secondary origin associated with other phosphates and iron oxides in iron deposits and iron-bearing pegmatites.

*Pima County:* Silver Bell district, in the Oxide Pit of the Silver Bell mine as golden to brownish yellow acicular crystals to 1 mm clustered in sheaths and forming undulatory mats; associated with turquoise(?) crystals and sharp torbernite crystals in the contact zone of an andesite dike (Kenneth W. Bladh, pers. comm., 1974).

## CALCIOVOLBORTHITE

Calcium copper vanadate hydroxide, CaCu-$(VO_4)(OH)$. A rare secondary mineral found in small quantities in association with other oxidized vanadium minerals.

*Apache County:* Monument Valley, Garnet Ridge, found in vein fillings and in the cement of the Navajo Sandstone, both along the contact with a breccia dike and nearby, associated with malachite, chrysocolla, tyuyamunite, limonite, pyrite, and chalcopyrite (Gavasci and Kerr, 1968).

*Cochise County:* Reported from the Gallagher Vanadium property, near Tombstone.

*Navajo County:* Monument Valley, at the Monument No. 1 mine.

*Yavapai County:* Big Bug district, NW ¼, Sec. 18, T.8N, R.1W. (Robert O'Haire, pers. comm., 1972).

## CALCITE

Calcium carbonate, $CaCO_3$. A very broadly distributed mineral formed under a wide range of conditions. A widespread sedimentary rock-forming mineral as limestone and chalk; as a metamorphic rock-former in marble derived from limestone and dolomite; as travertine precipitated from thermal springs and other surface waters. Calcite is sometimes at its spectacular best in its occurrence as dripstone (stalactites and stalagmites) formed by the evaporation of underground waters carrying calcium and carbonate ions in aqueous solution. Some limestone caverns in Arizona contain excellent examples of dripstone development. The rock materials popularly termed *onyx marble* and *Mexican onyx* are distinctly banded and variously colored travertines. These materials are not related to onyx, which is banded, colored silica.

By far the most abundant occurrence of calcite is as the sedimentary rock limestone. Widely distributed throughout the sedimentary rock sequence of the state, particularly those of the Paleozoic era. The most extensive limestones of northern Arizona are the Red Wall and Kaibab Formations; in south central and southeastern Arizona, the Mississippian Escabrosa Limestone, and a number of formations of Pennsylvanian and Permian ages. Calcite also is commonly formed in veins as a result of hydrothermal processes. It is in this mode of formation that calcite attains its best morphological development.

The examples cited here are intended to be representative of the occurrences of this very abundant mineral.

*Cochise County:* Warren district, as remarkable masses of fine crystals in oxidized ore, as scalenohedra stained bright red by chalcotrichite (H 70306) and green by included malachite (Hovey, 1900); Copper Queen mine (Guild, 1911); stalactites were abundant in the oxidized-zone workings of the Copper Queen mine and these were locally colored by copper salts. A cavern in limestone about 340 feet in diameter and 80 feet high was encountered on the 300 level of the Shattuck mine in 1914. Some of the magnificent cavestone material from Bisbee, found during the early days of mining, has fortunately found its way into collections in many parts of the United States. Tombstone district, as coarsely crystalline aggregates along the flanks of the "roll" deposits, and as snow-white linings of caverns in manganese mineral-bearing ore bodies; Lucky Cuss mine, blue (UA 9739), Empire mine, black (UA 9747). Chiricahua Mountains, as crystal aggregates at Crystal Cave, also as extensive marble deposits near Fort Bowie in Immigrant Canyon and at the bend of Whitetail Creek. Little Dragoon Mountains, northwest of Manzora, as marble; marble was quarried to some extent a few miles southeast of Dragoon station (*Engr. Mining Jour.*, 1926). Turquoise district, as crystals (UA 1046).

*Coconino County:* Near Cameron, replacing aggregates of lenticular gypsum crystals (AEC). Supai, Grand Canyon, as scalenohedral crystals twinned on the basal pinacoid (S 94328); as travertine at Havasupai Falls; in Havasu Canyon, and at Mooney, Bridal Veil, and other falls. As a brown variety of travertine shading into

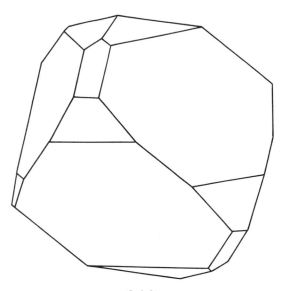

**Calcite.**
Copper Queen mine, Bisbee (Guild, 1911).

amber ("onyx marble"), 20 miles south of Canyon Diablo (*Engr. Mining Jour.*, 1922).

*Gila County:* Globe-Miami district, Old Dominion mine, as fine groups of scalenohedral crystals in cavities in limestone; quarried as marble at the foot of Sleeping Beauty Mountain, about 10 miles west of Globe; large deposits of multicolored and variegated travertine are said to occur about 20 miles north of Globe in T.4N, R.16E (Funnell and Wolfe, 1964). Banner district, 79 mine, sparingly as "butterfly" habit twins on descloizite, smithsonite, and rosasite; also as ⅛-inch, frosty-white plates having well-developed basal pinacoids, perched on aurichalcite (Keith, 1972).

*Graham County:* As "sand crystals," consisting of abundant quartz sand grains incorporated in calcite crystals, from the Safford area (UA 6474, 4873).

*Maricopa County:* As travertine at Cave Creek (Anon., 1892a); on Camp Creek, west of Cave Creek, a deposit of soft travertine contains boulders variegated in greens and yellows and veined in browns and reds (Bowles, 1940).

*Mohave County:* "Sand crystals" occur in the sediments of the Big Sandy Valley (UA 7548) and in the Kingman area (UA 8921). Moss mine area, as excellent large, clear, tabular crystals (UA 880). Abundant at the mouth of the Grand Canyon.

*Pima County:* As fine groups of large scalenohedra, Quijotoa Mountains (UA 7559, 9351). Santa Rita Mountains, at the King mine, Helvetia (UA 9350); at Greaterville as brown travertine unusually free of cracks; quarried as marble 6 miles north of Helvetia. Pima district, Twin Buttes open pit mine, as creamy white crusts several inches thick, composed of simply-terminated rhombohedral crystals; the crusts are often several square feet in extent (Stanley B. Keith, pers. comm., 1973) (UA 10083). Both stalactites and stalagmites are abundant in Colossal Cave, near the southwestern slopes of the Rincon Mountains. Dust layers are often intercalated with bands of calcite, indicative of dry periods of nondeposition. Ajo district, as crystals having inclusions of chalcotrichite.

*Pinal County:* Superior district, as delicate, pink scalenohedral crystal groups at the Magma mine. With molybdenite at the Childs-Aldwinkle mine (UA 537). As dripstone in a cave in Peppersauce Canyon, northern flank, Santa Catalina Mountains; as marble of good quality near Condon Mountain, northern end of the Santa Catalina Mountains.

*Santa Cruz County:* Santa Rita Mountains, as magnificent groups of scalenohedral crystals in Onyx Cave (UA 2024–2027) where dripstone deposits also occur. Tyndall district, Cottonwood Canyon, as yellow-stained clusters of scalenohedral crystals at the Glove mine (UA 9346) where exceptionally attractive coralloidal material occurs in naturally cavernous areas. Pata-

gonia Mountains, Holland mine (UA 6707); Washington Camp, as groups of large, pink, equant crystals on quartz crystals.

*Yavapai County:* As beautiful examples of crystallized calcite, quartz, adularia, and ore minerals from the Cash mine, Hassayampa district, Bradshaw Mountains; also as deposits of banded travertine which have yielded much decorative stone from the Big Bug Creek area near Mayer. As manganiferous banded travertine from near Cordes (UA 882). Other travertine deposits are found in the Eureka district and at Ash Fork (DeKalb, 1895); travertine production is reported from near the Montezuma Castle National Monument (Funnell and Wolfe, 1964). Calcite occurs as pseudomorphous replacements of glauberite crystals from the lake beds of Verde Valley (Snyder, 1971) (UA 9871, 9874). Calcite of optical quality has been reported from the Castle Hot Springs area.

*Yuma County:* Gila Mountains, where very pure marble deposits are reported at a locality south of Dome station. Marble of a variety of colors is reported to occur south of the Harquahala Mountains, about 9 miles east of Wenden (Strong, 1962).

## CALEDONITE

Lead copper carbonate sulfate hydroxide, $Pb_5Cu_2(CO_3)(SO_4)_3(OH)_6$. An uncommon secondary mineral found in small amounts in some oxidized copper-lead deposits.

**Caledonite.**
Mammoth-St. Anthony mine,
Pinal County. University of Arizona
collection. Jeff Kurtzeman.

*Maricopa County:* Painted Rock Mountains, at the Rowley mine near Theba, as fine, clear blue-green crystals up to several millimeters in length in small veins in the Jobes shaft; associated with cerussite, linarite, and leadhillite (Wilson and Miller, 1974) (UA 4066).

*Pima County:* Cababi district, south Comobabi Mountains, Mildren mine, as single anhedra up to 25 mm in diameter, surrounded by cerussite altering to malachite (Williams, 1962, 1963).

*Pinal County:* Mammoth district, Mammoth–St. Anthony mine, as excellent crystals from the 400 foot level, Collins vein. Forms which have been observed are {100}, {120}, {010}, {101}, {111}, {131}, {011}. An exceptionally large specimen, 3½ x 3½ x 2 inches in size, consists of a solid mass of interlocking acicular crystals of a fine, deep blue color (UA 9675).

## *CALOMEL

Mercurous chloride, $Hg_2Cl_2$. A secondary mineral formed by the alteration of other mercury minerals, principally cinnabar but also eglestonite.

*Maricopa County:* Sunflower district, Mazatzal Mountains, with eglestonite and metacinnabar (UA 217, specimen collected by Carl Lausen).

## *CANNIZZARITE

A rare lead bismuth sulfide whose properties and composition are poorly known.

*Graham County:* Aravaipa district, Landsman Camp, as tinfoil-like microcrystals on and imbedded in quartz veinlets in hedenbergite tactite.

## CARBONATE-CYANOTRICHITE

Copper aluminum carbonate sulfate hydroxide hydrate, $Cu_4Al_2(CO_3,SO_4)(OH)_{13} \cdot 2H_2O$. A very rare secondary mineral found associated with other oxidized copper minerals in some deposits.

*Cochise County:* Warren district, as fibrous radiating needles in tiny spherules on silicified shaly limestones from the Holbrook pit; associated with antlerite.

## *CARNALLITE

Potassium magnesium chloride hydrate, $KMgCl_3 \cdot 6H_2O$. Found in sedimentary salt deposits deposited from oceanic waters, with halite, sylvite, polyhalite, anhydrite, and gypsum.

*Apache and Navajo Counties:* In the subsurface of the southern portions of east central Arizona, in a northeast-trending zone of Permian evaporite deposits (Peirce, 1969). The log of a hole drilled at Sec. 24, T.18N, R.25E, showed, in addition to carnallite, halite, sylvite, polyhalite, anhydrite, and gypsum (Peirce, pers. comm., 1972).

## CARNOTITE

Potassium uranyl vanadate hydrate, $K_2(UO_2)_2(VO_4)_2 \cdot 3H_2O$. A secondary mineral of wide distribution on the Colorado Plateau where it is an important ore of uranium and vanadium. Formed by the action of meteoric waters on preexisting uranium and vanadium minerals, including uraninite and montroseite; disseminated or locally concentrated in sandstone, and associated with fossilized tree trunks or other vegetal matter. Commonly associated with tyuyamunite and metatyuyamunite and other oxidized uranium and vanadium minerals.

*Apache County:* At numerous places in the Salt Wash Member of the Morrison Formation throughout the Carrizo Mountains (UA 7748, 534) (Isachsen et al., 1955). Lukachukai Mountains (UA 7141), and Chuska Mountains (Joralemon, 1952; Masters, 1955; Wright, 1955; Lowell, 1955; Garrels and Larsen, 1959). Monument Valley, Monument No. 2 and Cato Sells mines, mixed with tyuyamunite (Witkind et al., 1963; Isachsen et al., 1955; Wright, 1955; Mitcham and Evensen, 1955; Finnell, 1957; Evensen and Gray, 1958; Young, 1964).

*Coconino County:* Cameron district, Huskon No. 10 mine, associated with schroeckingerite (Austin, 1964). Vermilion Cliffs, in petrified wood.

*Maricopa County:* Vulture Mountains, southeast of Aguila, on minor fractures in tuff (Hewett, 1925) (H 106863).

*Navajo County:* Monument Valley, at the Monument No. 1 and Mitten No. 2 mines and other prospects (Holland et al., 1958; Evensen and Gray, 1958; Witkind, 1961; Witkind and Thaden, 1963).

*Pima County:* Cienega Wash, near Vail where the road crosses the Southern Pacific tracks (Robert O'Haire, pers. comm., 1972).

*Yavapai County:* Anderson mine, sparse on sandstone (AEC). Eureka district, Hillside mine (Axelrod et al., 1951) (S 117681).

*Yuma County:* Muggins Mountains, Red Knob claims, associated with weeksite, opal, vanadinite, gypsum, calcite, and azurite (Outerbridge et al., 1960).

**Carnotite and weeksite.**
Anderson mine, Yavapai County. Julius Weber.

## CASSITERITE

Tin oxide, $SnO_2$. The most important and widely distributed tin mineral. Found in high-temperature veins and pyrometasomatic deposits associated with felsic granitic igneous rocks; also in pegmatites and in rhyolites.

*Graham County:* Found in spherulitic rhyolite and in associated placers, 25 miles east of Safford on highway 70, Apache Tin claims, as pebbles (UA 7045), and in rhyolite (UA 7028).

*Maricopa County:* In the Outpost pegmatite, as tabular crystals having well-defined faces, honey-yellow to very dark brown, some zoned, up to 1.5 inches in diameter, associated with copper and bismuth minerals.

*Pinal County:* A single small piece of stream tin was identified from an area of rhyolite flows east of the Tablelands.

*Yavapai County:* A rare constituent of the pegmatites of the White Picacho district (Jahns, 1952).

## *CELADONITE

Potassium magnesium iron aluminum silicate hydroxide, $K(Mg,Fe^{2+})(Fe^{3+},Al)Si_4O_{10}(OH)_2$. A member of the mica group, having composition and properties much like those of glauconite, but occurring in vesicles in basaltic rocks. It is common throughout the state but seldom mentioned in the literature.

*Cochise County:* Steele Hills, in seams and vesicles in a basalt within the Threelinks Conglomerate (Cooper and Silver, 1964). Also reported from the Warren district.

*Navajo County:* Hopi Buttes volcanic field, Seth-La-Kai diatreme, in volcanic sandstone and tuffs, with limonite, laumontite, gypsum, and montmorillonite, associated with weak uranium mineralization (Lowell, 1956).

*Pinal County:* Midway Station, as vesicle fillings (Bideaux et al., 1960).

## CELESTITE

Strontium sulfate, $SrSO_4$. In veins, beds, or lenticular masses in sedimentary rocks; also as a gangue mineral with lead-zinc ores.

*Cochise County:* One mile northwest of Portal, as sharp, well-formed crystals up to 5 mm in length, free in vesicles in basalt; occurs with prehnite, pyrite, and pumpellyite.

*Maricopa County:* Occurs with gypsum, sandstone, and conglomerate as beds in sandy tuff on the northwest side of a low range of mountains about 15 miles from Gila Bend and 3 miles east of the Black Rock railroad siding (Phalen, 1914; Moore, 1935, 1936) (UA 907). As beds in shaly tuff on the northeast slope of the Vulture Mountains, 15 miles southeast of Aguila in the NW ¼, Sec. 20, T.6N, R.7W (Butler, 1929; Moore, 1935, 1936; Hewett et al., 1936; Harness, 1942) (UA 1154).

*Mohave County:* Artillery Mountains, Graham prospect, as nodules scattered about on the surface (Lasky and Webber, 1949).

*Yuma County:* Plomosa district, with lead and silver ores.

## CELSIAN

Barium aluminum silicate, $BaAl_2Si_2O_8$. An uncommon monoclinic member of the feldspar group. Usually associated with manganese deposits.

*Yavapai County:* Reported from an unspecified locality near Yarnell.

## CERUSSITE

Lead carbonate, $PbCO_3$. A very common secondary mineral of oxidized lead deposits, formed by reaction between carbonated waters and lead minerals or solutions containing lead. Frequently found as concentric layers about the lead sulfate, anglesite, which surrounds a core of unaltered galena. Cerussite is so widespread in Arizona that only a few localities can be noted.

*Cochise County:* Tombstone district, the most abundant lead mineral of the district; Toughnut mine (UA 9737) (Butler et al., 1938; Rasor, 1938; Frondel and Pough, 1944). Warren district, as sharp, pale gray, nearly perfect, tabular "sixling" twinned crystals up to two inches in diameter, on psilomelane (Sinkankas, 1964); found in Hendricks Gulch as impure "sand car-

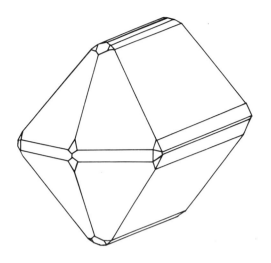

**Cerussite.**
C and B Vanadium mine (Trebisky and Keith, 1975).

bonate" near a fissure in limestone (Ransome, 1904; Guild, 1910; Schwartz and Park, 1932). Chiricahua Mountains, as large twinned crystals at the Hilltop mine (Pough, 1941). Outlook mine, Huachuca Mountains (UA 7642). The principal lead ore mineral of the Turquoise district. Gunnison Hills, Texas-Arizona mine (Cooper and Silver, 1964).

*Gila County:* Dripping Spring Mountains, at the London-Arizona mine, where it is the principal ore mineral of lead; Banner district, at the 79 mine, as the most abundant supergene mineral, in the form of "sand carbonate," and as beautifully crystallized material in which are "sixling" and V-shaped twinned crystals; associated with anglesite, wulfenite, and vanadinite (Kiersch, 1949; Keith, 1972). Globe-Miami district, the principal ore mineral at the Defiance mine (Wilson et al., 1950); in large masses associated with anglesite and galena at the Apache mine (Wilson et al., 1951).

*Graham County:* Aravaipa district (Denton, 1947b; Simons, 1964). Stanley district (Ross, 1925a).

*Greenlee County:* Clifton-Morenci district (Reber, 1916), Hormeyer mine, with gold ore.

*Maricopa County:* Painted Rock Mountains, at the Rowley mine near Theba, as sharp, clear, tiny crystals, also massive (Wilson and Miller, 1974). Wickenburg district, at the Moon Anchor, Potter-Cramer property, and the Rat Tail claim, where it is associated with galena, sphalerite, and a variety of lead and chromium oxi-

dized minerals (Williams, 1968; Williams and Anthony, 1970; Williams et al., 1970).

*Mohave County:* Cerbat Mountains, Chloride, Mineral Park, and Gold Basin districts; at some properties with free gold; Tennessee-Schuylkill mine, Wallapai district (Schrader, 1909; Thomas, 1949). At the McCracken mine, McCracken Peak.

*Pima County:* Santa Rita Mountains, Golden Gate and Blue Jay mines, Helvetia district; Greaterville district. Empire Mountains, Total Wreck, Chief, and Hilton mines (Schrader and Hill, 1915), and at the 49 mine as trillings to ¾ inch, free in finer "sand carbonate," associated locally with aurichalcite and turquoise (Gene D. Schlepp, pers. comm., 1974); C and A lease, southwest of Pantano (UA 7680). Sierrita Mountains, Pima district, as "sand carbonate"; Paymaster mine, Olive Camp, as massive and crystallized material (Ransome, 1922; Nye, 1961); Twin Buttes open pit mine (Stanley B. Keith, pers. comm., 1973). Silver Bell Mountains, as silky crystals and as earthy mixtures with smithsonite, El Tiro and other properties. Quijotoa Mountains, Morgan mine, with chlorargyrite. Tucson Mountains, at the Old Yuma mine, with wulfenite and vanadinite (Guild, 1911).

*Pinal County:* Mammoth district, Mammoth-St. Anthony mine, as single crystals and magnificent twinned and reticulated crystal aggregates from the Collins vein (Pogue, 1913; Peterson, 1938; Palache, 1941b; Fahey et al., 1950) (UA 850). A complex single crystal in the collection

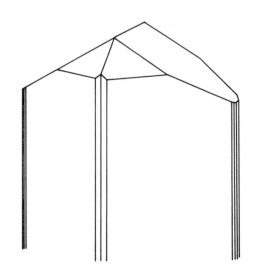

**Cerussite.**
Mildren mine (Williams, 1962).

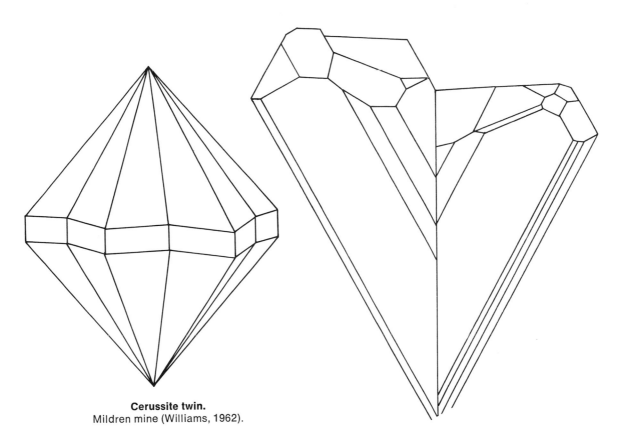

**Cerussite twin.**
Mildren mine (Williams, 1962).

**Cerussite twin.**
Mammoth–St. Anthony mine (Pogue, 1913).

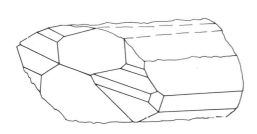

**Cerussite.**
Old Yuma mine (Guild, 1911).

of the U.S. National Museum exhibits the following crystallographic forms: {001}, {010}, {100}, {110}, {130}, {012}, {023}, {011}, {032}, {021}, {031}, {041}, {051}, {102}, {302}, {111}, {112}, and {211}. Galiuro Mountains, at the Blue Bird and other lead deposits of the Copper Creek district; Saddle Mountain district. Tortilla Mountains, Florence Lead-Silver mine, formed by the alteration of primary ores containing galena, pyrite, sphalerite, and tennantite, and associated with hemihed-

rite, wulfenite, willemite, vauquelinite, minium, and mimetite (Williams and Anthony, 1970).

*Santa Cruz County:* Santa Rita Mountains, Tyndall district, at the Victor, Ivanhoe, and Rosario properties; Cottonwood Canyon, Glove mine as large crystalline masses (Olson, 1966); Montosa Canyon, Isàbella mine, as "sand carbonate," along a mineralized fault in limestone, associated with specular hematite (Anthony, 1951); American Boy mine, Wrightson district. Patagonia Mountains, at the Flux and Domino mines, as magnificent groupings of needle- and pencil-like crystals and as massive material (Kunz, 1885) (UA 213, 7134, 9727), and at several other properties (Kunz, 1885; Schrader, 1917).

*Yavapai County:* Black Hills, Verde district, Copper Chief mine. Bradshaw Mountains, Silver Belt mine, with chlorargyrite, in ancient mine workings (Lindgren, 1926). Eureka district, Hillside mine (Axelrod et al., 1951). Bella mine (Kunz, 1885). Big Bug district, Iron King mine

**Cerussite.**
Flux mine, Patagonia Mountains,
Santa Cruz County. Tom McKee
collection. Jeff Kurtzeman.

(Creasey, 1952; Anderson and Creasey, 1958).

*Yuma County:* Trigo Mountains, Silver district, Red Cloud mine, as single crystals, twinned crystals, and as reticulated masses of substantial size; many of the crystals are arrowhead-shaped twins (Pirsson, 1891; Guild, 1910; Foshag, 1919). Castle Dome Mountains, associated with anglesite and yellow and red lead oxides (Wilson, 1933; Batty et al., 1947).

## CERVANTITE

Antimony oxide, $Sb_2O_4$. A secondary mineral formed through the oxidation of stibnite and other antimony minerals.

*Yuma County:* Dome Rock Mountains, in veins, as radiating blades of stibnite partly altered to cervantite and stibiconite.

## *CESAROLITE

Basic lead manganese oxide; formula uncertain, perhaps $PbMn_3O_7 \cdot H_2O$ or $H_2PbMn_3O_8$. A rare secondary mineral sometimes associated with the oxidation of galena.

*Cochise County:* Warren district, where it is probably of supergene origin, found in upper Paleozoic limestones (Hewett and Fleischer, 1960; Hewett et al., 1963).

*Maricopa County:* As dull black crystalline films on vein quartz which carries cerussite and minium in cavities.

## *CHABAZITE

Calcium aluminum silicate hydrate, $CaAl_2Si_4O_{12} \cdot 6H_2O$. A member of the zeolite group. Typically found in amygdules and fissures in mafic volcanic rocks; also from the alteration of volcanic glass in tuffs.

*Cochise County:* San Simon Basin, 7 miles northwest of Bowie, in bedded lake deposits with analcime, herschelite, erionite, clinoptilolite, halite, and thenardite (Regis and Sand, 1967).

*Coconino County:* Found in mafic dikes associated with the breccia pipe at Black Peak, near Cameron; localized in the centers of dolomite replacements of olivine, and as small veins (Barrington and Kerr, 1961).

*Gila County:* Christmas mine, as rhombohedral crystals in vugs in hydrothermally altered andesite porphyry where it appears to be a late-stage mineral (D. Perry, pers. comm., 1967); also as chalky, pinkish-white material (F. A. Mumpton, pers. comm.). Near Bowie, as chalky, pale buff-yellow material (F. A. Mumpton, pers. comm.).

*Mohave County:* East of Big Sandy Wash, eastern half of T.16N, R.13W, in Pliocene tuff with analcime, clinoptilolite, erionite, and phillipsite (Ross, 1928, 1941).

*Navajo County:* Coliseum diatreme, near Indian Wells, filling or lining vesicles.

*Pinal County:* In a railway cut at Malpais Hill, north of Mammoth, as crystals intergrown with calcite (Bideaux et al., 1960).

**Chabazite and mordenite.**
Superior, Pinal County. Julius Weber.

*Yavapai County:* Found in basalts west of Perkinsville, with phillipsite (?) (McKee and Anderson, 1971).

## CHALCANTHITE

Copper sulfate pentahydrate, $CuSO_4 \cdot 5H_2O$. A water-soluble secondary mineral found in the oxidized zone of sulfide copper deposits; not uncommonly found as crusts and stalactites in mine workings. Decomposes in dry atmospheres.

*Cochise County:* Warren district, Copper Queen mine, as stalactites and irregular, porous excrescences several inches thick on mine walls (Merwin and Posnjak, 1937); in the Lavender open pit mine (UA 1651); as stalactites at the Calumet and Arizona mine (Mitchell, 1921).

*Coconino County:* Grand Canyon National Park, Horseshoe Mesa, Grandview (Last Chance) mine, as cross-fiber veinlets associated with clay; some clearly of post-mine origin (Leicht, 1971).

*Gila County:* Globe-Miami district, as stalactites and as coatings on floors of old openings, Old Dominion mine; Castle Dome mine, largely as a post mine-opening mineral (Peterson et al., 1951; Peterson, 1962); Inspiration ore body (Olmstead and Johnson, 1966). Dripping Spring Mountains, Banner district, 79 mine, with olivenite, sphalerite, galena, siderite, anglesite, and brochantite (Kiersch, 1949; Thomas Trebisky, pers. comm., 1972). Cherry Creek area, Donna Lee mine; Sierra Ancha Mountains, First Chance, Little Joe, and Shipp No. 2 mines, as efflorescences on mine walls (Granger and Raup, 1969).

*Greenlee County:* Clifton-Morenci district, as small bodies in the oxidized ores of Copper Mountain, and as stalactites in one of the upper drifts of the Jay shaft (Lindgren, 1905; Guild, 1910; Moolick and Durek, 1966).

*Mohave County:* Cerbat Mountains, as a common secondary mineral in the Wallapai district (Thomas, 1949).

*Navajo County:* Monument Valley, in the Mitten No. 2 mine (Witkind and Thaden, 1963).

*Pima County:* Silver Bell Mountains, as thick coatings on the walls of old mine workings, Silver Bell district. Santa Rita Mountains, Helvetia-Rosemont district, as fibrous veins (Schrader and Hill, 1915; Creasey and Quick, 1955). Pima district, San Xavier West mine (Arnold, 1964).

*Pinal County:* Galiuro Mountains, as coatings on the walls of drifts and in fractures, Old Reliable, Copper Giant, Glory Hole, and Copper Prince mines, Copper Creek district (Simons, 1964). Mammoth district, Mammoth–St. Anthony mine area (UA 4585).

*Santa Cruz County:* Patagonia Mountains, Duquesne and Washington Camp (Schrader, 1917).

*Yavapai County:* Black Hills, as stalactites up to 2 feet in length, United Verde mine (Guild, 1910) (UA 5569). Crown King district, on timbers in the dump of the Springfield mine, as fine crystals. Also found in the Copper Basin district. Abundant in the old mine dumps at Bagdad (Therese Murchison, pers. comm., 1972).

## Chalcedony (see QUARTZ)

## CHALCOALUMITE

Copper aluminum sulfate hydroxide hydrate, $CuAl_4(SO_4)(OH)_{12} \cdot 3H_2O$. A very rare secondary mineral known from only a few oxidized copper deposits. The Bisbee occurrence is the type locality.

*Cochise County:* The species was described by Larsen and Vassar (1925), based on material from an unspecified locality in Bisbee. An analysis by H. E. Vassar is as follows:

| | | | |
|---|---|---|---|
| Insol. | 0.09 | $Na_2O$ | $H_2O(-)$ 28.95 |
| $Al_2O_3$ | 38.71 | $+ K_2O$ 0.50 | $H_2O(+)$ 1.90 |
| MgO | 0.05 | | $SO_3$ 14.80 |
| CaO | 0.01 | | CuO 14.78 |
| | | | TOTAL 99.79% |

Specific gravity: 2.29.

Sacramento pit, in vugs in a dense quartz-geothite gossan, associated with cuprite and malachite, as tiny, beautifully formed, single crystals and twins; commonly altered to pale blue, mas-

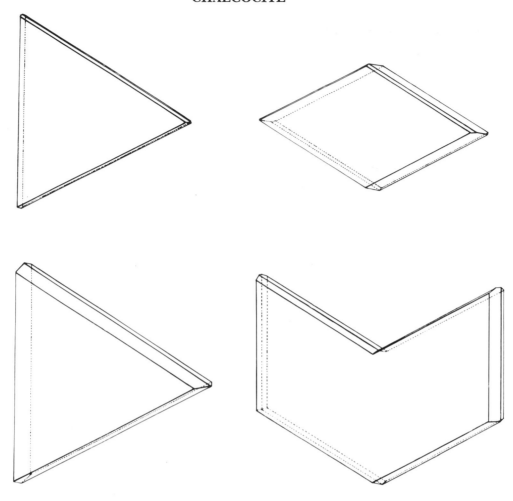

**Chalcoalumite.**
Bisbee (Williams & Khin, 1971).

**Chalcoalumite.**
Bisbee, Cochise County. S. A. Williams.

sive, botryoidal gibbsite. The crystals are equilateral-triangular in shape; the flat sides are bounded by $\{100\}$, one edge by $\{010\}$, and the other two edges by $\{012\}$, and $\{11\bar{2}\}$ (Williams and Khin, 1971).

*Coconino County:* Grand Canyon National Park, Horseshoe Mesa, Grandview (Last Chance) mine, sky-blue to greenish in color, as botryoidal crusts on limonite coating earlier zeunerite crystals; also associated with scorodite, olivenite, and brochantite (Leicht, 1971).

## CHALCOCITE

Copper sulfide, $Cu_2S$. An important ore mineral of copper, of wide distribution. Uncommon as a primary mineral, but important as a secondary mineral in the zone of secondary enrichment where it may replace other sulfides. Of commercial importance in Arizona where many large, low-grade, copper orebodies owe much of their value to secondary chalcocite.

*Apache and Navajo Counties:* Monument Valley district, where it is associated with other sulfide minerals in uranium-vanadium ores in sandstone (Mitcham and Evensen, 1955; Evensen and Gray, 1958).

*Cochise County:* Warren district, Campbell and Shattuck mines, where it is locally abundant as a secondary mineral in limestone replacements; in the Sacramento pit mine, as disseminated ores (Ransome, 1904; Guild, 1910; Trischka et al., 1929; Schwartz and Park, 1932; Schwartz, 1934, 1939; Grout, 1946; Hutton, 1957; Bain, 1952); the principal ore mineral of the Lavender open pit mine (Bryant and Metz, 1966). Cochise district, Johnson Camp area, in small amounts in quartz veins cutting granite, schist, and limestone (Kellogg, 1906). Tombstone district, with acanthite and stromeyerite (Butler et al., 1938); Empire mine (UA 9731). Turquoise district, an important constituent of the enriched ores.

*Coconino County:* Grand Canyon National Park, Horseshoe Mesa, Grandview (Last Chance) mine (Rogers, 1922; Leicht, 1971); Orphan mine (UA 2368).

*Gila County:* Globe-Miami district, an important constituent of the disseminated sulfide deposits of the region (Ransome, 1903; Schwartz, 1921, 1928, 1934, 1939, 1947, 1958; Peterson, 1962). Old Dominion mine, as compact massive bodies (Ransome, 1903); Castle Dome mine, replacing chalcopyrite in the upper part of the deposit (Peterson, 1947); Copper Cities deposit (Peterson, 1954; Simmons and Fowells, 1966). Banner district, Christmas mine, locally associated with bornite in the chalcopyrite-bornite inner ore zone of the ore bodies, replacing dolomite (Perry, 1969); Dripping Spring Mountains, 79 mine (Kiersch, 1949; Keith, 1972).

*Graham County:* Aravaipa district (Denton, 1947b; Simons, 1964); Ten Strike group of claims. Lone Star district, Gila Mountains, as both the steely and the sooty varieties, replacing pyrite, chalcopyrite, and bornite in the Safford porphyry copper deposit, where it forms a blanket-like deposit on the primary ore zone (Robinson and Cook, 1966).

*Greenlee County:* Clifton-Morenci district, in a thick blanket at the Morenci open pit mine (Moolick and Durek, 1966), and the principal ore mineral of the disseminated and vein deposits, in places as solid seams 2 to 3 feet thick (Lindgren, 1903, 1904, 1905; Guild, 1910); Ryerson mine (S 86029), (sample collected by W. Lindgren); Montezuma mine (S 86042).

*Maricopa County:* Cave Creek district, Red Rover mine, with argentiferous tetrahedrite.

*Mohave County:* Noted in the vicinity of Bill Williams Fork by Genth (1868). Grand Wash Cliffs, Grand Gulch, Bronze L, and Copper King mines (Hill, 1914). Cerbat Range, Wallapai district, Mineral Park mine (Thomas, 1949; Field, 1966); also in other districts of the region.

*Pima County:* Ajo district, disseminated in a narrow band bordering the New Cornelia ore body on the south (Joralemon, 1914). Santa Rita Mountains, Helvetia and other districts of the range (Schrader, 1917). Sierrita Mountains, Pima district, Esperanza mine, in an extensive enriched blanket (Schmidt et al., 1959; Lynch, 1967). Pima mine, as a product of secondary enrichment (Journeay, 1959), and as large, nearly pure masses at the Copper Glance and Queen mines (Ransome, 1922), in the secondarily enriched ores of the Twin Buttes mine (Stanley B. Keith, pers. comm., 1973). Silver Bell Mountains, as an important secondarily enriched ore mineral at the El Tiro and Oxide open pit mines, Silver Bell district (Richard and Courtright, 1966). Cababi district, southern Comobabi Mountains, Mildren and Steppe claims (Williams, 1963). Cerro Colorado Mountains, Heintzelman mine (S 534).

*Pinal County:* Pioneer district, as large, nearly pure secondary bodies in the Magma mine, and an important constituent of the primary ores of

**Chalcocite in gypsum.**
Mission mine, Pima County. Julius Weber.

the deeper levels, associated with digenite, bornite, and djurleite; at the Belmont mine, as fine-grained, sooty material (Ransome, 1914; Guild, 1917; Bateman, 1929; Harcourt, 1942; Short et al., 1943; Morimoto and Gyobu, 1971). The essential mineral of the disseminated ores at Ray (Ransome, 1919; Schwartz, 1947; Clarke, 1953; Lewis, 1955; Metz and Rose, 1966). Galiuro Mountains, at several properties including the Childs-Aldwinkle mine where primary chalcocite occurs in the deeper levels (Kuhn, 1941; Denton, 1947a). Mammoth district, Mammoth–St. Anthony mine, as thin films and as replacements of chalcopyrite (Peterson, 1938; Fahey et al., 1950); San Manuel mine, in hydrothermally altered monzonite and quartz monzonite porphyry, replacing chalcopyrite in the secondary sulfide zone; associated with bornite, chalcopyrite, cuprite, chrysocolla, and native copper (Chapman, 1947; Schwartz, 1947, 1949; Lovering, 1948; Thomas, 1966; Lowell, 1968).

*Santa Cruz County:* Santa Rita Mountains, at a number of properties including the Ivanhoe mine, Tyndall district, where it occurs in rather large bodies (Schrader and Hill, 1915; Schrader, 1917). Palmetto district, as an ore body of secondary origin at the 3R mine (Handverger, 1963).

*Yavapai County:* Black Hills, in the oxidized zone of the United Verde mine, and in pure, massive bodies of exceptional size in the United Verde Extension mine (Fearing, 1926; Schwartz, 1938; Anderson and Creasey, 1958). Eureka district, Bagdad mine (Anderson, 1950; Moxham et al., 1965). Copper Basin district, in the oxidation zone of sulfide deposits in breccia pipes (Johnston and Lowell, 1961).

*Yuma County:* In prospects at Cinnabar, 8 miles southwest of Quartzsite. Castle Dome Mountains, with malachite (S 65493).

## *CHALCOPHANITE

Zinc iron manganese oxide hydrate, (Zn,Fe,-Mn)Mn$_2$O$_5$·nH$_2$O. A secondary mineral which occurs with other manganese and iron oxides.

*Cochise County:* Warren district, Cole shaft, as crystalline material perched on gossany, geothite-rich material (AM 34586).

*Pinal County:* In a railway cut between San Manuel and Mammoth (UA 9726).

## CHALCOPHYLLITE

Copper aluminum arsenate hydroxide hydrate, Cu$_{18}$Al$_2$(AsO$_4$)$_3$(SO$_4$)$_3$(OH)$_{27}$·33H$_2$O. A rare secondary mineral found in the oxidized zone of copper deposits with other oxidized copper minerals.

*Cochise County:* Warren district, at the Calumet and Arizona mine, where it is associated with cuprite and connellite (Palache and Merwin, 1909).

*Graham County:* Turtle Mountain area, east of Bonito Creek, as crystals in vesicles in basalt (Bideaux et al., 1960).

## CHALCOPYRITE

Copper iron sulfide, CuFeS$_2$. The most important ore mineral of copper. Occurs in appreciable quantity in nearly all copper sulfide deposits. Predominantly of primary origin in veins and replacement bodies, as disseminated particles in a variety of rock types, and in contact metamorphic zones.

*Apache County:* Monument Valley, Garnet Ridge, found in vein fillings and in the cement of the Navajo Sandstone both along its contact with a breccia dike and nearby (Gavasci and Kerr, 1968).

*Apache and Navajo Counties:* Monument Valley, associated with chalcocite, bornite, and native copper in the uranium-vanadium ores in Shinarump Conglomerate, Moenkopi Formation, and DeChelly Member (Mitcham and Evensen, 1955; Evensen and Gray, 1958).

*Cochise County:* Warren district, as massive ore bodies; mined at the Copper Queen, Calumet and Arizona, Junction (Mitchell, 1920), and other properties. Tombstone district, the most abundant copper mineral of the district. Turquoise district, the principal ore mineral in pyritic bodies. Little Dragoon Mountains, Cochise district, an important ore mineral at several properties (Cooper, 1957); Keystone and St. George properties (Romslo, 1949). Dragoon Mountains, Primos mine, near Dragoon (Palache, 1941a). Reef mine, Huachuca Mountains (Palache, 1941). Chiricahua Mountains, Humboldt mine (Hewett and Rove, 1930).

*Gila County:* Globe-Miami district, as large masses in Mescal Limestone, Old Dominion mine (Woodbridge, 1906), and forming the bulk of the ore at the Summit mine; in the protore of the Miami and Inspiration mines, and the principal ore mineral at the Castle Dome mine; Copper Cities deposit (Peterson, 1954). The consistent occurrence of a chalcopyrite enriched zone beneath the supergene chalcocite blanket in the Inspiration mine has led some workers to believe the chalcopyrite to be of supergene origin; occurs as thin films on the surfaces of pyrite crystals (Olmstead and Johnson, 1966). Payson district, as the chief ore mineral of the copper deposits. Dripping Spring Mountains, Banner district, Christmas mine, where it is the most abundant ore mineral (Tainter, 1948; Knoerr and Eigo, 1963; Perry, 1969); 79 mine, rarely as disphenoidal crystals up to ½ inch, showing {112}, coated by smithsonite (Keith, 1972).

*Graham County:* Present in ores of the Stanley (Ross, 1925a) and Aravaipa (Denton, 1947b; Simons, 1964) districts. Lone Star district, San Juan mine, in a low grade sulfide deposit (Rose, 1970), and in the Safford porphyry copper deposit (Robinson and Cook, 1966).

*Greenlee County:* Clifton-Morenci district, in the lower levels of the veins and disseminated in limestone near porphyry contacts (Lindgren, 1903, 1904; Guild, 1910); it is the only primary copper mineral of importance in the Morenci open pit mine where it is associated with pyrite, molybdenite, and sphalerite as disseminations and small veinlets (Moolick and Durek, 1966).

*Mohave County:* Cerbat Range, common in nearly all the copper mines and prospects of the region; Johnny Bull-Silver King property (Tainter, 1947c). Grand Wash Cliffs, the main ore mineral at the Bronze L and Copper King mines. Ithaca Peak, present in veins with pyrite, sphalerite, and galena in a quartz monzonite stock (Eidel, 1966). In 1868, Genth noted the mineral, associated with native copper, cuprite, chrysocolla, malachite, brochantite, chalcocite, covellite, and pyrite, at Bill Williams Fork. Copper World mine, near Yucca, associated with sphalerite, pyrrhotite, and very minor loellingite (Rasor, 1946).

*Pima County:* The most important copper mineral in the mines and prospects of the Santa Rita Mountains (Schrader and Hill, 1915); Helvetia district, the principal ore mineral at a number of properties including the Rosemont lease, Copper World, and Leader mines (Creasey and Quick, 1955). At several properties in the Pima district (Webber, 1929; Eckel, 1930; Guild, 1934) including the Esperanza mine (Tainter, 1947b; Lynch, 1967); the Pima mine, as a primary ore mineral closely associated with grossular hornfels (Journeay, 1959; Himes, 1972); the Mission mine where it is the principal ore mineral (Richard and Courtright, 1959);

and the Twin Buttes open pit mine (Stanley B. Keith, pers. comm., 1973). Catalina district, Pontotoc mine, with epidote and hematite in gneiss associated with limestone (Guild, 1934); also in contact deposits with garnet and epidote near Marble Peak (Dale et al., 1960). Ajo district, as scattered grains in the New Cornelia Quartz Monzonite (Joralemon, 1914; Gilluly, 1937). Cababi district, South Comobabi Mountains, Little Mary mine (Williams, 1963). Silver Bell Mountains, El Tiro and Oxide open pit mines (Kerr, 1951; Richard and Courtright, 1966).

*Pinal County:* Pioneer district, Magma mine as massive replacements of limestone (Sell, 1961; Short et al., 1943) and Belmont mine. Galiuro Mountains, Copper Creek district, Copper Prince mine, where several thousand tons of nearly pure chalcopyrite were mined (Joralemon, 1952), the Childs-Aldwinkle mine (Kuhn, 1941), and the Old Reliable mine (Denton, 1947a). In protore at the Ray district. Pioneer district, Silver King mine (Guild, 1917). Mammoth district, San Manuel mine, of hypogene origin; the most important ore mineral (Chapman, 1947; Thomas, 1966).

*Santa Cruz County:* Santa Rita Mountains, the most abundant copper ore mineral in most of the districts; in the Tyndall district, American Boy mine. Patagonia Mountains, Santo Niño mine, with large bodies of massive molybdenite (Blanchard and Boswell, 1930); Indiana mine, with pyrite, sphalerite, and chalcopyrite (Stanley B. Keith, pers. comm., 1973). Oro Blanco district, Idaho and Montana mines, Annie Laurie prospect (Warren and Loofburrow, 1932; Anderson and Kurtz, 1955).

*Yavapai County:* The principal ore mineral in the pyritic ore body of the United Verde mine, Black Hills (Anderson and Creasey, 1958); also abundant at the Copper Chief and Shea properties. Bradshaw Mountains, found in the Big Bug, Agua Fria, Black Canyon, and Pine Grove districts; in the Big Bug district at the Iron King mine (Creasey, 1952). Monte Cristo mine, Wickenburg Mountains, with native silver and nickel arsenides (Bastin, 1922).

*Yuma County:* Buckskin Mountains, Planet mine (Bancroft, 1911). Harquahala Mountains, Golden Eagle mine.

## CHALCOSIDERITE

Copper iron aluminum phosphate hydroxide hydrate, $Cu(Fe,Al)_6(PO_4)_4(OH)_8 \cdot 4H_2O$. The iron-bearing analogue of turquoise (which see), but much less common in occurrence than that mineral.

*Cochise County:* Warren district, reported in small quantities at the Shattuck mine (UA 1468).

## *CHENEVIXITE

Copper iron arsenate hydroxide hydrate, $Cu_2Fe_2(AsO_4)_2(OH)_4 \cdot H_2O$ (?). A widespread but uncommon mineral; of secondary origin in the oxidized portion of copper deposits. A relatively common mineral in leached cappings over enargite-pyrite mineralization.

*Santa Cruz County:* Patagonia Mountains, Alum Gulch, as minute green spherules in porous quartzose rock with contact silicates and specular hematite (UA 8218) (Loghry, 1972).

## Chert (see QUARTZ)

## CHEVKINITE

Calcium cerium iron magnesium titanium silicate, $(Ca,Ce)_4(Fe,Mg)_2(Ti,Fe)_3Si_4O_{22}$. A rare mineral found in granite pegmatites, certain granites, and volcanic ash deposits.

*Mohave County:* Aquarius Range, in a vein traversing a granitic dike; intimately associated with sphene, monazite, cronstedtite, and quartz (Kauffman and Jaffe, 1946). An analysis reported by these authors gave:

| | | | |
|---|---|---|---|
| $SiO_2$ | 12.04 | $Fe_2O_3$ | 9.56 |
| $TiO_2$ | 17.08 | FeO | 7.76 |
| $ThO_2$ | 0.82 | MnO | 0.50 |
| $Ce_2O_3$ | 25.29 | CaO | 3.35 |
| $(La,Di)_2O_3$ | 18.35 | MgO | 0.74 |
| $Y_2O_3$ | 1.50 | $P_2O_5$ | 0.38 |
| $Al_2O_3$ | 0.93 | $H_2O$ | 1.50 |

TOTAL 99.80%

## Chillagite (tungstenian WULFENITE)

## CHLOANTHITE (see SKUTTERUDITE)

## CHLORARGYRITE (cerargyrite)

Silver chloride, AgCl. A complete substitutional series extends from chlorargyrite to bromargyrite (AgBr). Formed in the oxidized zones of silver deposits from primary silver minerals.

*Cochise County:* In the oxidized ores of the Tombstone district (Butler et al., 1938; Rasor, 1938; Romslo and Ravitz, 1947); Santa Anna mine (UA 9162); also at the Bradshaw mine, near Charleston, in granular aggregates. Pearce district, Pearce Hills, at the Commonwealth mine with bromargyrite, embolite, iodargyrite, and acanthite in quartz veins (Endlich, 1897). Dragoon Mountains, in the oxidized lead-silver ores of the Turquoise district. Warren district, Cole mine, in pockets in massive sulfides (Hutton, 1957) and on malachite; Campbell mine (Schwartz and Park, 1932); cementing silica breccia at the Shattuck mine (UA 5566).

*Gila County:* Globe-Miami district, in many of the surficial ores (Peterson, 1962); Old Dominion mine, with manganese oxides. Richmond Basin, where massive chlorargyrite forms plates one half inch thick and several inches in diameter. Payson district, with native silver at the Silver Butte mine (Lausen and Wilson, 1927). A mass of pure chlorargyrite weighing more than 50 pounds was discovered loose in soil in the Globe region by a local cowboy in the 1950s.

*Graham County:* Aravaipa district, Orejana adit, with bromargyrite; Windsor shaft (Ross, C. P., 1925a). Copper Creek district, Blue Bird mine (Kuhn, 1951).

*Greenlee County:* Clifton area, Hargo mine (UA 5523).

*Mohave County:* Hualpai Mountains, at several properties in the Maynard district (Schrader, 1907). Cerbat Range, at several properties, principally in the Mineral Park, Stockton Hills, White Hills (*Engr. Min. Jour.*, 1892c), and Wallapai districts; also in the Chloride district at the Distaff mine where, with native silver, it was the principal silver mineral (Bastin, 1925).

*Pima County:* Santa Rita Mountains, Blue Jay mine, Helvetia-Rosemont district; also in the Greaterville district. Empire Mountains, at the Total Wreck mine (Schrader and Hill, 1915; Schrader, 1917). Quijotoa Mountains, at the Morgan mine. Cerro Colorado Mountains, Cerro Colorado and other mines (Guild, 1910). Papago district, Sunshine-Sunrise group, with galena, chalcocite, cerussite, and anglesite (Ransome, 1922). Cababi district, South Comobabi Mountains, Mildren and Steppe claims (Williams, 1963). Prince Rupert mine, Crittenden, with wire silver (*Engr. Min. Journ.*, 1892b).

*Pinal County:* Pioneer district, as the chief near-surface silver ore mineral in the Belmont area (Romslo and Ravitz, 1947). Mammoth district, Mammoth–St. Anthony mine, as tiny, yellowish, cubo-octahedral crystals implanted on caledonite, from the Collins vein. Vekol mine, Vekol Mountains, as millimeter-size crystals on jarosite (AM 24967). Tortolita Mountains, Owl Head district, at the Apache prospect.

*Santa Cruz County:* Ivanhoe mine, Tyndall district (Guild, 1910); Anaconda group, Wrightson district. Patagonia Mountains, La Plata and Meadow Valley mines, Redrock district; Hermosa and American mines, Harshaw district; Palmetto mine, Palmetto district (Schrader and Hill, 1915; Schrader, 1917).

*Yavapai County:* Bradshaw Mountains, Dos Oris mine, Hassayampa district, with acanthite and native silver; Thunderbolt mine, Black Canyon district, with proustite and native silver; Tuscumbria mine, Bradshaw district, with stephanite; with ruby silver at the Tip Top mine, Tip Top district. At the Silver Belt mine, Big Bug district, where the presence of stone hammers and grads in ancient workings indicates that the deposit was mined in ancient times. Peck district, Swastika mine, as fine crystals. Eureka district, Hillside mine, associated with anglesite, cerussite, smithsonite, native silver, and hemimorphite (Axelrod et al., 1951). Prescott area, Stonewall Jackson mine, with acanthite, embedded in siderite gangue; formed by alteration of native silver with which it is associated (Guild, 1917); Lida mine (*Engr. Min. Jour.,* 1892d).

*Yuma County:* Trigo Mountains, Silver district, Silver Clip and Red Cloud mines, as the principal silver mineral in oxidized lead ores (Wilson, 1933).

## CHLORITE

A mineral group name, members of which have compositions that can be expressed by the general formula $(Mg,Fe^{2+},Fe^{3+},Mn)(AlSi_3)O_{10}$-$(OH)_8$. The chlorite minerals are very abundant in Arizona and are found primarily in thermally metamorphosed rocks and as a product of the hydrothermal alteration of ferromagnesian silicates such as biotite, pyroxene, and amphibole. A few localities, usually of the less common chlorite species, are listed under the individual species name.

*Colorado Plateau Region:* Present with illite and montmorillonite in several members of the Morrison Formation (Keller, 1962). Abundant locally in the Moenkopi Formation (Schultz, 1963).

*Pima County:* Santa Catalina Mountains, clinochlore is present in limestone at the contact

with the Leatherwood Quartz Diorite (Wood, 1963).

*Pinal County:* Mammoth district, Mammoth–St. Anthony mine, as green felted masses in the lower levels of the Collins vein (Peterson, 1938).

*Yavapai County:* Verde district, United Verde mine, where it is an abundant alteration mineral. The term "black schist" refers to chlorite-rich aggregates which have replaced rocks in the region. The chlorite of the United Verde mine has been classified in the clinochlore-prochlorite group, and rumpfite, ripidolite, diabantite, aphrosiderite, brunsvigite, and thuringite have been recognized locally by Anderson and Creasey (1958).

## *CHLORITOID

Iron magnesium manganese aluminum silicate hydroxide, $(Fe,Mg,Mn)_2Al_4Si_2O_{10}(OH)_4$. A fairly common product of intermediate grade thermal metamorphism of aluminum- and iron-rich pelitic sedimentary rocks.

*Cochise County:* Chiricahua Mountains, as crystals in phyllite (Bideaux et al., 1960).

## *CHONDRODITE

Magnesium iron silicate hydroxide fluoride, $(Mg,Fe)_3SiO_4(OH,F)_2$. A member of the humite group; an uncommon mineral found in contact metamorphosed limestones and dolomites and in skarns related to ore deposits.

*Cochise County:* Cochise district, Johnson Camp area, in tactite formed in the Martin Formation, in bright pink lenses with an unidentified silicate mineral (Cooper and Silver, 1964).

*Pinal County:* Locally abundant in forsterite marble at the Lakeshore mine.

## CHROMITE

Iron chromium oxide, $FeCr_2O_4$. A member of the spinel group. Commonly associated with peridotites or serpentines in which it may occur in veins or as segregations; common associates include magnetite, olivine, garnet, vesuvianite, ilmentite, and pyroxene.

*Apache County:* Monument Valley, Garnet Ridge, as fragments in a breccia dike piercing sedimentary rocks (Gavasci and Kerr, 1968).

*Coconino County:* Canyon Diablo area, associated with krinovite, roedderite, high albite, richterite, and ureyite in the Canyon Diablo meteorite (Olsen and Fuchs, 1968).

*Yuma County:* Trigo Mountains, Silver district, as disseminated grains and small masses with mariposite (a variety of muscovite) in mica schist.

## *CHRYSOBERYL

Beryllium aluminum oxide, $BeAl_2O_4$. An uncommon mineral found in granite pegmatites, certain metamorphic rocks, and as detrital grains.

*Mohave County:* Virgin Mountains, with beryl crystals up to 15 inches long in pegmatite dikes on the Hummingbird claims (Olson and Hinrichs, 1960; Meeves, 1966).

## CHRYSOCOLLA

Copper acid silicate hydroxide, $Cu_2H_2Si_2O_5$-$(OH)_4$. A widespread and locally abundant secondary mineral found in practically all oxidized copper deposits of the state. Commonly intimately associated with tenorite and malachite.

*Apache and Navajo Counties:* Monument Valley, associated with the oxidized uranium-vanadium ores in sandstone (Mitcham and Evensen, 1955).

*Apache County:* Garnet Ridge, in veins cutting Navajo Sandstone and as cement along the contacts with a nearby breccia dike; associated with malachite, calciovolborthite, tyuyamunite, limonite, chalcopyrite, and pyrite (Gavasci and Kerr, 1968).

*Cochise County:* Warren district, with cuprite, disseminated flakes of copper, azurite, malachite, small crystals of brochantite, and, locally, tufts of connellite (Ransome, 1904; Holden, 1922; Schwartz, 1934). Tombstone district, widely distributed but not abundant in the properties of the district (Butler et al., 1938). Cochise district, Johnson Camp area, associated with the copper sulfide ore bodies of the district, commonly associated with other oxidized copper minerals (Kellogg, 1906; Cooper and Huff, 1951). Sulfur Springs Valley, in small amounts in numerous veinlets cutting porphyritic andesite flows (Lausen, 1927).

*Coconino County:* Jacobs Canyon district, near Jacobs Lake, where it is one of the principal constituents of low-grade copper deposits in the Red Wall and Kaibab Limestones, associated with malachite and azurite (Fischer, 1937). White Mesa district, with malachite cementing sandstone (Hill, 1912).

*Gila County:* Globe-Miami district (Woodbridge, 1906), abundant in the Live Oak, Keystone, Black Warrior, and Geneva mines (Wells, 1937). An important ore mineral at the Old Dominion mine (Schwartz, 1921, 1934); at the Keystone Copper mine, located 5 miles west of Globe (Wells, 1937). In the Bulldog tunnel of the Inspiration mine, with malachite, chalcedony, and quartz, as aggregates of great beauty (Ransome, 1919; Peterson, 1962; Sun, 1963); Black Copper portion of the Inspiration ore body, forming thin coatings on fractures in Pinal Schist, occasionally as veins up to 5 cm wide, also as matrix cementing sub-angular schist pebbles in an old stream channel with White Tail Conglomerate, interlayered with heulandite in some outcroppings. An analysis of the black variety shows it to contain 6.7% $MnO_2$; the blue-green variety contains 0.06% $MnO_2$ (Throop, 1970; Throop and Buseck, 1971; see also Kemp, 1905). Banner district, 79 mine, as pseudomorphs after hemimorphite (Keith, 1972) (UA 7587); Christmas mine, as a moderately common supergene mineral associated with andradite-bearing skarns in the Naco Formation (David Perry, pers. comm., 1967).

*Graham County:* Lone Star district, Gila Mountains, near Safford, in metamorphosed latites and andesites, with a variety of other copper minerals (Hutton, 1959b); Safford porphyry copper deposit, where it is the most abundant copper mineral in the deposit, making up 50 to 85% of all the oxide mineralization; as veins, blebs and coatings, and commonly intimately associated with kaolinite and halloysite (Robinson and Cook, 1966).

*Greenlee County:* Clifton-Morenci district, common in the district but of little commercial interest; as fine, glassy-green specimens (Kunz,

**Chrysocolla.**
79 mine, Gila County. Julius Weber.

1885; Lindgren and Hillebrand, 1904; Lindgren, 1905; Reber, 1916), frequently pseudomorphic after malachite and brochantite (Moolick and Durek, 1966).

*Maricopa County:* Painted Rock Mountains, at the Rowley mine, near Theba, where small masses and stringers, locally associated with malachite, have yielded quartz-rich material of gemmy quality; most, however, is friable to powdery (Wilson and Miller, 1974).

*Mohave County:* Cerbat Mountains, Wallapai district, as fine specimen material at the Emerald Isle mine where it occurs in both vein and blanket deposits in granite porphyry and in alluvium; associated with dioptase and "copper pitch" which cements alluvial detritus (Thomas, 1949; Searls, 1950; Newberg, 1967); a chemical analysis on material from the Emerald Isle mine by L. T. Richardson (Wells, 1937) is as follows:

| | | | | | |
|---|---|---|---|---|---|
| $SiO_2$ | 40.30 | CaO | 1.05 | $H_2O(+)$ | 8.38 |
| FeO | — | $H_2O(-)$ | 16.92 | CuO | 28.90 |
| MgO | 0.40 | | | TOTAL | 95.95% |

Genth observed chrysocolla at Bill Williams Fork in 1868. Planet mine on Bill Williams Fork, as beautifully banded material associated with cuprite in limestone replacement ore bodies (McCarn, 1904).

*Navajo County:* White Mesa district, as cementing material in sandstone beds (Mayo, 1955).

*Pima County:* Pima district, reported from the Sierrita Mountains (Eckel, 1930), and in large masses with brochantite and cuprite at the Twin Buttes mine (Henry Worsley, pers. comm., 1972) (UA 9608). Silver Bell Mountains, as clear emerald-green material at the El Tiro mine (UA 8869). Santa Rita Mountains, Helvetia area (UA 3571). South Comobabi Mountains, Cababi district, Mildren and Steppe claims (Williams, 1963). Ajo district, widespread in the ore body of the New Cornelia mine (Joralemon, 1914; Gilluly, 1937). Santa Catalina Mountains, at Pusch Ridge (UA 9328).

*Pinal County:* Mammoth district, San Manuel mine, the most abundant copper mineral in the oxidized zone of the ore body (Chapman, 1947; Schwartz, 1947, 1949, 1953; Thomas, 1966; Lowell, 1968; Throop and Buseck, 1971); Mammoth–St. Anthony mine, in places as material of gem quality (Peterson, 1938; Galbraith and Kuhn, 1940; Fahey et al., 1950). Near Florence (UA 6049). Virgus Canyon, Galiuro Mountains, Table Mountain mine, as

the only abundant copper mineral (Simons, 1964). Mineral Creek district, Ray mine, as spherulitic aggregations of highly birefringent material, as green and black varieties, with tenorite, malachite, halloysite, and heulandite (Schwartz, 1934; Clarke, 1953; Stephens and Metz, 1967; Throop and Buseck, 1971); Copper Butte mine (Phelps, 1946).

*Santa Cruz County:* Santa Rita Mountains, Tyndall district, Cottonwood Canyon, Glove mine (Olson, 1966). Patagonia Mountains, Duquesne and Washington Camp (Schrader, 1917); Helvetia, with limonite, forming warty masses and incrustation pseudomorphs after gypsum crystals.

*Yavapai County:* Black Hills, as mammillary fillings in Tertiary conglomerate, Arizona-Dundee property (Anderson and Creasey, 1958). Bradshaw Mountains, where material of a bright blue color occurs at the Whipsaw and Copperopolis properties, Castle Creek district (Lindgren, 1926).

*Yuma County:* Buckskin Mountains, Planet mine (Bancroft, 1911; Cummings, 1946b). Betty Lee group and other properties, Copper Mountain (Wilson, 1933). Harquahala Mountains, near Salome (BM 1966, 113).

Chrysoprase (see QUARTZ)

CHRYSOTILE (see Serpentine)

## CINNABAR

Mercury sulfide, HgS. Of near-surface origin as veins, replacement deposits, or impregnations. Commonly associated with rocks and hotspring deposits of recent volcanic origin.

*Gila and Maricopa Counties:* Mazatzal Mountains, where it occurs in a number of properties, usually as veinlets or as thin films on fracture surfaces and as discontinuous, more or less definite ore shoots in a belt of "schist" which consists of a variety of rock types. Cinnabar is the only important ore mineral in the deposits, but the chloride (calomel), mercury, and metacinnabar are present in small amounts as are pyrite, chalcopyrite, azurite, and malachite (Lausen, 1926); a number of producers are situated on Alder and Sycamore Creeks (Ransome, 1916; von Bernewitz, 1937; Dreyer, 1939; Beckman and Kerns, 1965; Faick, 1958). Phoenix Mountains, at the Rico, Mercury, and Eureka groups of claims. Also at the Sam Hughes claims, north of Phoenix (UA 1391).

*Mohave County:* Northern Black Mountains (River Range), Fry mine, Gold Basin district.

*Pima County:* Roskruge Mountains, Roadside mine (Beckman and Kerns, 1965). Cerro Colorado Mountains, with malachite (UA 3095). As tiny crystals and scales lining cavities in quartz at the Mary G mine, Cerro Colorado district (Davis, 1955).

*Pinal County:* At the Mickey Welch claims, south of Casa Grande.

*Yavapai County:* Copper Basin district (Beckman and Kerns, 1965), Mercury, Cinnabar, Queen, Zero Hour, and Shylock properties (Johnston and Lowell, 1961). White Picacho district, Westerdahl claims.

*Yuma County:* Mined on a small scale near Ehrenberg where it is sparsely distributed in veins. Dome Rock Mountains, Plomosa district, French, American and Colonial properties, 8 miles southwest of Quartzsite (Bancroft, 1909; von Bernewitz, 1937; Beckman and Kerns, 1965).

## CLAUDETITE

Arsenic oxide, $As_2O_3$. A secondary mineral formed from the oxidation of other arsenic minerals; also as a sublimation product of mine fires.

*Yavapai County:* Black Hills, Jerome, United Verde mine, as silky crystals filling a small cavity above the burned pyritic ore body (Palache, 1934; Buerger, 1942) (H 92682).

## *CLAUSTHALITE

Lead selenide, PbSe. A rare lead mineral sometimes found in complex ores with other sulfides and selenides.

*Greenlee County:* Found in a deep drill hole in the Blue Range wilderness area. A sample from 2807 feet showed silvery films of clausthalite lining fractures in a calcite-epidote marble. Traces of pyrite and sphalerite were also present.

## *Cliftonite

Carbon, C. A rare form of carbon occurring as tiny cubic polycrystalline pseudo-crystals in metallic meteorites. Controversy surrounds the origins of this material: some believe it to be pseudomorphic after diamond; others feel that it formed by the diffusion of carbon, which was a product of the decomposition of cohenite, $(Fe,Ni)_3C$, to nucleation sites where subsequent growth produced the cubic morphology.

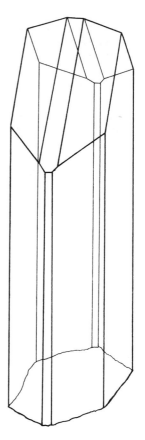

**Claudetite.**
United Verde mine (Palache, 1934).

*Coconino County:* Meteor Crater area, found in kamacite (alpha-Fe,Ni) which surrounds troilite (FeS)-graphite-silicate nodules in the Canyon Diablo meteorite; as tiny cube-shaped masses composed of a large number of crystallites (El Goresy, 1965; Brett and Higgins, 1967).

## *CLINOCHLORE

Magnesium iron aluminum silicate hydroxide, $(Mg,Fe^{2+},Al)_6(Si,Al)_4O_{10}(OH)_8$. A member of the chlorite group formed during thermal metamorphism and by hydrothermal processes.

*Cochise County:* Warren district, as coarse crystals formed during retrograde metamorphism of garnet-epidote tactites, commonly found in the deeper levels of the Cole and Dallas mines.

## *CLINOCHRYSOTILE

Magnesium silicate hydroxide, $Mg_3Si_2O_5(OH)_4$. A member of the serpentine group (which see); the dimorph of chrysotile.

*Cochise County:* Warren district, as scaly micaceous intergrowths with stevensite ("deweylite") and chrysocolla, from the Holbrook pit.

## *CLINOCLASE

Copper arsenate hydroxide, $Cu_3(AsO_4)(OH)_3$. An uncommon mineral of supergene origin, formed in the oxidized portion of copper deposits.

*Gila County:* Banner district, reported from the 6th level of the 79 mine (Keith, 1972).

*Pinal County:* Hull claims, south of Ray, in thin purple crusts and veinlets (AM 35561).

*Santa Cruz County:* Temporal Gulch, St. Louis mine, as druses on barite, derived from the alteration of tennantite (Bideaux et al., 1960).

*Yavapai County:* Copper Mountain, Old Robertson claim 4 miles east-northeast of Mayer, with olivenite (R. O'Haire, pers. comm., 1972); with cornwallite (UA 2211, 5758).

## *CLINOHUMITE

Magnesium silicate fluoride hydroxide, $Mg_9Si_4O_{16}(F,OH)_2$. A member of the humite group, it usually occurs in contact metamorphosed limestones and dolomites and in skarns related to mineral deposits.

*Apache County:* As the variety titanclinohumite, a prominent constituent of a kimberlite tuff plug in the Buell Park diatreme, located about 15 miles north of Fort Defiance (Sun, 1954; Balk, 1954; Bideaux et al., 1960; McGetchin et al., 1970). A chemical analysis by H. B. Wiik is reported by Balk (1954) as follows:

| | | | | | |
|---|---|---|---|---|---|
| $SiO_2$ | 35.34 | $Fe_2O_3$ | 1.23 | $H_2O+$ | 1.14 |
| $TiO_2$ | 5.11 | MgO | 46.45 | $H_2O-$ | 0.02 |
| $Al_2O_3$ | 0.00 | MnO | 0.23 | $P_2O_5$ | 0.12 |
| $Cr_2O_3$ | 0.10 | CaO | 0.07 | F | .06 |
| FeO | 10.20 | | | $-O$ | 0.02 |

TOTAL 99.95%

The specific gravity was determined by Sun (1954) to be 3.364.

## *CLINOPTILOLITE

Sodium potassium calcium aluminum silicate hydrate, $(Na,K,Ca)_{2-3}Al_3(Al,Si)_2Si_{13}O_{36} \cdot 12H_2O$. A member of the zeolite group. A very abundant mineral in Arizona, formed by the devitrification and alteration of volcanic glass in tuffs. Remarkably enough, until recent years the mineral was practically unknown in the state because its presence in the tuffaceous volcanic rocks is not readily discernible without careful study by optical or x-ray diffraction techniques.

*Apache County:* Near Nutrioso, with analcite in a tuff and sandstone formation of Tertiary age (Wrucke, 1961; Sheppard, 1971).

*Cochise County:* San Simon Basin, 7 miles northwest of Bowie, in bedded lake deposits with analcime, chabazite, herschelite, erionite, thenardite, and halite (Regis and Sand, 1967).

*Greenlee County:* About 6 miles north of Morenci with mordenite in Tertiary age tuff (Sheppard, 1969, 1971).

*Maricopa County:* Near Horseshoe Reservoir, Sec. 3, T.7N, R.6E, in tuff of the Verde Formation (Sheppard, 1971; F. A. Mumpton, pers. comm.).

*Mohave County:* Near Wikieup, in a lacustrine formation of Pliocene age (Sheppard, 1971); east of Big Sandy Wash, eastern half of T.16N, R.13W, in Pliocene tuff with analcime, chabazite, erionite, and phillipsite (Ross, 1928, 1941).

*Pinal County:* Northern Tortilla Mountains, where it has formed by the alteration of feldspar fragments in abundant, large pumice masses contained in the middle tuff and tuffaceous sandstone member of the Ripsey Wash beds (Schmidt, 1971).

*Yuma County:* Near Dome, with bentonite and tuff in a lacustrine formation of late Tertiary age (Bramlette and Posnjak, 1933; Sheppard, 1971). An analysis by J. G. Fairchild (Wells, 1937) is as follows:

| | | | | | |
|---|---|---|---|---|---|
| $SiO_2$ | 64.30 | CaO | 2.42 | MgO | 0.62 |
| $Al_2O_3$ | 12.78 | $Na_2O$ | 3.96 | $H_2O(-)$ | 4.78 |
| $Fe_2O_3$ | 0.82 | $K_2O$ | 1.36 | $H_2O(+)$ | 9.50 |

TOTAL 100.54%

## CLINOZOISITE

Calcium aluminum silicate hydroxide, $Ca_2Al_3Si_3O_{12}(OH)$. A member of the epidote group. Typically a product of both regional and contact metamorphism; much like epidote in occurrence. Much of what has been called zoisite is probably this mineral.

*Cochise County:* Tombstone district, as small, vitreous green grains with vesuvianite, monticellite and thaumasite in the Lucky Cuss mine (Butler et al., 1938). Warren district, as pseudomorphs after biotite, with chlorite, sericite, sphene, quartz, and muscovite (Schwartz, 1958).

*Gila County:* Globe-Miami district, as stringers and scattered grains in sericitized plagioclase

**Conichalcite on plancheite.**
Table Mountain mine, Pinal County. R. Bideaux.

of the quartz monzonite, Castle Dome mine (Peterson et al., 1946, 1951; Creasey, 1959); a common mineral along the north side of the Miami-Inspiration ore body (Peterson, 1962); Copper Cities deposit, formed during hydrothermal alteration, associated with chlorite, epidote, pyrite, sericite, and calcite (Peterson, 1954).

*Pima County:* North end of the South Comobabi Mountains, and north of the Ko Vaya Hills, as the variety withamite, with manganaxinite and piemontite, as vesicle fillings in andesite flow boulders (Bideaux et al., 1960). Santa Catalina Mountains, with diopside, epidote, actinolite, tremolite, microcline, and plagioclase in meta-arkose at the contact with the Leatherwood Quartz diorite (Wood, 1963).

*Pinal County:* Mammoth district, San Manuel mine, as coarse grains with epidote, as a hydrothermal alteration product (Schwartz, 1953).

*Yavapai County:* Big Bug district, Iron King mine, in diorite, along the northwestern contact with the breccia facies, associated with epidote, chlorite, and hornblende (Creasey, 1952).

## *CLINTONITE

Calcium magnesium aluminum silicate hydroxide, $Ca(Mg,Al)_3(Al_3Si)O_{10}(OH)_2$. A member of the mica group which may be regarded as a calcium analogue of phlogopite (Deere et al., 1962). Occurs in chlorite schist with talc and in metamorphosed limestones.

*Gila County:* Banner district, Christmas mine, as the variety xanthophyllite, associated with forsterite, garnet, idocrase, and calcite on the 800-foot level, northside, in contact metamorphosed diorite (David Perry, pers. comm., 1967; 1969).

## COBALTITE

Cobalt arsenic sulfide, $CoAsS$. A rare mineral most commonly encountered as disseminations in contact metamorphic rocks, but also in high-temperature veins. Associated with other cobalt and nickel arsenides and sulfides.

*Apache County:* Reported from the White Mountains, exact locality not known.
*Maricopa County:* Mazatzal Mountains, along the Apache Trail between Fish Creek and Roosevelt Dam.
*Pima County:* Cababi district, Comobabi Mountains, exact locality unknown.
*Yavapai County:* Black Hills, near the Old Prudential claim, along the contact between Bradshaw Granite and greenstone of the Yavapai Schist; altered at the surface to erythrite (Guild, 1910).

## *COCONINOITE

Iron aluminum uranyl phosphate sulfate hydroxide hydrate, $Fe_2Al_2(UO_2)_2(PO_4)_4(SO_4)$-$(OH)_2 \cdot 20H_2O$. An uncommon secondary mineral which occurs in the oxidized zones of Colorado Plateau-type uranium deposits.

*Apache County:* Black Water No. 4 mine, in the oxidized zone of a sandstone uranium deposit in Triassic strata, associated with gypsum, jarosite, limonite, and clay; as aggregates of microcrystalline grains of a light creamy yellow color (Young et al., 1966). A chemical analysis on 20 mg of this material by Robert Meyrowitz gave the following results:

| | | | | | |
|---|---|---|---|---|---|
| $Fe_2O_3$ | 9.7 | $SO_3$ | 5.4 | CaO | 0.1 |
| $Al_2O_3$ | 6.6 | $H_2O$ | 24.0 | $Na_2O$ | $<0.1$ |
| $UO_2$ | 34.9 | Acid | | $CO_2$ | $<0.1$ |
| $P_2O_5$ | 18.3 | Insol. | } 0.5 | | |

TOTAL 99.5%
$H_2O(-)$   18.2

Approximate specific gravity (Sun Valley mine material): 2.70.
*Coconino County:* Near Cameron, at the Sun Valley and Huskon No. 7 mines, "in seams 1 mm or less thick, predominantly along bedding planes of a light-colored arkosic sandstone that is fine-grained, poorly sorted, and thinly bedded" (Young et al., 1966). The Sun Valley mine is the co-type locality.

## *COESITE

Silicon dioxide, $SiO_2$. First known as a synthetic product created in the laboratory under extreme conditions of pressure. The only natural occurrences are associated with meteoric impacts in high silica rocks. The natural occurrence of coesite was predicted by geological inference before its actual discovery.

*Coconino County:* Meteor Crater, as an abundant constituent of sheared Coconino Sandstone in debris under the crater floor and at depth in breccia below the crater floor as well as beyond the crater rim. It also occurs in lechatelierite in water-lain beds within the crater. As irregular grains, 5 to greater than 50 microns in size. Meteor Crater is the type locality. (Chao et al., 1960).

## COFFINITE

Uranium silicate hydroxide, $U(SiO_4)_{1-x}(OH)_{4x}$. An important primary ore mineral of uranium, presumably of hydrothermal origin, which occurs with uraninite in the deeper unoxidized portions of sandstone uranium deposits.

*Apache County:* Monument Valley, Monument No. 2 mine, with uraninite, corvusite, montroseite, and doloresite, as constituents of the dark, unoxidized uranium-vanadium ores; found in tabular bodies in sandstone-filled paleo stream channels or scours (Young, 1964).

*Coconino County:* Cameron area, Yazzie No. 102 mine (Austin, 1964).

*Gila County:* At the Workman mine, in Dripping Spring Quartzite (Granger and Raup, 1962, 1969).

*Navajo County:* North of Holbrook, at the Ruth group of claims. Also on Stinking Spring Mountain. Monument Valley, Monument No. 1 mine, with uraninite, corvusite, and montroseite (Evensen and Gray, 1958).

## COHENITE

Iron nickel cobalt carbide, $(Fe,Ni,Co)_3C$. Found in some metallic meteorites of the state.

## COLEMANITE

Calcium borate hydrate, $Ca_2B_6O_{11} \cdot 5H_2O$. Most commonly formed by the evaporation of inland bodies of salt water.

*Maricopa County:* As the well-crystallized colorless mineral with bitumen in a fossil egg 62 x 40 mm in size, enclosed in limestone matrix; found in gravels on the Gila River. An analysis of the colemanite (Morgan and Tallmon, 1904) follows:

| | | |
|---|---|---|
| CaO 27.07 | $B_2O_3$ 51.00 | $H_2O$ 22.01 |
| | | TOTAL 100.08% |

## COLUMBITE-TANTALITE

Iron manganese columbium-tantalum oxides, $(Fe,Mn)(Nb,Ta)_2O_6$. A complete (ionic substitutional) (?) series exists between the end members columbite (columbium) and tantalite (tantalum). The minerals typically occur in lithium-bearing granite pegmatites.

*Maricopa County:* White Picacho district, as crystals up to 5 inches in diameter in the quartz-rich zone of the Midnight Owl pegmatite (Jahns, 1952). The ferberite deposits of the Cave Creek district carry 2.19% combined columbium-tantalum, probably present as columbite-tantalite.

*Yavapai County:* In small quantities in several pegmatites of the White Picacho district (Jahns, 1952). A three-inch twinned crystal was recovered from pegmatites near Crown King.

## *COLUSITE

Copper arsenic sulfide with tin, vanadium, and iron, $Cu_3(As,Sn,V,Fe)S_4$. An uncommon mineral associated with bornite, chalcocite, tetrahedrite, tennantite, and enargite in some copper deposits.

*Pinal County:* Pioneer district, where it occurs in the Magma mine (Hammer and Peterson, 1968).

## CONICHALCITE

Calcium copper arsenate hydroxide, $CaCu(AsO_4)(OH)$. A member of the adelite group. Found in the oxidized portions of copper deposits; associated with other secondary minerals including austinite, olivenite, limonite, brochantite, malachite, azurite, and jarosite.

*Cochise County:* Warren district, a mineral named higginsite from the Higgins mine (Palache and Shannon, 1920) was later shown to be identical with conichalcite; it occurs as excellent crystals and small masses in manganese oxides and in limonite (Taber and Schaller, 1930; Richmond, 1940; Berry, 1951; Qurashi and Barnes, 1954; Radcliffe and Simmons, 1971). An analysis by E. V. Shannon gave:

| | | |
|---|---|---|
| CuO 28.67 | $As_2O_5$ 41.23 | $Fe_2O_3$ 0.48 |
| CaO 20.83 | $H_2O$ 3.49 | MnO 2.84 |
| $V_2O_5$ 1.97 | | Insol. 0.86 |
| | | TOTAL 100.37% |

Tombstone district, Little Joe mine (UA 43).

*Gila County:* Globe district, Inspiration Consolidated Copper Company's Thornton Pit, as "spheroidal" alteration areas in masses of chrysocolla 2 to 3 inches thick associated with faults

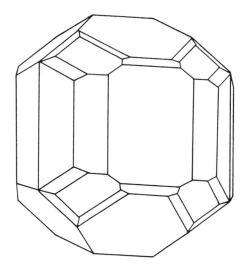

**Conichalcite (higginsite).**
Bisbee (Palache, 1920).

(Kenneth W. Bladh, pers. comm., 1974); Globe Hills, Copper Hill mine (UA 6425).

*Pima County:* As small green crystals in vugs associated with shattuckite at the New Cornelia mine, Ajo (Schaller and Vlisidis, 1958) (UA 8987). Also from near Rosemont in the Santa Rita Mountains (S 112807).

*Pinal County:* Galiuro Mountains, Table Mountain mine, as good quality crystals in vugs with willemite, plancheite, and malachite (Bideaux et al., 1960). Sawtooth Mountains (Robert O'Haire, pers. comm., 1972). Tortolita Mountains, Azurite mine, with malachite, plancheite, bisbeeite, and shattuckite (Bideaux et al., 1960).

*Santa Cruz County:* North end of the Tumacacori Mountains (UA 9265).

*Yavapai County:* In a shallow pit 13 miles west of Congress, just off State road 93. Also in the Bagdad open pit mine, lining vugs (S 114678).

## CONNELLITE

Copper sulfate chloride hydroxide hydrate, $Cu_{27}(SO_4)_2Cl_8(OH)_{62} \cdot 8H_2O$. A secondary mineral of great rarity from oxidized copper deposits; usually found in small cavities in cuprite, with brochantite and malachite. One "fist-sized" specimen from Bisbee contained probably as much connellite as is otherwise known in the entire world.

*Cochise County:* Warren district, Copper Queen (UA 163; BM 55921), Calumet and Arizona, and Czar mines, as small, radiating aggregates of slender crystals (Ransome, 1904; Palache and Merwin, 1909; Holden, 1922, 1924; Frondel, 1941). An analysis by W. E. Ford (Ford and Beadley, 1915) on material from the Czar mine gave:

| | | |
|---|---|---|
| CuO 73.38 | SO₃ 3.15 | N₂O₅ 0.72 |
| Cl 6.82 | | H₂O 17.13 |

$$CuO\ 73.38 \quad SO_3\ 3.15 \quad N_2O_5\ 0.72$$
$$Cl\ \ \ \ 6.82 \quad\quad\quad\quad\quad\quad H_2O\ 17.13$$

TOTAL 101.20%

$(-)O = Cl$ 99.67%

Crystals of "footeite" (= connellite) on the type specimens of paramelaconite from the Copper Queen mine attain the size of matchsticks. Tombstone district, Toughnut mine, as slender needle-like crystals and aggregates in cavities in cuprite (McLean and Anthony, 1972) (UA 26).

*Pima County:* At the New Cornelia mine, Ajo district, in vugs with cuprite (BM 1965, 257). Daisy shaft, Mineral Hill mine, Pima district, with pseudoboleite, gerhardtite, and atacamite in cuprite (Williams, 1961).

*Pinal County:* Mammoth district, Mammoth–St. Anthony mine, rarely as microcrystals associated with caledonite.

## COOKEITE

Lithium aluminum silicate hydroxide, $LiAl_4$-$(AlSi_3)O_{10}(OH)_8$. A rare member of the chlorite group. Found in granite pegmatite associated with beryl, lepidolite, and spodumene.

*Maricopa County:* White Picacho district, as rare coatings of pale pink flakes and foils commonly grouped in felt-like aggregates on lepidolite and spodumene and filling fractures in the interior parts of several lithium-bearing pegmatites (Jahns, 1952).

## COPIAPITE

Iron magnesium sulfate hydroxide hydrate, $(Fe,Mg)Fe_4^{3+}(SO_4)_6(OH)_2 \cdot 20H_2O$. A fairly common secondary mineral formed, with other sulfates such as melanterite, alunogen, and halotrichite, from the alteration of sulfide minerals, principally pyrite.

*Cochise County:* Warren district, Copper Queen mine, in crusts several inches thick, with coquimbite and voltaite (Merwin and Posnjak, 1937); Campbell mine, 2100 level, as bright yellow crystals.

*Coconino County:* Cameron area, a fairly common mineral associated with jarosite and other sulfates in the uranium deposits (Austin, 1964); Yazzie No. 102 mine, in large, pure masses of yellow scales.

*Gila County:* Sierra Ancha Mountains, First Chance mine (Granger and Raup, 1969).

*Pima County:* Sierrita Mountains, as silky fibers and foliated masses, Pima district.

*Yavapai County:* Black Hills, United Verde mine, as incrustations nearly 1 cm thick, crystals and crystalline masses, formed by burning of sulfide ore (Anderson, 1927; Lausen, 1928) (H 90538). An analysis by T. F. Buehrer (Lausen, 1928) gave:

| | | |
|---|---|---|
| $H_2O$ 31.03 | CuO 2.26 | $K_2O$ 0.24 |
| $SO_3$ 38.45 | FeO 0.38 | $Na_2O$ 1.74 |
| $Fe_2O_3$ 27.12 | | TOTAL 101.22% |

Specific gravity: 2.09.

## COPPER

Copper, Cu. Most importantly and abundantly of secondary origin, widely distributed in the oxidized zones of many sulfide copper deposits, where it may be accompanied by cuprite, malachite, azurite, tenorite, and limonite. Also in sedimentary rocks and in cavities in volcanic rocks. May pseudomorphically replace other minerals such as malachite, cuprite, azurite, and chalcopyrite. Very abundant in Arizona, usually in small amounts in the oxidized portions of copper deposits.

*Apache and Navajo Counties:* Monument Valley, where it occurs as a primary mineral associated with copper sulfides in the uranium-vanadium deposits of the area (Evensen and Gray, 1958).

*Cochise County:* Abundant at several mines of the Warren district, locally as pseudomorphs after cuprite crystals; in the oxidized ore of the Copper Queen mine above the third level, as masses of several hundred pounds weight (Petereit, 1907; Guild, 1910). Fine specimens of crystallized material, some coated with native silver, were recovered from a single pocket at the Campbell mine (Schwartz and Park, 1932; Schwartz, 1934). Calumet and Arizona mine, as small crystals and irregular networks throughout cuprite and in earthy mixtures of cuprite, limonite, and kaolinite (Ransome, 1904; Holden, 1922; Papish, 1928; Schwartz, 1934; Frondel, 1941). Junction mine (Mitchell, 1920). Tombstone district, as microscopic particles in cuprite (Butler et al., 1938). Turquoise district, as large, arborescent masses. Cochise district, Johnson Camp area, in trace amounts in the oxidized ores of the Mammoth, Republic, and Copper Chief mines (Kellogg, 1906; Cooper and Silver, 1964). Huachuca Mountains, in dark basaltic rock on claims of the Jack Wakefield

Mining Company (Robert Davis, pers. comm.).

*Gila County:* Globe-Miami district, in quartzite, Old Dominion mine, as small, hackly particles (Ransome, 1903; Peterson, 1962). Uncommon as a secondary mineral at the Castle Dome mine (Peterson et al., 1946, 1951). Banner district, associated with tenorite, chrysocolla, and cerussite at the 79 mine (Stanley B. Keith, pers. comm., 1973).

*Graham County:* Lone Star district, Gila Mountains, of erratic occurrence along thin veinlets or fractures, usually associated with cuprite or chalcotrichite in the oxidized zone of the Safford porphyry copper deposit (Robinson and Cook, 1966).

*Greenlee County:* Clifton-Morenci district, common in the upper parts of the veins of the district, in part as branching coralloid forms and groups of indistinct crystals (Kunz, 1885), mostly with cuprite, at the upper limits of the chalcocite zone; Williams vein, as solid copper up to 8 inches thick, with fibrous structure and probably pseudomorphous after chalcocite (Lindgren, 1904, 1905; Reber, 1916).

*Maricopa County:* Cave Creek district, Red Rover mine, as tiny scales impregnating schist (Lewis, 1920).

*Mohave County:* In small amounts as a secondary mineral in the Wallapai district, Cerbat Mountains (Thomas, 1949). Observed at Bill Williams Fork by Genth (1868). Noted in small amounts by Bastin (1925) in thin, plate-like masses, associated with chalcocite at the King claim, Mineral Park district (see also Eidel et al., 1968).

*Pima County:* Santa Rita Mountains, at several properties in the Helvetia-Rosemont district (Schrader and Hill, 1915); found in exploratory drill core with kinoite, apophyllite, djurleite, bornite, and chalcopyrite, between Helvetia and Rosemont (Anthony and Laughon, 1970). Ajo district, sparingly distributed as an oxidation product in the New Cornelia mine (Schwartz, 1934; Gilluly, 1937). Pima district, as a secondary mineral at the Pima mine (Journeay, 1959); the New Years mine, with malachite and cuprite (Stanley B. Keith, pers. comm., 1973); Mission mine in gypsum (UA 5560) and with cuprite (Stanley B. Keith, pers. comm., 1973). W. P. Blake (1855) observed "the pure metal" at a locality he described as "near Altar," in a vein containing "the red oxyd of copper . . . and green crusts of carbonate." South Comobabi Mountains, Cababi district, Mildren and Steppe claims (Williams, 1963). Silver Bell Mountains, El Tiro pit (Kerr, 1951).

**Copper.**
Ajo, Pima County. Harry Roberson collection.
Jeff Kurtzeman.

*Pinal County:* Ray district, Ray Central mine, as large masses (Ransome, 1919) and as unusually large sawtooth-shaped single crystals and aggregates up to 5½ inches long by 1 inch wide, loose in clay (White, 1974); also associated with cuprite and malachite in the Pearl Handle pit (Stanley B. Keith, pers. comm., 1973). Galiuro Mountains, as twisted and wire-like masses in the oxidized ore at the Copper Princes mine. Mammoth district, San Manuel mine, generally in the lower part of the supergene sulfide zone (Schwartz, 1949, 1953) (UA 6769, a tetrahexahedral crystal). Pioneer district, Magma mine, in small amounts (Short et al., 1943). Copper Prince mine, Copper Creek (UA 536).

*Santa Cruz County:* Patagonia Mountains, Three R mine, as thin sheets and films apparently derived from chalcocite (Schrader and Hill, 1915; Schrader, 1917).

*Yavapai County:* Verde district, United Verde Extension mine, locally abundant with cuprite; as fine specimens from near Walker (Lindgren, 1926; Schwartz, 1938; Anderson and Creasey, 1958). Copper Basin district (Johnston and Lowell, 1961). Eureka district, Bagdad mine (Anderson, 1950).

## COQUIMBITE

Iron sulfate hydrate, $(Fe_{2-x},Al_x)(SO_4)_3 \cdot 9H_2O$.
A not uncommon mineral found in the oxidized portions of base metal deposits with other secondary sulfates.

*Cochise County:* Warren district, as porous crusts several inches thick at the Copper Queen mine (Merwin and Posnjak, 1937).

*Pima County:* Pima district (UA 5905), San Xavier West mine (Arnold, 1964).

*Yavapai County:* Black Hills, Jerome, United Verde mine as an aluminous variety, with copiapite and other sulfate minerals, formed as a result of the burning of pyritic ore (Anderson, 1927; Lausen, 1928). (UA 54; H 91845; H 90623.) An analysis by T. F. Buehrer (Lausen, 1928) gave:

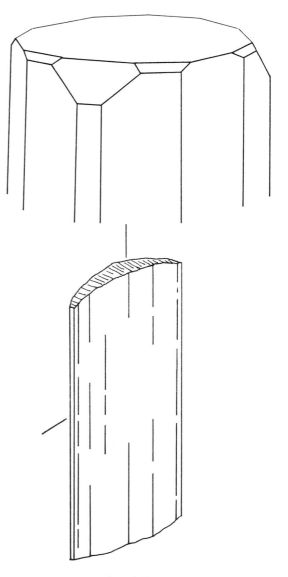

**Coquimbite.**
United Verde mine (Lausen, 1928).

$H_2O$ 31.72    $Fe_2O_3$ 14.69    $Al_2O_3$ 6.93
$SO_3$ 44.05                        $Na_2O$ 2.13

TOTAL 99.52%

Specific gravity: 2.07.

**Cornetite.**
Saginaw Hill mine, Pima County. Julius Weber.

## *CORDIERITE

Aluminum magnesium iron silicate, $(Mg,-Fe^{2+})_2Al_3(Si_5Al)O_{18}$. An early formed product of the metamorphism of argilliaceous sedimentary rocks which persists into conditions of higher-grade metamorphism.

*Graham County:* Graham Mountains, as dark gray, slightly greenish, large (up to ½ inch in diameter) crystals which have been completely replaced by muscovite (variety pinite), in graphic granite (John S. White, pers. comm.; locality observed by John W. Donowick).

*Mohave County:* From an unspecified locality in Mohave County, as sharp, 1 inch crystals (H 91900).

*Pinal County:* Sacaton Mountains, Gila River Indian Reservation, 10 miles north of Casa Grande, variety iolite, as cyclic twins, with sillimanite, corundum, and titanandalusite (Bideaux et al., 1960).

## *CORKITE

Lead iron phosphate sulfate hydroxide, $PbFe_3(PO_4)(SO_4)(OH)_6$. An uncommon secondary mineral found in the oxidized cappings of certain base-metal deposits.

*Yavapai County:* As drusy crusts of brilliant yellow-green crystals on vein quartz from a prospect near the Old Dick mine.

## *CORNETITE

Copper phosphate hydroxide, $Cu_3(PO_4)-(OH)_3$. A rare mineral found in the oxidized portions of base-metal deposits at only a few localities in the world.

*Pima County:* Southern side of Saginaw Hill, about 7 miles southwest of Tucson, in association with other oxide zone minerals including brochantite, pseudomalachite, malachite, libethenite, atacamite, and chrysocolla; in well-crystallized, fine-grained, soft clusters plastering the fracture surfaces in chert or chert gangue; as clear light peacock-blue crystal aggregates and as darker deep blue or deep greenish-blue crystalline clots which may alter to pseudomalachite. Crystals are usually between 0.05 and 0.20 mm in size (Khin, 1970).

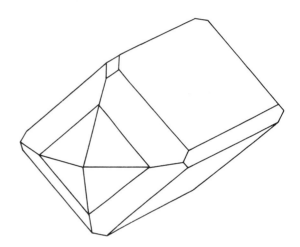

**Cornetite.**
Saginaw Hill (Khin, 1970).

## *CORNUBITE

Copper arsenate hydroxide, $Cu_5(AsO_4)_2-(OH)_2$. A dimorph of cornwallite and exceedingly rare in the oxide zone.

*Santa Cruz County:* At a prospect near the Humboldt mine as lovely pistachio-green crystals associated with luetheite and chenevixite. The crystals are measurable and clearly triclinic, verifying the symmetry proposed in the original description.

### *Cornuite (?)

Amorphous silicate of copper. An ill-defined material whose validity as a species is questioned by some authorities. Commonly found with chrysocolla.

*Pima County:* Cababi district, South Comobabi Mountains, Mildren and Steppe claims, a product of the oxidation of sulfide ores in quartz veins cutting andesite (Williams, 1962, 1963).

## *CORNWALLITE

Copper arsenate hydroxide hydrate, $Cu_5(AsO_4)_2(OH)_4 \cdot H_2O$. A rare secondary mineral first recognized from Cornwall and recently from several localities in the western United States. Found with other copper arsenates such as clinoclase, chenevixite, and duftite.

*Yavapai County:* Near Mayer, with clinoclase (UA 5758); Old Robertson claim, 4 miles east-northeast of Mayer, as films and coating (S 117492).

## CORONADITE

Lead manganese oxide, $Pb(Mn^{2+},Mn^{4+})_8O_{16}$. Related to cryptomelane. A rare mineral found in the oxidized portion of mineralized veins. First described in 1904 from the Coronado vein in the Clifton-Morenci district by Lindgren and Hillebrand.

*Greenlee County:* In fairly large amounts at the west end of the Coronado vein (Lindgren and Hillebrand, 1904; Lindgren, 1905; Guild, 1910; Fairbanks, 1923; Frondel and Heinrich, 1942; Fleischer and Richmond, 1943; Hewett and Fleischer, 1960) (H 83825). An analysis by W. F. Hillebrand gave:

| | | | | |
|---|---|---|---|---|
| $MnO_2$ | 56.13 | $MoO_3$ | 0.34 | CaO, |
| MnO | 6.56 | $Al_2O_3$ | 0.63 | MgO,+ |
| PbO | 26.48 | $Fe_2O_3$ | 1.01 | Loss 0.45 |
| ZnO | 0.10 | $H_2O$ | 1.03 | Insol. 7.22 |
| CuO | 0.05 | | | TOTAL 100.00% |

*Mohave County:* Artillery Mountains, in numerous veinlets of manganese oxides along fractures, joints, bedding planes, and breccia zones in Tertiary rocks; associated with cryptomelane, hollandite, psilomelane, pyrolusite, ramsdellite, and lithiophorite (Mouat, 1962).

*Pinal County:* Pioneer district, in small amounts near the lower limits of the oxidized zone, Magma mine, with sauconite (Short et al., 1943; Fleischer and Richmond, 1943; Hewett and Fleischer, 1960; Hewett et al., 1963).

## CORUNDUM

Aluminum oxide, $Al_2O_3$. A relatively common mineral which occurs in pegmatites and in contact metamorphic zones related to intrusions of silica-undersaturated rocks such as nepheline syenites.

*Mohave County:* Grand Wash Cliffs, Red Lake district, as blue, red, and white material in a pegmatite dike with andalusite.

*Pinal County:* From the Sacaton Mountains, Gila River Indian Reservation, with sillimanite, cordierite, and titanandalusite (Bideaux et al., 1960); also with rutile and quartz in felsite at Sec. 12, T.5S, R.5E (Wilson, 1969).

*Yavapai County:* An accessory mineral in trachyandesite in the Mingus Mountain quadrangle (McKee and Anderson, 1971).

## CORVUSITE

Vanadium oxide hydrate, $V_2O_4 \cdot 6V_2O_5 \cdot nH_2O$. A widespread mineral in the sandstone uranium-vanadium deposits of the Colorado Plateau; locally an abundant ore mineral of vanadium. Probably of primary origin.

*Apache County:* Monument Valley, abundant at the Monument No. 2 and Cato Sells mines where it is associated with coffinite, montroseite, uraninite, and doloresite (Weeks et al., 1955; Finnell, 1957; Witkind et al., 1963; Young, 1964). Lukachukai Mountains, in the Salt Wash Member of the Morrison Formation (Chenoweth, 1967).

*Navajo County:* Monument Valley, Monument No. 1 and Mitten No. 2 mines, in the dark uranium-vanadium ores, associated with a variety of uranium and vanadium minerals (Holland et al., 1958; Evensen and Gray, 1958; Witkind, 1961; Witkind and Thaden, 1963).

## COSALITE

Lead bismuth sulfide, $Pb_2Bi_2S_5$. A rare sulfosalt which occurs in moderate temperature veins in contact metamorphic deposits, and in pegmatites.

*Graham County:* With calcite and diopside on the Landsman claim, Aravaipa district.

## COTUNNITE

Lead chloride, $PbCl_2$. A rare mineral found in association with cerussite and anglesite in oxidized mineral deposits as an alteration product of galena.

*Mohave County:* Grand Wash Cliffs, where it is reported to occur as small veinlets replacing chalcocite at the Grand Gulch mine, Bentley district (Hill, 1914).

## COVELLITE

Copper sulfide, CuS. Occurs most commonly as a secondary mineral in the zone of oxidation and secondary enrichment of copper deposits, with chalcocite and other copper sulfides. Also occurs as a primary mineral. Widespread in small amounts in most copper deposits, usually as coatings and iridescent tarnish on other sulfides.

*Cochise County:* Tombstone district, lining boxwork structures formed by the removal of primary sulfide minerals (Butler et al., 1938). Warren district, Campbell mine, in massive sulfide ore bodies in limestone, believed to be of supergene origin; at the Cole mine, in veins in limestone with chalcocite (Schwartz and Park, 1932; Bain, 1952; Hutton, 1957); Junction mine (Mitchell, 1920). Cochise district, Little Dragoon Mountains, Johnson Camp area, where it is a common mineral in the upper portions of cupriferous veins and pyrometasomatic deposits (Cooper and Silver, 1964). Primos mine, near Dragoon (Palache, 1941a).

*Gila County:* Globe-Miami district, where it is a widespread but minor constituent of the disseminated copper deposits (Peterson, 1962); Roseboom (1966) notes that its occurrence with djurleite in the district should constitute an unstable assemblage; Castle Dome mine (Peterson, 1947; Peterson et al., 1951); Copper Cities mine (Peterson, 1954). Payson area, in small amounts replacing chalcopyrite and bornite. Dripping Spring Mountains, Banner district, as a replacement of bornite, with secondary chalcocite at the Christmas mine (Peterson and Swanson, 1956; Perry, 1969). 79 mine, in minor amounts, associated with oxidation products of galena (Kiersch, 1949).

*Graham County:* As films and blebs in enriched ores of the Aravaipa district (Ross, C. P., 1925a; Denton, 1947b).

*Greenlee County:* Clifton-Morenci district, Ryerson and Montezuma mines (Lindgren, 1905; Guild, 1910).

*Mohave County:* Common as an accessory mineral in sulfide deposits of the Wallapai district, Cerbat Mountains (Thomas, 1949).

*Pima County:* Ajo district, as minute blebs and coatings on other copper minerals in the New Cornelia open pit mine (Gilluly, 1937). Helmet Peak area (UA 7159). La Coronado mine, Pima district, Sierrita Mountains.

*Pinal County:* Mammoth district, widespread but sparse at the San Manuel mine (Chapman,

1947; Schwartz, 1949, 1953); Mammoth–St. Anthony mine, replacing chalcopyrite. Pioneer district, sparingly distributed in the Magma mine (Bateman, 1929; Short et al., 1943).

*Santa Cruz County:* Patagonia Mountains, as films on other sulfides at the Three R mine and other properties (Schrader and Hill, 1915). Oro Blanco district, near Ruby, Idaho and Montana mines (Warren and Loofburrow, 1932).

*Yavapai County:* As fine specimens from the United Verde Extension mine (Lindgren, 1926; Schwartz, 1938). Eureka district, Bagdad mine, found as films on chalcopyrite in the chalcocite zone (Anderson, 1950).

## *COWLESITE

Calcium sodium aluminum silicate hydrate, $(Ca,Na)Al_2Si_3O_{10} \cdot 6H_2O$. An uncommon member of the zeolite group of minerals.

*Pinal County:* Found as tiny, white, bladed crystals up to 1 mm long in vesicles in olivine bombs and scoria associated with cinder cone material of middle Tertiary age, $5\frac{1}{2}$ miles south of Superior. Associated with calcite and the zeolites thompsonite, chabazite, analcime, and mordenite (Wise and Tschernich, 1975). Chemical analysis of the anhydrous constituents of Superior material by electron microprobe gave the following result:

| | | |
|---|---|---|
| $SiO_2$  46.37 | CaO 15.12 | $Na_2O$ 0.80 |
| $Al_2O_3$ 25.87 | | $K_2O$  0.10 |
| $Fe_2O_3$  0.08 | | TOTAL 88.34% |

Specific gravity: 2.12±

## *CREASEYITE

Copper lead aluminum iron silicate hydrate, $Cu_2Pb_2(Fe,Al)_2Si_5O_{17} \cdot 6H_2O$. A rare mineral known only from three localities; two of these are in Arizona, one is in Sonora, Mexico. Found in partially oxidized lead-copper ores. The Mammoth–St. Anthony mine is the type locality.

*Maricopa County:* Potter-Cramer mine in vuggy, leached fluorite gangue with willemite, mimetite, wickenburgite, and ajoite (Williams and Bideaux, 1975).

*Pinal County:* Mammoth district, Mammoth–St. Anthony mine, as pale-green fibrous tufts in or on cerussite, wulfenite, and fluorite (Williams and Bideaux, 1975). The average of several analyses by M. Duggan gave:

CuO 13.5    Fe$_2$O$_3$ 12.3    Al$_2$O$_3$ 2.1
PbO 37.0    SiO$_2$ 25.5       H$_2$O  8.8
ZnO  1.2                        TOTAL 100.4%

## *CREEDITE

Calcium aluminum sulfate fluoride hydroxide hydrate, Ca$_3$Al$_2$(SO$_4$)(F,OH)$_{10}$·2H$_2$O. A rare mineral found in hydrothermal veins.

*Graham County:* Aravaipa district, Grand Reef mine, with linarite and anglesite in quartz-lined veins (Richard L. Jones, pers. comm., 1969).

## CRISTOBALITE

Silicon dioxide, SiO$_2$. A high temperature polymorph of SiO$_2$. Typically associated with volcanic rocks where it occurs in the metastable condition. Also the major constituent of opal.

*Pima County:* Roskruge Mountains, in cavities in andesite, associated with tridymite and clay (anauxite).

## Opal

Hydrous silica, SiO$_2$·nH$_2$O. Usually regarded as amorphous but actually a submicrocrystalline aggregate of crystallites of cristobalite containing much water (Frondel, 1962). Most opal in nature results from the accumulation of the tests of silica-secreting sea animals; also formed by deposition from hot spring waters, as a hydrothermal product in mineral deposits, deposition from underground and surface waters, and as a product of the weathering of silicate minerals. Distinct from chalcedonic silica (which see).

*Cochise County:* Tombstone district, Lucky Cuss mine, as small seams in ore, presumed to be of late stage hydrothermal origin (Butler et al., 1938).

*Gila County:* Globe-Miami district, Inspiration mine, in the oxidized zone as thin layers and in vugs, with chrysocolla and malachite (Sun, 1963). Northwestern Gila County, as fluorescent hyalite, at a number of uranium deposits in the Dripping Spring Quartzite (Granger and Raup, 1969).

*Greenlee County:* Clifton-Morenci district, in monzonite porphyry where it forms in cavities left after leaching of pyrite (Reber, 1916), especially around the periphery of the Morenci ore body (Moolick and Durek, 1966).

*Maricopa County:* On the north slope of Saddle Mountain, west of Hassayampa. In pink rhyolite, near Castle Hot Springs.

*Mohave County:* As fire opal, in tiny specks at Union Pass, west of Kingman. Also from the eastern slopes of the Black Mountains, northwest of Kingman.

*Pima County:* Ajo district, associated with shattuckite, ajoite, and quartz (Sun, 1961). Silver Bell Mountains, as veinlets in the oxidized zone of the Silver Bell mine with jarosite (Kerr, 1951). As common opal formed in the old dump at the Copper Glance mine, Pima district (T. V. Murchison, pers. comm., 1972).

*Pinal County:* In cavities in dacite at Picket Post Mountain, near Superior (UA 6879).

*Santa Cruz County:* Grosvenor Hills, near Santa Cruz.

*Yavapai County:* Bradshaw Mountains, with chalcedony, 14 miles from Mayer on the Agua Fria River (Hewett et al., 1963). Eureka district, Bagdad mine, in quartz veins. East of the Black Canyon Road, near Moore Wash.

*Yuma County:* Muggins Mountains, Red Knob claims, associated with secondary uranium minerals (Outerbridge et al., 1960).

## Crocidolite
### (a variety of RIEBECKITE)

## CROCOITE

Lead chromate, PbCrO$_4$. An unusual secondary mineral formed in oxidized lead deposits where there is a source of available chromium; commonly associated with galena, sphalerite, cerussite, and exotic secondary lead chromate minerals.

*Gila County:* Workmans Creek, Sierra Ancha Mountains, as smears in the upper Dripping Spring Quartzite, with metatorbernite (Robert O'Haire, pers. comm., 1972).

*Maricopa County:* Reported by Silliman (1881) in the "Vulture region" at the Collateral, Chromate, Blue Jay, and Phoenix properties. South of Wickenburg, at the Moon Anchor mine, Potter-Cramer property, and the Rat Tail claim, as a minor oxide zone mineral derived from lead-zinc ores; associated with a variety of secondary lead, zinc, and chromium minerals (Williams, 1968; Williams et al., 1970).

*Pima County:* Cababi district, South Comobabi Mountains, Mildren and Steppe claims (Williams, 1963).

## CRONSTEDTITE

Iron silicate hydroxide, Fe$_4^{2+}$Fe$_2^{3+}$(Fe$_2^{3+}$Si$_2$)O$_{10}$(OH)$_8$. An uncommon member of the septachlorite group, related chemically to the chlorites and structurally to the kaolinite group of minerals. Occurs under conditions similar to those of the chlorites.

*Mohave County:* Aquarius Range, in pegmatite at the Rare Metals mine with sphene, monazite, apatite, and chevkinite (Kauffman and Jaffe, 1946; Heinrich, 1960).

## CRYPTOMELANE

Potassium manganese oxide, $K(Mn^{2+},Mn^{4+})_8O_{16}$. Commonly associated with the psilomelane manganese ores; probably most commonly of secondary origin. The Tombstone occurrence is a co-type locality.

*Cochise County:* Material collected at Tombstone by A. E. Granger exhibited fine-grained, cleavable, and fibrous varieties in the same specimen (Richmond and Fleischer, 1942). An analysis of massive cleavable material by Fleischer gave the following result:

| | | | | | |
|---|---|---|---|---|---|
| $MnO_2$ | 83.13 | BaO | 0.13 | $H_2O+$ | 2.58 |
| MnO | 2.08 | SrO | none | $Al_2O_3$ | 0.37 |
| CuO | 0.12 | CaO | 0.27 | $Fe_2O_3$ | 0.46 |
| NiO | none | $Na_2O$ | 0.44 | $SiO_2$ | 0.58 |
| CoO | none | $K_2O$ | 3.50 | $TiO_2$ | 0.01 |
| ZnO | 5.23 | $H_2O-$ | 0.81 | $P_2O_5$ | 0.07 |
| MgO | 0.05 | | | TOTAL | 99.83% |

Specific gravity: 4.33.

*Coconino County:* Peach Springs district, as fine-grained massive material intimately associated with hollandite. Contains 0.23% thallium and 5.5% BaO (Crittenden et al., 1962). Adams-Woodie prospect, along the Aubrey Cliffs, about 22 miles northeast of Peach Springs, cementing rock fragments in veins cutting Kaibab Limestone; associated with psilomelane (Hewett et al., 1963).

*Gila County:* Sierra Ancha district, Apache mine, as botryoidal material, intimately associated with hollandite; contains 0.34% thallium, about 9% BaO, and about 2% $K_2O$ (Crittenden et al., 1962; Hewett et al., 1963).

*Mohave County:* Artillery Mountains, Black Jack, Price, and Priceless veins, and the Plancha and Maggie Canyon bedded deposits (Mouat, 1962).

*Pinal County:* Galiuro Mountains, Blake Place (UA 7037). Pioneer district, Magma mine (Hewett and Fleischer, 1960).

*Santa Cruz County:* Patagonia Mountains, Patagonia district, at the Mowry mine (Fleischer and Richmond, 1943); Harshaw district, where

argentian cryptomelane is present in significant quantities (Davis, 1975).

*Yavapai County:* Burmeister mine, near Mayer, Sec. 17, T.11N, R.3W, with other manganese oxides interlayered with volcanic ash, clastic sediments, and a basalt flow; also found in and near mounds of opalized dolomite deposited after travertine and silica by an extinct spring (Hewett and Fleischer, 1960; Hewett et al., 1963).

## CUBANITE

Copper iron sulfide, $CuFe_2S_3$. Typically a high-temperature sulfide, commonly associated with chalcopyrite, with which it may be intimately intergrown, pyrrhotite, and pentlandite.

*Gila County:* Banner district, Christmas mine, as lamellae of probable exsolution origin in chalcopyrite, in the pyrrhotite-chalcopyrite zone of the lower Martin ore body (Perry, 1969; McCurry, 1971). Workman Creek area, Workman Adit No. 1, in chalcopyrite (Granger and Raup, 1969).

## CUMMINGTONITE

Magnesium iron silicate hydroxide, $(Mg,Fe,Mn)_7Si_8O_{22}(OH)_2$. A monoclinic amphibole which is typically a product of regional metamorphism, although it is known to form as a primary igneous mineral in certain mafic rocks.

*Mohave County:* Reported from the Antler mine.

## CUPRITE

Copper oxide, $Cu_2O$. A secondary copper mineral of widespread occurrence; found in many of the oxidized copper mines and prospects of Arizona where it may be associated with malachite, azurite, tenorite, limonite, or, locally, native copper.

*Cochise County:* Warren district, where it has been an important constituent of the ores. Magnificent cuprite specimens from Bisbee are to be found in mineral collections throughout the world. At the Copper Queen mine, mainly as earthy material mixed with limonite but also as crystals and as the variety chalcotrichite (H 97648). At the Calumet and Arizona mine, as large crystalline masses associated with native copper, and in beautiful druses of ruby-red crystals which are mostly simple cubes. (Douglas, 1899; Ransome, 1903, 1904; Mitchell, 1920; Holden, 1922; Schwartz and Park, 1932; Schwartz, 1934; Frondel, 1941). Tombstone

**Cuprite (chalcotrichite) on calcite.**
Bisbee, Cochise County. Julius Weber.

**Cuprite.**
Morenci, Greenlee County. Julius Weber.

district, as bright red cubic crystals associated with malachite, brochantite, and locally, connellite, which lines small cavities in cuprite at the Toughnut mine (Butler et al., 1938). Dragoon Mountains, Turquoise district, as aggregates of octahedral crystals. Cochise district, as deep-red splendent crystals lining pockets in quartz veins in the Texas Canyon Quartz Monzonite (Cooper and Silver, 1964).

*Gila County:* Globe-Miami district, Castle Dome and Iron Cap mines, where it occurs sparsely throughout the leached and chalcocite zones of porphyry copper deposits; associated with malachite, azurite, native copper, and turquoise (Schwartz, 1921, 1934; Peterson, 1947); as massive material and as chalcotrichite at the Buffalo and Continental mines, and as large, dull octahedra at the Old Dominion mine (Ransome, 1903; Peterson, 1962). Dripping Spring Mountains, Banner district, 79 mine (Keith, 1972).

*Greenlee County:* At several mines in the Clifton-Morenci district, at the upper limit of the chalcocite zone; as cubic crystals and as chalcotrichite (Kunz, 1885; Lindgren, 1903, 1904, 1905; Reber, 1916).

*Maricopa County:* White Picacho district, as a supergene mineral in some of the pegmatites.

*Mohave County:* Bill Williams Fork (Genth, 1868); Planet mine, with chrysocolla (McCarn, 1904). Cerbat Mountains, Wallapai district (Thomas, 1949), Altata mine, as "rich and beautifully crystalline" material.

*Pima County:* Santa Rita Mountains, as crystal aggregates lining cavities at Rosemont (Schrader and Hill, 1915). Silver Bell Mountains, as cubic crystals and as chalcotrichite in small fractures (*Engr. Mining Jour.,* 1904; Kerr, 1951). Waterman Mountains, Silver Hill mine,

with azurite and malachite (UA 6719). Pima district, Copper Glance mine, where it occurs with native copper which has replaced an iron pipe in the sump (UA 6386); present in the oxidized ores of the Twin Buttes open pit mine (Stanley B. Keith, pers. comm., 1973). Tucson Mountains, disseminated in porphyry over a considerable area on the Saginaw and Arizona Tucson properties, Amole district. Ajo district, where it occurs sparingly in the New Cornelia open pit mine (Schwartz, 1934; Gilluly, 1937). Cababi district, South Comobabi Mountains, Mildren and Steppe claims (Williams, 1963).

*Pinal County:* Copper Creek district, Childs-Aldwinkle mine (Kuhn, 1941). Mammoth district, San Manuel mine, near the base of the oxide zone (Schwartz, 1949, 1953; Thomas, 1966); Mammoth–St. Anthony mine, rarely as crystalline masses associated with chalcotrichite. Mineral creek property, Mineral Creek district, as slender ruby-red crystals up to 1 cm long, in a Holocene gravel deposit, with jarosite and goethite (Phillips et al., 1971). Ray district, as sparkling ruby-red aggregates in one of the stopes worked from the old Ray shaft (Ransome, 1919), and as crystalline material and chalcotrichite in the oxidized zone of the Ray deposit (Metz and Rose, 1966).

*Santa Cruz County:* Patagonia Mountains, Patagonia district, Westinghouse property. Oro Blanco Mountains, Montana mine, as fine bright crystals in vugs (Schrader, 1917).

*Yavapai County:* Black Hills, United Verde Extension mine, where it is locally abundant and commonly accompanied by native copper; as beautiful druses of crystallized material and as chalcotrichite (Lindgren, 1926; Schwartz, 1938; Anderson and Creasey, 1958). White Picacho district, as a supergene mineral in some of the pegmatites (Jahns, 1952). Copper Basin district, as massive crystalline material at Sec. 20 and 21, T.13N, R.3W (David Shannon, pers. comm., 1971). Walnut Grove district, Zonia Copper mine, with malachite and chrysocolla (Kumke, 1947).

*Yuma County:* Muggins Mountains, Red Knob mine, associated with wulfenite, vanadinite, chalcedony, and limonite (Honea, 1959).

## Cuprogoslarite
### (a variety of GOSLARITE)

## CUPROTUNGSTITE

Copper tungstate hydroxide, $Cu_2WO_4(OH)_2$. A rare secondary mineral formed by the alteration of scheelite; generally in concentric layers about a scheelite core.

*Cochise County:* Little Dragoon Mountains, Burrell claim, 1.5 miles west of Dragoon; altering from scheelite (Dale et al., 1960).

*Maricopa County:* Cave Creek district, where it occurs with ferberite (UA 1192). Recalculated analyses by W. T. Schaller (1932) gave:

(1)

| | | |
|---|---|---|
| $WO_3$ 55.36 | CaO 4.12 | MgO 0.67 |
| CuO 32.66 | | $H_2O$ 7.19 |
| | | TOTAL 100.00% |

(2)

| | | |
|---|---|---|
| $WO_3$ 59.04 | | MgO 0.45 |
| CuO 32.68 | CaO 2.89 | $H_2O$ 4.94 |
| | | TOTAL 100.00% |

*Pima County:* From the 200 level of the Helvetia mine, Santa Rita Mountains.

*Yuma County:* Reported from the Livingston claims, south of Quartzsite.

## CYANOTRICHITE

Copper aluminum sulfate hydroxide hydrate, $Cu_4Al_2(SO_4)(OH)_{12} \cdot 2H_2O$. A rare secondary mineral found with other oxidized copper minerals in some deposits.

*Cochise County:* Warren district, with azurite and malachite (UA 3107).

*Coconino County:* Grand Canyon National Park, Horseshoe Mesa, Grandview (Last Chance) mine, as radiating crystals of exceptional beauty and massive nodules and veins, closely associated with brochantite (Rogers, 1922; Gordon, 1923; Leicht, 1971).

**Cyanotrichite.**
Grandview mine, Grand Canyon, Coconino County.
Julius Weber.

*Greenlee County:* Copper Mountain mine near Morenci, as narrow seams in a siliceous gangue, coated with earthy hematite; as incrustations up to 2 mm thick, and as thin fibers and small tufts in cavities (Genth, 1890). The mean of three analyses by Genth gave:

| | | |
|---|---|---|
| CuO 46.71 | $SO_3$ 12.49 | $H_2O$ 21.89 |
| $Al_2O_3$ 16.47 | $Fe_2O_3$ 1.34 | Insol. 0.44 |
| | | TOTAL 99.34% |

*Yavapai County:* Jerome, in the lower part of the oxidized zone in chlorite schist which contains chalcopyrite; with antlerite, and brochanite.

# D

## Danaite
### (cobaltian ARSENOPYRITE)

### *DANALITE

Iron zinc manganese beryllium silicate sulfide, $(Fe,Zn,Mn)_4Be_3(SiO_4)_3S$. Occurs as a rare accessory mineral in granite and greisens.

*Yavapai County:* Black Hills, south of Jerome (UA 7970).

### *DATOLITE

Calcium borosilicate hydroxide, $CaBSiO_4(OH)$. An uncommon mineral found in metamorphosed limestones or basaltic rocks; it may occur in amygdules in basalts. Found with prehnite, diopside, and grossularite.

*Pima County:* Pima district, in thin veins in diopside-garnet tactites at the Pima mine; the material found was visible only microscopically.

# DAVIDITE

Iron lanthanum uranium calcium rare earth titanium oxide hydroxide, $(Fe^{2+},La,U,Ca,RE)_6$-$(Ti,Fe^{3+})_{15}(O,OH)_{36}$. A rare mineral which, in Arizona, occurs in a transitional contact zone between a metaspessartite dike and quartz monzonite intrusive.

*Pima County:* Quijotoa Mountains, at the Pandora prospect located about 5 miles west of Covered Wells, as dark brown, pitchy-lustered masses in a matrix of sphene, epidote, and feldspar, in a transition zone between a metaspessartine dike and a quartz monzonite intrusive; nearly entirely metamict (non-crystalline, as a result of $\alpha$-particle bombardment) (Pabst and Thomssen, 1959; Pabst, 1961).

# DELAFOSSITE

Copper iron oxide, $CuFeO_2$. A rare secondary mineral found in the oxidized portions of copper deposits with hematite, cuprite, tenorite, and native copper.

*Cochise County:* Warren district, Cole shaft, with cuprite (UA 7124); Calumet and Arizona mine, where it was formed in a stope above the 14th level of the Hoatson shaft, in "kaolin and ferruginous clay at about the lowest zone of oxidation" (Rogers, 1913; Pabst, 1938). The average of two analyses by G. S. Bohart follows:

| | | |
|---|---|---|
| Cu 41.32 | Insol. 0.21 | O 21.21 |
| Fe 37.26 | TOTAL (by difference) 100.00% | |

(see also Frondel, 1935).

*Pinal County:* Found at Ray, with chalcotrichite (UA 3925).

**Delafossite.**
Bisbee, Cochise County. Julius Weber.

*Yavapai County:* Verde district, at the United Verde mine, as crusts of black tabular crystals, up to 8 mm on an edge, perched on milky quartz.

# *DELESSITE

Magnesium iron aluminum silicate hydroxide, $(Mg,Fe^{2+},Fe^{3+},Al)_6(Si,Al)_4O_{10}(OH)_8$. A member of the chlorite group, known to occur as amygdules in volcanic rocks.

*Cochise County:* Reported from the Warren district.
*Mohave County:* At the Antler mine, located on the southwest flank of the Hualpai Mountains, Secs. 3, 4, and 9, T.17N, R.16W, in schist (Romslo, 1948).
*Pima County:* With epidote and pumpellyite as amygdules in altered andesite at the Chicago mine, near Sells.

# DESCLOIZITE

Lead zinc vanadate hydroxide, $Pb(Zn,Cu)$-$VO_4(OH)$. A secondary mineral which occurs in small amounts in some oxidized lead-zinc or copper deposits; associated with cerussite, vanadinite, and other secondary minerals. See also the related species mottramite.

*Cochise County:* Warren district, Shattuck mine (Harcourt, 1942), where it occurs in some quantity as the cuprian variety; reported to have been the first vanadium mineral found in the district. As small stalactites several centimeters long and up to about 8 mm across at the base. The mean of three analyses made by Wells (1913) follows:

| | | | | | |
|---|---|---|---|---|---|
| Insol. | 0.17 | ZnO | 0.31 | $P_2O_5$ | 0.24 |
| PbO | 55.64 | $V_2O_5$ | 21.21 | $CrO_3$ | 0.50 |
| CuO | 17.05 | $As_2O_5$ | 1.33 | $H_2O$ | 3.57 |

TOTAL 100.02%

Tombstone district, Lucky Cuss, Toughnut, and Tombstone Extension mines (Butler et al., 1938). An analysis by Hillebrand (1889b) on material from the Lucky Cuss mine (S 48406) is as follows:

| | | | | | |
|---|---|---|---|---|---|
| PbO | 57.00 | $As_2O_5$ | 1.10 | CaO | 1.01 |
| CuO | 11.21 | $P_2O_5$ | 0.19 | MgO | 0.04 |
| FeO | tr. | $H_2O$ | 2.50 | $K_2O$ | 0.10 |
| ZnO | 4.19 | Cl | 0.07 | $Na_2O$ | 0.17 |
| $V_2O_5$ | 19.79 | $SiO_2$ | 0.80 | $CO_2$ | 0.82 |

TOTAL 98.99%

*Coconino County:* Havasu Canyon, as fine stalactitic crystal groups.

**Descloizite.**
C and B Vanadium mine (Trebisky and Keith, 1975).

*Gila County:* Payson district, in small amounts at the Ox Bow and Zulu mines (Lausen and Wilson, 1927). Globe district, from a locality 2 miles north of the Old Dominion mine; found on the 400 foot level of the Comstock Extension mine; disseminated in the Defiance mine (Peterson, 1962). Apache mine, brown to reddish in poorly crystallized masses and coatings, often associated with vanadinite (Wilson, 1971). Banner district, 79 mine, as tiny wedge-shaped crystals on hemimorphite; locally, crystals up to ¼ inch encrust wulfenite and line cavities (Keith, 1972). C and B Vanadium mine, Dripping Spring Mountains 10 miles northwest of Christmas, as crystal aggregates and sharp, brown to brown-black crystals up to 5 mm, associated with vanadinite and mimetite (Trebisky and Keith, 1975).

*Maricopa County:* At several localities south of Wickenburg, including the Moon Anchor mine, the Potter-Cramer property, and the Rat Tail claim; associated with a suite of exotic secondary minerals (Williams, 1968). Vulture Mountains, at the Black Hawk property one mile south of the Vulture mine, as a velvety coating of fine crystals. Also from the pegmatites of the White Picacho district. Painted Rock Mountains, Rowley mine, near Theba, where it is sparse as black coatings of minute crystals associated with vanadinite (Wilson and Miller, 1974).

*Mohave County:* Grand Wash Cliffs, Bentley district, Grand Gulch mine (Hill, 1914).

*Pima County:* Tucson Mountains, Amole district, Old Yuma mine, rarely as brownish-orange crystals to 5 mm.

*Pinal County:* Mammoth district, Mammoth–St. Anthony mine, with mottramite, forming crusts of small pointed crystals (Galbraith and Kuhn, 1940); the mineral was first noted at the mine by Genth in 1887. Dripping Spring Mountains, Bywater mine (S 16420). Copper Creek district, Blue Bird mine (Kuhn, 1951). Mineral Creek district, Ray (S 93866). Slate Moun-

tains, as tiny red crystals coating vanadinite at the Turning Point mine (Hammer, 1961).

*Yuma County:* Near Radium Hot Springs (UA 7894). Reported from the Castle Dome district (Guild, 1910).

## *DEVILLINE

Calcium copper sulfate hydroxide hydrate, $CaCu_4(SO_4)_2(OH)_6 \cdot 3H_2O$. A rare species found in the oxidized portion of copper ore deposits.

**Devilline.**
Little Mary Mine (Williams, 1962).

*Cochise County:* Warren district.

*Coconino County:* Horseshoe Mesa, Grand Canyon National Park, found as bluish-green crusts of lath-like crystals with gypsum at the Grandview mine (Leicht, 1971).

*Pima County:* South Comobabi Mountains, Cababi district, Little Mary mine, as a zincian variety associated with brochantite, anglesite, zincian dolomite, and gypsum; in quartz veins cutting andesite (Williams, 1962, 1963) (UA 6737).

*Pinal County:* Mammoth–St. Anthony mine, Mammoth district, rarely as a scaly alteration of powdery djurleite.

## *DIABANTITE (a member of the CHLORITE group)

## DIABOLEITE

Lead copper chloride hydroxide, $Pb_2CuCl_2(OH)_4$. A rare secondary mineral previously known only from Somerset, England. Occurs in oxidized lead ores with boleite, cerussite, hemimorphite, and other secondary minerals. From a study of diaboleite and its associated minerals, Wenden (Winchell and Wenden, 1968) concluded that diaboleite formed at low temperatures consistent with data from their experiments on synthesis of the species under hydrothermal conditions (between $100°$ and $170°$ C).

*Maricopa County:* Painted Rock Mountains, at the Rowley mine, near Theba, where it occurs sparingly in association with linarite and anglesite. Distinct crystals were not observed; rather, it is present as sandy aggregates (Wilson and Miller, 1974).

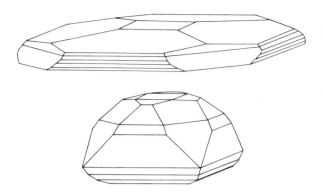

**Diaboleite.**
Mammoth–St. Anthony mine (Palache, 1941).

*Pinal County:* Mammoth district, Mammoth–St. Anthony mine, as small crystals of superb quality; associated with cerussite, wulfenite, phosgenite, and boleite, from the 400-foot level of the Collins vein (Palache, 1941b; Fahey et al., 1950) (UA 6473, 5168, 6202; BM 1947, 91). Crystallographic forms recognized by Palache include $\{001\}$, $\{010\}$, $\{0.1.10\}$, $\{110\}$, $\{101\}$, $\{012\}$, $\{011\}$, $\{021\}$, $\{201\}$, $\{112\}$, $\{111\}$. An analysis by F. A. Gonyer gave:

| | | | |
|---|---|---|---|
| PbO | 72.01 | $Cl_2$ | 11.42 | Insol. | 0.19 |
| CuO | 12.68 | $H_2O$ | 6.03 | $O = Cl_2$ | 2.57 |

TOTAL 99.76%

Specific gravity: 5.42.

## DIADOCHITE

Iron phosphate sulfate hydroxide hydrate, $Fe_2(PO_4)(SO_4)(OH) \cdot 5H_2O$. A secondary mineral found in gossans and formed in mine openings.

*Santa Cruz County:* Santa Rita Mountains, in fist-sized masses of fine-grained crystalline material, from Gringo Gulch, about 3 miles north-northwest of Patagonia. (Sample collected by James A. Yanez.)

## DIAMOND

Carbon, C. The only known Arizona occurrence of diamond is in meteorites. Evidence has been put forward to support the concept that meteoric diamonds were formed by shock on impact with the Earth (Carter and Kennedy, 1964); arguments refuting this idea were advanced by Anders and Lipschutz (1966).

*Coconino County:* In 1891 a 40-pound mass of the Canyon Diablo meteorite was dissolved and found to contain tiny black diamonds. Sub-sequently, small diamonds embedded in graphite, and in places associated with lonsdaleite, have been found in other fragments from the same fall (Foote, 1891; Kunz and Huntington, 1893; Ksanda and Henderson, 1939; Frondel and Marvin, 1967).

## *DIASPORE

Aluminum oxide hydroxide, AlO(OH). The dimorph of boehmite. A constituent of bauxite ores formed by extensive weathering and leaching of aluminous rocks, and common in solfatarically altered volcanics with alunite.

*Cochise County:* From the Warren district, as a microscopic constituent of sericitized quartz monzonite.
*Greenlee County:* Associated with widespread nacrite which formed as a result of solfataric alteration of dacites and quartz latites, in the Steeple Rock area northwest of Duncan.
*Maricopa County:* From an unspecified locality near Tempe (UA 7451).
*Santa Cruz County:* Abundant in hydrothermally altered andesites at Red Mountain, Patagonia. Occurs with quartz, pyrite, and alunite.

## *DICKITE

Aluminum silicate hydroxide, $Al_2Si_2O_5(OH)_4$. A member of the kaolinite group. Of hydrothermal origin, usually found in mineral deposits with sulfides.

*Cochise County:* Warren district, Lavender pit, as dense, white, earthy material cementing massive pyrite.
*Pinal County:* Mammoth district, San Manuel mine, associated with alunite in the most intensely altered rocks (Schwartz, 1953).

## DIGENITE

Copper sulfide, $Cu_9S_5$. Not uncommon in copper deposits where it usually is associated with chalcocite; known to form in both hypogene and supergene environments.

*Cochise County:* Warren district, where it has been observed in association with djurleite (Roseboom, 1966). Cochise district, Johnson Camp, Black Prince adit, intergrown with chalcopyrite (Cooper and Silver, 1964).
*Pinal County:* Pioneer district, forming a part of all chalcocite-bornite intergrowths on and below the 3,400-foot level of the Magma mine (Short et al., 1943; Morimoto and Gyobu, 1971).

*Yavapai County:* Black Mountains, United Verde mine, in the fire zone as distinct crystals (Harcourt, 1942).

## DIOPSIDE

Calcium magnesium silicate, $CaMgSi_2O_6$. A member of the pyroxene group. Generally of metamorphic origin; most abundant in crystalline limestones with other contact metamorphic calcium silicate minerals.

*Apache County:* Buell Park, a chromian variety found as detrital grains (Bideaux et al., 1960). Monument Valley, Garnet Ridge, emerald green diopside occurs in a breccia dike, associated with garnet (Gavasci and Kerr, 1968).

*Cochise County:* Warren district, as grains in unoxidized pyritic ores. Tombstone district, as small pale-green crystals in the contact zone at Comstock Hill (Butler et al., 1938). Little Dragoon Mountains, Johnson area, Republic and Moore mines, as one of the abundant contact metamorphic silicate minerals in limestone (Cooper and Huff, 1951; Baker, 1960); abundant in contact metamorphosed limestones at the April mine (Perry, 1964).

*Gila County:* With tremolite, a prominent gangue mineral at the Christmas mine, Banner district, as a common constituent of hornfels and an important skarn mineral (Peterson and Swanson, 1956; Perry, 1969); 79 mine, in contact metamorphosed limestones with andradite, tremolite, and epidote (Keith, 1972).

*Greenlee County:* Clifton-Morenci district, abundant in contact metamorphosed limestones of the Longfellow Formation (Lindgren, 1905; Moolick and Durek, 1966).

*Pima County:* Santa Rita Mountains, in the wall rocks of ore bodies in metamorphic limestones, Helvetia-Rosemont district (Schrader and Hill, 1915; Schrader, 1917); Peach-Elgin copper deposit, with almandine, and grossular, as bedded replacements (Heyman, 1958). Pima district, in contact metamorphosed limestones of the Sierrita Mountains; present in the skarns of the Twin Buttes mine (Stanley B. Keith, pers. comm., 1973); abundant in limestone hornfels with grossular and tremolite at the Pima mine (Journeay, 1959); and as the most abundant mineral in the hornfels host rock at the Mission mine (Richard and Courtright, 1959; Kinnison, 1966).

*Santa Cruz County:* Patagonia Mountains, Westinghouse property, Duquesne area (Schrader and Hill, 1915).

*Yavapai County:* Bradshaw Mountains, Big Bug district, with magnetite at the Henrietta mine.

*Yuma County:* Harquahala Mountains, Sec. 14, T.5N, R.10W, in coarse masses with sparse crystals of very light-colored material, along contact zones (Funnell and Wolfe, 1964; David Shannon, pers. comm., 1972).

## DIOPTASE

Copper silicate hydroxide, $CuSiO_2(OH)_2$. A rare mineral of secondary or mesogene origin found with other oxidized copper minerals.

*Gila County:* Globe-Miami district, Inspiration mine, as clusters of radiating crystals and sheaves coated in part by later chalcotrichite, with cubic pseudomorphs after fluorite(?). Payson district, as small, prismatic crystals at the Ox Bow and Summit mines (Lausen and Wilson, 1927).

*Greenlee County:* From the Bon Ton group, near the head of Chase Creek, about 9 miles from Clifton, "as brilliant crystals lining cavities in what is called locally 'mahogany ore'. . ." (Hills, 1882) (see also Kunz, 1885; Lindgren and Hillebrand, 1904; Lindgren, 1905; Guild, 1910).

*Mohave County:* Cerbat Mountains, Wallapai district, in minor amounts with chrysocolla in both blanket and vein deposits (Thomas, 1949).

*Pima County:* Santa Rita Mountains, Helvetia-Rosemont district (UA 8075). Papago Indian Reservation, Lake Shore copper deposit, associated with chrysocolla (Romslo, 1950).

*Pinal County:* Mammoth district, Mammoth–St. Anthony mine, as aggregates of deep emerald-green small crystals and as druses of stout crystals in chrysocolla (Galbraith and Kuhn, 1940; Palache, Fig., 1941b) (UA 7921; BM 1966, 112; H 104226, 104227). Pioneer district, Magma mine, as deep green crystal incrustations, partly coated with small olivenite crystals, from the upper levels, particularly from the outcrop of the No. 1 Glory Hole (Short et al., 1943). At an unspecified locality near Riverside on the Gila River (Smith, 1887). Galiuro Mountains, Virgus Canyon, Table Mountain mine, where it is fairly common in vugs in jasperoid of the Escabrosa Limestone (Bideaux et al., 1960; Simons, 1964) (UA 548, 673). Banner district, Christmas mine (UA 2549). Ray area (UA 1276, 2876).

*Yavapai County:* From Amazon Wash, near the Gold Bar mine, Black Rock district, 15 miles northwest of Wickenburg.

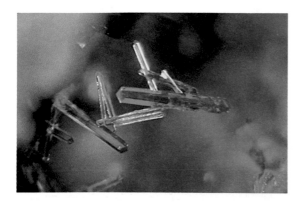

**Dioptase.**
Table Mountain mine, Pinal County. R. W. Thomssen collection. R. A. Bideaux.

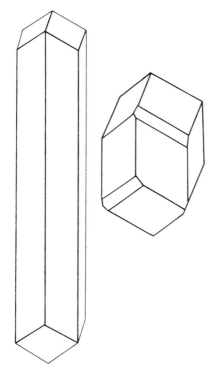

**Dioptase.**
Riverside (Kelvin) (Smith, 1887).

*Yuma County:* Harquahala Mountains, Harquahala mine, with crystalline hematite and malachite on cherty-appearing brownish rock (John S. White, Jr., pers. comm., 1972). Buckskin Mountains, in small quantities at the Chicago prospect.

## *DJURLEITE

Copper sulfide, $Cu_{1.96}S$. Djurleite was described by Roseboom in 1966, and his predic-

tion that it should prove to be a relatively common supergene mineral has been borne out. It occurs in intimate association with the more common chalcocite, and the distinction between the two usually rests on x-ray diffraction analysis. It is highly probable that as the details of the sulfide mineralogy of the copper mines of Arizona are worked out, many more occurrences of djurleite will come to light.

*Cochise County:* Warren district, observed in association with digenite (Roseboom, 1966).

*Gila County:* Globe-Miami district, where it occurs in association with covellite (Roseboom, 1966). (AM 34772, a pure, massive lump which constitutes part of the type material.)

*Pima County:* Near Helvetia and Rosemont, in drill core which penetrated contact metamorphosed Paleozoic limestones and dolomites, with chalcopyrite, bornite, apophyllite, copper, and kinoite (Anthony and Laughon, 1970).

*Pinal County:* Pioneer district, Magma mine, with digenite and bornite (Morimoto and Gyobu, 1971). At a small prospect in the Copper Creek area with vesignieite (UA 9607). Mammoth district, Mammoth–St. Anthony mine, replacing galena.

*Yuma County:* Dome Rock Mountains, as large, pure masses in vein quartz; partial oxidation has produced graemite, teineite, and brochantite (Williams and Matter, 1975).

## DOLOMITE

Calcium magnesium carbonate, $CaMg(CO_3)_2$. Most common as a sedimentary rock-forming mineral in a rock of the same name, and abundant in dolomitic limestones; also as a gangue mineral in hydrothermal mineral deposits.

*Cochise County:* Tombstone district, as massive beds interbedded with limestones and shales in the Naco Formation (Butler et al., 1938). Cochise district, Johnson Camp area (Cooper and Huff, 1951). Warren district in several beds of the Martin Formation (Hewett and Rove, 1930).

*Coconino County:* Portions of the Kaibab Limestone in the Flagstaff and Grand Canyon areas are highly dolomitic, especially beds west of El Tovar (Weitz, 1942).

*Gila County:* Christmas mine area, in many of the units in the Martin and Escabrosa Formations (Perry, 1969).

*Greenlee County:* Clifton-Morenci district, in beds in the lowest part of the Modoc Formation and in the upper part of the Morenci Shale;

sparingly in the Longfellow Limestone (Lindgren, 1905).

*Maricopa County:* In the Agua Fria-Humbug area, 38 miles north of Marinette; also east of Agua Fria, near Castle Hot Springs (Weitz, 1942).

*Mohave County:* As thick beds on Tassai Ridge, below Pierce Ferry, Lake Mead area; also in flat lying beds in the Peach Springs area (Weitz, 1942).

*Pima County:* Sierrita Mountains, Pima district, as coarsely crystallized material in fissures. Cababi district, South Comobabi Mountains, Mildren and Steppe claims (Williams, 1963).

*Pinal County:* Vekol mine, Casa Grande area (UA 7661).

*Santa Cruz and Pima Counties:* Santa Rita and Patagonia Mountains. Widely distributed in contact metamorphosed areas in limestones near contacts with intrusive rocks (Schrader, 1917).

*Yavapai County:* Black Hills, as a fairly abundant gangue mineral at the United Verde mine. Bradshaw Mountains, Tillie Starbuck mine. Hassayampa district, as small rhombohedral crystals coating walls of cavities. Burmeister mine, near Mayer, in mounds which are in part replaced by opal and chalcedony, deposited by an extinct spring (Hewett and Fleischer, 1960; Hewett et al., 1963) in beds up to 80 feet thick south and southwest of Seligman (Weitz, 1942).

*Yuma County:* Harquahala Mountains, as a bed several feet thick near the Bonanza mine.

## DOLORESITE

Acid vanadium oxide, $H_8V_6O_{16}$ or $3V_2O_4 \cdot 4H_2O$. Occurs intimately mixed with oxide minerals, especially paramontroseite which it replaces, in slightly oxidized uranium-vanadium ores in sandstone.

*Apache County:* Monument Valley, Monument No. 2 mine, as crystals and chocolate-brown bladed masses with satin-like cleavage surfaces; associated with a variety of other vanadium minerals (Stern et al., 1957; Jensen, 1958; Witkind and Thaden, 1963; Young, 1964).

## DOMEYKITE

Copper arsenide, $Cu_3As$. A rare mineral, probably of primary origin; locally associated with nickeline.

*Cochise County:* Specimens of this species were received by the University of Arizona many years ago, but the exact location and other data regarding them are not known (Guild, 1910).

## DUFRENOYSITE

Lead silver sulfide, $Pb_2Ag_2S_5$. A rare mineral known from only a few localities; associated with sphalerite and other sulfides and sulfosalts.

*Mohave County:* Reported from the Mineral Park district, but the exact locality is not known.

## *DUFTITE

Lead calcium copper zinc arsenate hydroxide, $(Pb,Ca)(Cu,Zn)(AsO_4)(OH)$. A rare mineral found in the oxidized zones of base metal deposits. The mineral noted below is believed to be the equivalent of the $\beta$-duftite of Guillemin (1956).

*Gila County:* Banner district, at the Christmas mine as brilliant pea-green crusts of microcrystals hidden beneath drusy quartz on which are diopside spherules.

*Maricopa County:* At several localities south of Wickenburg, including the Moon Anchor mine and Potter-Cramer property, as a secondary mineral in the oxidized portion of galena-sphalerite veins, associated with a variety of exotic secondary minerals of lead and chromium (see wickenburgite description, Williams, 1968).

*Mohave County:* Abundant at the Grand Deposit mine as massive material cementing sandstone.

## DUMONTITE

Lead uranyl phosphate hydroxide hydrate, $Pb_2(UO_2)_3(PO_4)_2(OH)_4 \cdot 3H_2O$. A rare secondary mineral associated with kasolite and other secondary uranium minerals.

*Santa Cruz County:* Nogales district, White Oak mine, with kasolite, uranophane, and autunite (Granger and Raup, 1962).

## DUMORTIERITE

Aluminum borate silicate oxide, $Al_7(BO_3)(SiO_4)_3O_3$. An uncommon mineral typically found in the same environments as kyanite, andalusite, staurolite, and sillimanite with which it may be associated in schists and gneisses.

*Cochise County:* A vein of purple dumortierite is reported to occur about 4 miles north northeast of Willcox, just east of U.S. Highway 666 (Duke, 1960; Funnell and Wolfe, 1964).

*Santa Cruz County:* Patagonia district, 12 miles northeast of Nogales (UA 6268).

*Yuma County:* Fine fibrous dumortierite altering to pyrophyllite and associated with kyanite occurs in boulders along the Colorado River between Yuma and Ehrenberg (Wilson,

1929). Material from the vicinity of Clip has been described by Diller and Whitfield (1889), Ford (1902), Schaller (1905), Bowen and Wyckoff (1926), and by Wilson (1933). An analysis by W. E. Ford on material from the Clip occurrence follows:

| | | |
|---|---|---|
| $Al_2O_3$ 63.56 | $SiO_2$ 29.86 | $Fe_2O_3$ 0.23 |
| $B_2O_3$ 5.26 | | $H_2O$ 1.41 |

TOTAL 100.32%

## DYSCRASITE

Silver antimonide, $Ag_3Sb$. A rare mineral found in silver vein deposits with galena, silver, and silver sulfosalts.

*Pima and Santa Cruz Counties:* Santa Rita and Patagonia Mountains, as a secondary mineral in quartz diorite, mainly in the Tyndall and Old Baldy districts (Schrader, 1917).

# E

## *EGLESTONITE

Mercury oxychloride, $Hg_4OCl_2$. Commonly associated with other mercury minerals in ore deposits of that metal; closely associated with calomel from which it may be derived by oxidation.

*Maricopa County:* Sunflower district, Mazatzal Mountains, with calomel and metacinnabar (UA 217; specimen collected by Carl Lausen).

## ELBAITE (see TOURMALINE)

## EMBOLITE

Silver chloride-bromide, $Ag(Cl,Br)$. An intermediate member of the chlorargyrite-bromargyrite series formed, like them, from primary silver minerals in the oxidized portions of silver deposits.

*Cochise County:* Tombstone district, Toughnut mine; State of Maine mine, with chlorargyrite, as the primary silver ore mineral (Rasor, 1938). Especially abundant at the Commonwealth mine at Pearce where it occurs with chlorargyrite, bromargyrite, iodargyrite, and acanthite in quartz veins (Endlich, 1897; Guild, 1910) (UA 3567, 7527). Warren district, with malachite (H 94744).

*Pinal County:* Mammoth district, Mammoth–St. Anthony mine, as octahedral crystals on caledonite.

*Santa Cruz County:* Noon Camp, Nogales area (S 48704).

## EMMONSITE

Iron tellurium oxide hydrate, $Fe_2Te_3O_9 \cdot 2H_2O$. A rare secondary mineral formed by the alteration of tellurides· and tellurium. The occurrence near Tombstone is the type locality.

*Cochise County:* As "yellowish green, translucent, crystalline scales and patches throughout a rather hard brownish gangue composed of lead carbonate, quartz, and a brown substance containing oxidized iron and tellurium plus water" (Hillebrand, 1885), from an unknown locality near Tombstone. One of several partial analyses by Hillebrand on impure type material gave:

| | | |
|---|---|---|
| Te(Se) 59.14 | Zno 1.94 | CaO 0.56 |
| Fe 14.20 | | |

The specific gravity was determined as being "not less than 5." Toughnut-Empire mine, with mackayite, and native tellurium (Bideaux et al., 1960); Frondel and Pough (1944) described the mineral from Tombstone as occurring in a hard brownish gangue composed of an intimate mixture of cerussite, quartz, and a brownish oxygenated compound of iron and tellurium.

## ENARGITE

Copper arsenic sulfide, $Cu_3AsS_4$. Occurs in veins and replacement deposits formed under moderate temperature conditions; associated with other sulfides.

*Cochise County:* Warren district, Campbell mine, as rounded grains and blades, primarily in chalcocite but in bornite as well; also associated with tetrahedrite, tennantite, and famatinite (Schwartz and Park, 1932).

*Gila County:* Globe-Miami district, Miami mine, with tennantite and aikenite in veins cutting chalcopyrite (Legge, 1939). Superior district, where it is the most important ore mineral of copper in the lowest levels of the Magma mine (Short et al., 1943). Galiuro Mountains, sparingly with tennantite at the Childs-Aldwinkle mine (Kuhn, 1941).

*Santa Cruz County:* Patagonia district, Volcano mine (UA 5858).

*Yavapai County:* Wickenburg Mountains, Monte Cristo mine, where it is associated with tennantite, nickeline, and native silver (Bastin, 1922).

## ENSTATITE

Magnesium silicate, $MgSiO_3$. A member of the pyroxene group of rock forming minerals. Common in mafic igneous rocks such as gabbro, norite, and peridotite and their extrusive equivalents. Associated with calcic plagioclase feldspars.

*Gila County:* Dripping Spring Mountains, Banner district, on the Reagan claims near the 79 mine.

*Pima County:* Ajo district, in the basal facies of the Batamonte andesite series (Gilluly, 1937).

*Santa Cruz County:* Sparingly in some andesites of the Santa Rita and Patagonia Mountains.

## EPIDOTE

Calcium aluminum iron silicate hydroxide, $Ca_2(Al,Fe)_3Si_3O_{12}(OH)$. Occurs in a wide variety of rock types; characteristically a product of low to medium grade thermal metamorphism of igneous and sedimentary rocks. In the southwestern United States it is a product of propylitic alteration of country rock, associated with base metal mineralization. Also common in contact metamorphosed limestones with other calcium silicates. Widespread in the southern part of the state.

*Apache County:* Monument Valley, Garnet Ridge, present in the matrix of ejection boulders of garnet gneiss (Gavasci and Kerr, 1968).

*Cochise County:* Tombstone district, in shale and quartzite (Butler et al., 1938). Chiricahua Mountains, common in the California district where a copper-bearing epidote vein up to 5 feet in width extends for over one mile (Dale et al., 1960). Turquoise district, in the wall rocks of pyritic deposits in Abrigo Limestone. Warren district, Sacramento Hill, in hydrothermally altered porphyry dikes in limestone (Schwartz, 1947, 1958, 1959). Cochise district, abundant in metamorphosed limestones and shales near the Texas Canyon stock (Kellogg, 1906; Cooper and Silver, 1964).

*Gila County:* Globe-Miami district, found in the marginal parts of the mineralized area of the Castle Dome mine (Peterson et al., 1946; Creasey, 1959); also in the Copper Cities deposit (Peterson, 1954). Payson district, with chalcopyrite on the Harrington claims. Dripping Spring Mountains, Banner district, Christmas mine (Ross, C. P., 1925a; Perry, 1969); 79 mine, in contact metamorphosed limestones

with diopside, andradite, and tremolite (Keith, 1972).

*Graham County:* Widely distributed in contact metamorphic copper deposits, Aravaipa (Simons, 1964) and Stanley (Ross, C. P., 1925a) districts. Lone Star district, San Juan property, present in the strong propylitic alteration zone which is locally peripheral to a chlorite-pyrite zone, in andesite, associated with chlorite and carbonates (Rose, 1970).

*Greenlee County:* Clifton-Morenci district, in contact metamorphosed rocks, rarely as well-defined crystals (Lindgren, 1905; Reber, 1916).

*Mohave County:* Cerbat Mountains, Wallapai district, common as an alteration product of wall rocks in sulfide vein deposits in gneisses (Thomas, 1949).

*Pima County:* Santa Rita Mountains, where it is widespread in metamorphosed limestones in the wall rocks of copper deposits (Schrader and Hill, 1915; Schrader, 1917). Sierrita Mountains, common as a metamorphic mineral and as an alteration product in igneous dikes; Sierrita mine, as the variety pistacite, a product of the hydrothermal alteration of dioritic rocks (Roger Lainé, written comm., 1973); Pima district, in contact metamorphic deposits in limestones, in considerable amounts with magnetite, garnet, wollastonite, and hedenbergite (Webber, 1929; Eckel, 1930); in the tactites and skarns of the Twin Buttes mine (Stanley B. Keith, pers. comm., 1973). Santa Catalina Mountains, in contact metamorphic copper deposits near Marble Peak, in places as splendid crystals (Dale et al., 1960); Pontotoc mine (Guild, 1934). Ajo district, as a widespread but sparse mineral formed as an alteration product of dark silicate minerals (Gilluly, 1937; Schwartz, 1947, 1958; Hutton and Vlisidis, 1960). Tucson Mountains (UA 7139). Silver Bell Mountains, where it is an alteration product in dacite porphyry (Kerr, 1951).

*Pinal County:* Mammoth district, San Manuel and Kalamazoo ore bodies, in hydrothermally altered quartz monzonite and monzonite porphyries, associated with the less intensely altered areas; associated with zoisite, chlorite, hydrobiotite, and secondary biotite (Schwartz, 1947; Creasey, 1959; Lowell, 1968).

*Santa Cruz County:* Abundant in the metamorphosed limestones of the Santa Rita and Patagonia Mountains (Schrader and Hill, 1915); Tyndall district, Glove mine (Olson, 1966).

*Yavapai County:* Bradshaw Mountains, in lenses in schist; Weaver district, in dikes at Rich

Hill. Crystals with an indicated prism diameter of 5 inches have been reported from Pylan Creek, 12 miles southeast of Wagoner. White Picacho district, as small, widely dispersed crystals (Jahns, 1952). Big Bug district, Iron King mine, where it is found in gabbro and diorite, associated with clinozoisite (Creasey, 1952).

*Yuma County:* Abundant in metamorphosed limestones at a number of localities. Dome Rock Mountains, in wall rocks of cinnabar veins.

## EPSOMITE

Magnesium sulfate hydrate, $MgSO_4 \cdot 7H_2O$. A secondary mineral usually encountered as efflorescences in old mine workings or caves.

*Cochise County:* As a late secondary mineral in some of the mines of the Warren district (Richard Graeme, pers. comm., 1974).

*Greenlee County:* Clifton-Morenci district, as delicate efflorescences on the walls of mine openings (Lindgren, 1905; Guild, 1910).

*Pima County:* Silver Bell Mountains, as capillary, hair-like crystals at the El Tiro mine; also reported from the Pima district. Ajo district, in the oxidized portion of the New Cornelia ore body, possibly of post-mine origin (Gilluly, 1937).

*Pinal County:* Mammoth district, San Manuel mine, with other secondary sulfate minerals coating mine openings.

*Santa Cruz County:* From south of Patagonia (UA 6732).

## *ERIONITE

Calcium sodium potassium aluminum silicate hydrate, $(Ca,Na_2,K_2)Al_9Si_{27}O_{72} \cdot 27H_2O$. A member of the zeolite group. Disagreement exists in the literature as to whether or not this species is actually offretite. Hey and Fejer (1962) maintain that the species are identical and that offretite has priority; however, Fleischer (1966) states that a vote on the matter by the Commission on New Minerals and Mineral Names of the International Mineralogical Association was indecisive.

*Cochise County:* San Simon Basin, 7 miles northeast of Bowie, in bedded lake deposits associated with analcime, herschelite, chabazite, clinoptilolite, thenardite, and halite (Regis and Sand, 1967).

*Mohave County:* East of Big Sandy Wash, eastern half of T.16N, R.13W, in Pliocene tuff with analcime, chabazite, clinoptilolite, and phillipsite (Ross, 1928, 1941).

*Pinal County:* In a railway cut one mile north of Malpais Hill on the west side of the San Pedro River, with chabazite, phillipsite, heulandite, and calcite, on celadonite (?) (UA 9221).

## ERYTHRITE

Cobalt arsenate hydrate, $Co_3(AsO_4)_2 \cdot 8H_2O$. Commonly containing nickel; a complete substitutional series extends to annabergite, $Ni_3(AsO_4)_2 \cdot 8H_2O$. A secondary mineral usually formed by the oxidation of cobalt and nickel arsenides.

*Apache County:* Reported to occur in the White Mountains with cobaltite, exact locality not known.

*Gila County:* From a locality one half mile northeast of the Mule Shoe Bend of the Salt River.

*Navajo County:* Near the Salt River at Showlow (UA 1996).

*Yavapai County:* Black Hills, as powdery incrustations from alteration of cobaltite, near claims of the Old Prudential Copper Company (Guild, 1910).

## ETTRINGITE

Calcium aluminum sulfate hydroxide hydrate, $Ca_6Al_2(SO_4)_3(OH)_{12} \cdot 26H_2O$. Typically formed by the alteration of contact metamorphosed limestones.

*Cochise County:* As an alteration of calcium and aluminum silicates in the Lucky Cuss mine, Tombstone district (Butler et al., 1938) (H 68800). Material discovered by W. F. Stanton and analyzed by A. J. Moses (1893) gave the result shown below. This analysis is of interest because it permitted a formula to be assigned to the species for the first time.

| | | | |
|---|---|---|---|
| CaO | 25.615 | $H_2O(115°)$ | 33.109 |
| $Al_2O_3$ | 10.157 | Loss (red heat) | 10.872 |
| $SO_3$ | 17.675 | $SiO_2$ | 1.901 |

TOTAL 99.329%

## *EUCRYPTITE

Lithium aluminum silicate, $LiAlSiO_4$. Formed from the alteration of spodumene, commonly intergrown with albite in pegmatites.

*Maricopa County:* McDowell Mountains, with spodumene (AM 35337).

## EUXENITE

Yttrium calcium cerium uranium thorium niobium tantalum titanium oxide, $(Y,Ca,Ce,U,Th)(Nb,Ta,Ti)_2O_6$. A member of the euxenite-polycrase series. A rare mineral which occurs in granite pegmatites.

*Mohave County:* Cerbat Range, as scattered masses weighing up to 50 pounds in pegmatite of the Kingman Feldspar mine. In pegmatites as small pockets or kidneys, east of the Big Sandy River, south of Burro Creek and near the Aquarius Mountains (Shaw, 1959).

*Navajo County:* Hugh Baron claim, Holbrook district.

*Pima County:* Sierrita Mountains, New Year's Eve mine, associated with molybdenite and chalcopyrite.

# F

## *FAIRCHILDITE

Potassium calcium carbonate, $K_2Ca(CO_3)_2$. Found in clinkers formed in partly burned trees with buetschliite and calcite. Since the characterization of the species was made on material from Arizona as well as from Idaho, the occurrence noted below constitutes a co-type locality.

*Coconino County:* Grand Canyon National Park. Discovered by ranger William J. Kennedy in an unspecified, partly burned tree "at a fire on the north side of Kanobownits Canyon one-quarter mile from the Point Sublime road and one-half mile from the North Entrance road" (Milton, 1944; Milton and Axelrod, 1947; Mrose et al., 1966).

## FAMATINITE

Copper antimony sulfide, $Cu_3SbS_4$. Less common than enargite with which it commonly occurs and to which it is similar in structure. Typically formed in moderate-temperature replacement deposits and veins with other sulfides.

*Cochise County:* Tombstone district, in small amounts in the Ingersol and Toughnut mines (Butler et al., 1938). Warren district, Campbell mine, as rounded grains and blades, largely confined to chalcocite and bornite but also associated with tetrahedrite, tennantite, and enargite; may form coarse, graphic intergrowths with chalcocite (Schwartz and Park, 1932).

*Pinal County:* Pioneer district, reported with

the hypogene ores of the Magma mine (Hammer and Peterson, 1968).

## FAYALITE (see also OLIVINE)

Iron magnesium olivine, $(Fe,Mg)_2SiO_4$. Present in small amounts in certain felsic and alkaline volcanic and plutonic igneous rocks; also formed during the regional metamorphism of iron-rich sedimentary rocks.

*Gila County:* A minor constituent of volcanic bombs and stream gravels in the vicinity of Peridot and Tolklai (Mason, 1968).

## FERBERITE (see WOLFRAMITE)

## FERGUSONITE

Yttrium columbium tantalum oxide, $Y(Nb,Ta)O_4$. A rare mineral found in granite pegmatites with other rare earth minerals, columbates, and tantalates.

*Mohave County:* Aquarius Mountains, Rare Metals mine, in pegmatite (sample gave an x-ray diffraction pattern resembling fergusonite upon ignition) (Heinrich, 1960).

*Yavapai County:* Reported to occur in the Mica-Feldspar quarry northwest of the highway maintenance camp on White Spar Road, near Yarnell.

## FERNANDINITE

Calcium vanadium oxide hydrate, $CaV_2^{4+}\cdot V_{10}^{5+}O_{30}\cdot14H_2O$ (?). A rare vanadium mineral which may be of primary origin.

*Apache County:* Monument Valley, Monument No. 2 mine, with doloresite and other oxidized vanadium minerals (Witkind and Thaden, 1963).

## FERRIMOLYBDITE

Iron molybdate hydrate, $Fe_2(MoO_4)_3\cdot8H_2O$ (?). A secondary mineral typically formed by alteration of molybdenite.

*Gila County:* Globe-Miami district, Castle Dome and Copper Cities mines (Peterson et al., 1951; Peterson, 1962).

*Mohave County:* Kingman area as short fibrous material encrusting quartz (H 95978; also UA 8207–8211). Mineral Park district, in the oxide zone at Ithaca Peak as a common replacement product of molybdenite (Eidel et al., 1968).

*Pima County:* Cababi district, South Comobabi Mountains, Mildren and Steppe claims (Williams, 1963). Santa Rita Mountains, as hair-like crystals and tufts near Madera Canyon. The average of two analyses by Guild (1907) gave:

$Fe_2O_3$ 21.835  $MoO_3$ 60.805  $H_2O$ 17.355

TOTAL 99.995%

which yielded the formula $Fe_2(MoO_4)_3 \cdot 7H_2O$.

*Pinal County:* Galiuro Mountains, as yellow powder and radiating crystal aggregates in the Childs-Aldwinkle mine, Copper Creek district (Kuhn, 1941). Rare Metals mine, Tortolita Mountains (UA 9488).

*Santa Cruz County:* Patagonia Mountains, Red Mountain mine (Schrader and Hill, 1915).

*Yavapai County:* Found in the oxidized portions of sulfide deposits in the Copper Basin district (Johnston and Lowell, 1961).

## *FLAGSTAFFITE

Cis-terpin monohydrate, $C_{10}H_{18}(OH)_2 \cdot H_2O$. A mineral formed in fossil logs.

*Coconino County:* In debris washed down from the San Francisco Mountains, a few miles north of Flagstaff, as a filling in radial cracks of certain buried tree trunks; a yellowish resinous material probably derived from natural resins of the tree by processes of hydration or oxidation; as orthorhombic crystals up to 1 mm long in drusy cavities (Guild, 1920, 1921, 1922; Strunz and Contag, 1965). This is the type and only reported locality.

## Flint (see QUARTZ)

## FLUORAPATITE (see APATITE)

## FLUORITE

Calcium fluoride, $CaF_2$. A widespread and common mineral. Occurs most commonly as a primary mineral in veins of which it is the chief constituent, or in the gangue of lead, zinc, and silver ores. It is also found in sedimentary rocks such as limestones and dolomites, and in plutonic igneous rocks such as granites and monzonites.

*Cochise County:* Tombstone district, locally abundant in some silicified areas, particularly at the Empire mine (Butler et al., 1938). Purple crystals with quartz are reported from near Government Draw. Chiricahua Mountains, small quantities were mined from quartz veins near Paradise. Little Dragoon Mountains, in granite pegmatite with huebnerite (Guild, 1910; Palache, 1941a; Cooper and Huff, 1951; Cooper and Silver, 1964). Whetstone Mountains, in a vein west of San Juan siding (Hewett, 1964).

*Gila County:* Payson district, with epidote, Ox Bow mine (Lausen and Wilson, 1927). Eastern Tonto Basin, Packard claims, white, light blue, and purple, in veins (Batty et al., 1947). Globe district, in small amounts in open fractures at the Castle Dome mine, associated with barite (Peterson et al., 1951).

*Graham County:* Aravaipa district, Grand Reef mine and in veins of the Landsman group (Simons, 1964). As crystals in barite veins at Stanley Butte (Ross, C. P., 1925a).

*Greenlee County:* In gouge in veins in andesite at several properties near Duncan (Allen and Butler, 1921; Ladoo, 1923), especially the Luckie and Fourth of July mines where fluorite is covered by mammillary layers of psilomelane (Hewett et al., 1963; Hewett, 1964); also at the Ellis mine (Batty et al., 1947).

*Maricopa County:* Harquahala Mountains, in veins in Precambrian rocks at the Snowball property, with barite (Denton and Kumke, 1949; Hewett, 1964). Vulture Mountains, west of Morristown, in veins. Reported from the White Tank Mountains. In a quartz quarry at

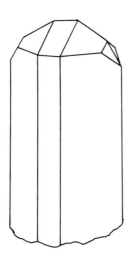

**Flagstaffite.**
(Guild, 1920).

Pinnacle Peak, east of Paradise Valley. Aguila district, Valley View mine, in layers consisting of fluorite crystals with barite which alternate with layers of black calcite, in a hydrothermal vein in an andesite flow (Hewett, 1964). Painted Rock Mountains, at the Rowley mine, near Theba, in small amounts, as colorless to violet masses in quartz veins; often as microcrystals showing the cube and dodecahedron and sometimes elongated on (111) (Wilson and Miller, 1974).

*Mohave County:* Black Mountains, as white to pale green bands or as linings of cavities in the northern part of the Oatman (San Francisco) district (Schrader, 1909); Skinner lode, south of Silver Creek "as beautiful octahedral crystals of green, white, and purple ... in a quartzose and feldspathic gangue with occasional gray spots of minutely diffused sulphide of silver"; also at other smaller properties of the district including the Caledonia, Dayton, Quackenbush, and Knickerbocker properties (Silliman, 1866). Cerbat Range, Wallapai district, Alta and Tintic mines, as a late gangue mineral known from two veins in a belt of sulfide-containing fissure vein deposits, associated with a granite porphyry stock (Thomas, 1949). Artillery Peak, Artillery Mountains district, with barite in veins containing manganese oxides which cut the Artillery Formation (Hewett and Fleischer, 1960; Hewett, 1964). Potts Mountain, Owens district, as purple material, cementing breccia (Robert O'Haire, pers. comm., 1972). Boriana mine, of a rich purple color, in quartz veins (Hobbs, 1944). Near Wright Creek ranch (about 15 miles south of Peach Springs), Blue Bird or Bountiful Beryl prospect, in several irregular pegmatite dikes in schist and granite, with sphene, garnet, schorl, and beryl, in microcline-albite-quartz pegmatite (Schaller et al., 1962).

*Pima County:* Santa Rita Mountains, New York mine, Helvetia-Rosemont district (Hewett, 1964). Silver Bell Mountains, at several properties (Guild, 1910). Sierrita Mountains, as veins from a few inches to 2 feet in width, Neptune property (Allen and Butler, 1921). As geodes, southern Santa Catalina Mountains (UA 1571).

*Pinal County:* Mammoth district, Mammoth–St. Anthony mine, abundant as microscopic crystals from the lower levels (Peterson, 1938; Fahey, 1955).

*Santa Cruz County:* Patagonia Mountains, as red material with embolite and chlorargyrite,

Alta mine, Harshaw district (Schrader and Hill, 1915; Schrader, 1917).

*Yavapai County:* Eureka district, Bagdad Copper mine, in small pockets in pegmatite, associated mainly with triplite and a green mica (Hurlbut, 1936). Bradshaw Mountains, Springfield group, Pine Grove district; Swallow mine, Castle Creek district. McCloud Mountains, at a prospect near the Leviathan mine from which fluorite has been shipped. As dodecahedral crystals 3 to 4 inches in diameter, northeast of Congress Junction, in pegmatite dikes (Jahns, 1953). Aguila district, Hatton mine, in hypogene veins as layers of crystals with barite which alternate with layers of black calcite (Hewett, 1964).

*Yuma County:* Castle Dome Mountains, Castle Dome district, as greenish, purple, and rose-colored crystals and clear cleavable masses up to several inches in diameter, locally associated with galena, barite, and wulfenite (Ladoo, 1923; Wilson, 1933; Batty et al., 1947; Peterson, 1947); Big Dome claim (Allen and Butler, 1921). Trigo Mountains, Silver district, as crystalline masses and dense, varicolored bands coating quartz, and as vein material with quartz and barite. Kofa district, in deposits in layered volcanic rocks (Hewett, 1964). Eastern slopes of the Plomosa Mountains where it is present in almost all barite veins in the area which form in association with manganese-oxide veins (Hewett, 1964). Buckskin Mountains, Chicago and Mammoth properties, with barite.

# *FORNACITE

Lead copper chromate arsenate hydroxide, $(Pb,Cu)_3[(Cr,As)O_4]_2(OH)$. A rare secondary mineral found at few localities in the world.

*Pima County:* Amole district, Old Yuma mine, on hematite and quartz (Bideaux et al., 1960); Gila Monster mine, with cerussite and wulfenite (Richard L. Jones, pers. comm.). With cerussite and chlorargyrite as films coating fracture surfaces in altered andesite, one half mile south of the New Cornelia pit, Ajo.

*Pinal County:* Mammoth district, at the Mammoth–St. Anthony mine, with wulfenite and fluorite, as crystals to several millimeters (Bideaux et al., 1960) (UA 3401). As sharply-formed crystals with shattuckite and cerussite in dumps near the Roadside mine, Slate Mountains.

*Yavapai County:* From Constellation, with crocoite (John S. White, Jr., written comm.).

## *FORSTERITE
### (see also OLIVINE)

Magnesium iron silicate, $(Mg,Fe)_2SiO_4$. The magnesium-rich portion of the olivine series of which fayalite is the iron-rich end-member. Forsterite is associated with dunites, peridotites, some basalts, and olivine-rich layered intrusive rocks. It is also formed during thermal and contact metamorphism of magnesium-bearing limestones and of dolomites.

*Cochise County:* Little Dragoon Mountains, near Johnson Camp, with epidote, zoisite, garnet, and other silicates, as a product of contact metamorphism of dolomitic rocks around the Texas Canyon stock (Cooper, 1957).

*Gila County:* Banner district, Christmas mine, where it is an important constituent of skarns formed in the Martin Formation and the Escabrosa Limestone; associated with garnet, idocrase, calcite, anhydrite, and xanthophyllite (Perry, 1969).

*Pima County:* Abundant in some portions of the meta-dolomite at the Lakeshore mine. Present in garnetite at the Twin Buttes mine, Pima district (Stanley B. Keith, pers. comm., 1973).

## FOURMARIERITE

Lead uranium oxide hydrate, $PbU_4O_{13} \cdot 4H_2O$. A secondary mineral formed from lead and uranium derived from the alteration of uraninite; also as a constituent of gummite formed by alteration of uraninite.

*Apache County:* Monument Valley, Monument No. 2 mine, as small, reddish grains in uraninite, with schoepite and becquerelite (Frondel, 1956; Witkind and Thaden, 1963).

## *FRANCOLITE

Calcium phosphate carbonate fluoride hydroxide, $2[Ca_5(PO_4,CO_3,OH)_3(F,OH)]$. The members of the apatite group are formed under a variety of conditions and are thus widespread in nature, in igneous, sedimentary, and metamorphic environments.

*Yavapai County:* Bagdad area, found rarely in the leached sulfide zone of the porphyry copper deposit in a quartz monzonite stock; as tiny cream-colored hexagonal plates deposited on chalcedony and in open cavities in the indigenous yellow-brown iron oxides (Anderson, 1950).

## FREIBERGITE
### (see TETRAHEDRITE)

## FREIESLEBENITE

Lead silver antimony sulfide, $AgPbSbS_3$. A rare medium to low temperature mineral which occurs in veins with acanthite, galena, siderite, and ruby silver.

*Yuma County:* Castle Dome Mountains, where a small amount was reported to have been mined and shipped with other argentiferous ores.

Fuchsite (chromian MUSCOVITE)

# G

## GADOLINITE

Beryllium iron yttrium silicate, $Be_2FeY_2Si_2O_{10}$. A rare mineral found principally in granites and granite pegmatites.

*Mohave County:* In the pegmatites of the Aquarius Range, 30 miles south of Hackberry, at the Rare Metals mine from which several tons of the mineral have been mined (Heinrich, 1960). Near Kingman, and, in the northern part of the county, as fragments in sand dunes. Cerbat Range, Chloride district, as black, vitreous, rough prismatic crystals a few inches long with beryl in pegmatite (Thomas, 1953).

*Pima County:* North side of the Rincon Mountains, with xenotime in biotite gneiss (Bideaux et al., 1960).

## *GAHNITE

Zinc aluminum oxide, $ZnAl_2O_4$. An uncommon member of the spinel group. Found in schists, high-temperature replacement bodies in metamorphic rocks, and in granite pegmatites.

*Maricopa and Yavapai Counties:* As small blue-gray to deep green crystals in the pegmatites of the White Picacho district which also contain greenish-black crystals of the ferroan variety of spinel, pleonaste (Jahns, 1952).

## GALENA

Lead sulfide, PbS. A widely distributed primary mineral commonly associated with zinc and copper sulfides and silver minerals. The most important ore mineral of lead; not uncommonly silver-bearing.

*Apache County:* Lukachukai Mountains, associated with secondary uranium and vanadium minerals in bedded deposits in the Salt Wash sandstone (Joralemon, 1952).

*Cochise County:* Warren district, Campbell mine, from which considerable quantities were mined during 1945–49 (Wilson et al., 1950); also at the Junction and Irish Mag mines. Tombstone district, in the ore bodies of both the "roll" deposits and the fissure veins. Extensively replaced by cerussite and to a lesser extent by anglesite (Hewett and Rove, 1930; Butler et al., 1938; Rasor, 1939). Dragoon Mountains, Turquoise district, as scattered bunches in copper sulfide ores; Mystery mine (UA 7188). Primos mine, Little Dragoon Mountains, Johnson Camp area (Wilson et al., 1950; Cooper and Huff, 1951; Cooper and Silver, 1964). Huachuca Mountains, at the Reef mine (Palache, 1941a). Chiricahua Mountains, Humboldt mine (Hewett and Rove, 1930); State of Texas and Panama mines (Wilson et al., 1951). Swisshelm Mountains, as replacements in Naco Limestone (Wilson et al., 1951).

*Coconino County:* Grand Canyon, on the south rim near Bright Angel Lodge, associated with the uranium component of copper-uranium-lead ores in Coconino Sandstone (Isachsen et al., 1955).

*Gila County:* Globe-Miami district, where it occurs sparingly in all of the deposits of the district (Peterson, 1962); Castle Dome mine (Brush, 1873; Peterson, 1947; Peterson et al., 1946); Defiance mine, as remnants enclosed in shells of anglesite and cerussite (Wilson et al., 1950). Dripping Spring Mountains, Banner district, in the 79 mine where it is the most abundant hypogene ore mineral, often coated by smithsonite (Kiersch, 1949; Wilson et al., 1951; Lewis, 1955; Keith, 1972); Christmas mine, in small amounts in ore bodies replacing dolomite (Perry, 1969).

*Graham County:* Aravaipa district (Denton, 1947b), Head Center, Iron Gap, and Grand Central mines (Ross, 1925a; Simons and Munson, 1963; Simons, 1964).

*Greenlee County:* Clifton-Morenci district, gold-bearing in the ores of the King mine; also at the Stevens group of claims.

*Maricopa County:* Painted Rock Mountains, Rowley mine (Wilson and Miller, 1974) (UA 4005). South of Wickenburg, at the Moon Anchor, Potter-Cramer property, and the Rat Tail claim, where oxidization has produced a variety of secondary lead and chromium minerals (Williams et al., 1970).

*Mohave County:* Common on many of the properties of the Cerbat Mountains (Haury, 1947; Tainter, 1947c) and Grand Wash Cliffs area (Schrader, 1909; Thomas, 1949). Sacramento district, from veins in metamorphic rocks (Silliman, 1866). Ithaca Peak area, with pyrite, chalcopyrite, and sphalerite in veins in a quartz monzonite stock (Eidel, 1966).

*Pima County:* Santa Rita Mountains, abundant at a number of properties including the Ridley mine near Helvetia. Empire Mountains, at the Chief, Prince, and other properties of the Hilton group (Schrader and Hill, 1915; Schrader, 1917; Wilson et al., 1950, 1951). Cerro Colorado Mountains, Cerro Colorado mine, with stromeyerite, tetrahedrite, and native silver. Sierrita Mountains, Pima district; the main ore mineral at the Sunshine mine, Papago district (Ransome, 1922). Olive Camp, Helmet Peak area. Cababi district, South Comobabi Mountains, Mildren and Steppe claims (Williams, 1963). Tucson Mountains, Old Yuma mine, with sparse tetrahedrite and supergene anglesite (Stanley B. Keith, pers. comm., 1973) (Guild, 1917).

*Pinal County:* Pioneer district, Silver King mine (Guild, 1917); in fairly large bodies in the Belmont mine and in the ores of the Magma mine (Short et al., 1943). Mammoth district, Mammoth-St. Anthony mine in the sulfide zone and altered to anglesite and cerussite in the oxidized zone (Peterson, 1938; Wilson et al., 1950; Fahey, 1955). Galiuro Mountains, Blue Bird mine, where it was the principal ore mineral (Simons, 1964); Saddle Mountain district, Adjust mine containing silver, and at the Saddle Mountain and Little Treasure properties. Dripping Spring Mountains, in several vanadium prospects near Kearny (Stanley B. Keith, pers. comm., 1973).

*Santa Cruz County:* Santa Rita Mountains, in many of the districts of the range; Tyndall district, Glove mine (Olson, 1966). Patagonia Mountains, where it is abundant in nearly all of the districts; the Mowry mine yielded ore having silver values up to 3800 ounces of silver per ton; Flux mine, as excellent specimens of cubo-octahedral crystals; Holland mine (Marshall and Joensuu, 1961); Harshaw district, Trench mine (Haggin shaft), with sphalerite, alabandite, and rhodochrosite (Hewett and Rove, 1930); Stella mine, near Duquesne, with chalcopyrite, diopside, and quartz (Stanley B. Keith, pers. comm., 1973). Oro Blanco Mountains, Montana mine, with sphalerite (Wilson et al., 1951).

*Yavapai County:* Bradshaw Mountains, at several properties in the Walker, Hassayampa, Big Bug, Turkey Creek, Peck, Pine Grove, Tiger, Tip Top, and Castle Creek districts, in many places with tetrahedrite; in the Big Bug district at the Iron King mine, as the most abundant ore mineral (Creasey, 1952; Anderson and Creasey, 1958). Eureka district, Hillside (Axelrod et al., 1951) and Bagdad mines (Anderson, 1950). Black Hills, Shea property, Verde district; Shylock mine, Black Hills district.

*Yuma County:* Trigo Mountains, Black Rock, Chloride, Silver King, and Silver Glance properties. The highly argentiferous galena at the Red Cloud mine, Silver district, probably contains acanthite (Hamilton, 1884; Foshag, 1919). Castle Dome Mountains, Flora Temple, Señora, Little Dome, Hull, Lincoln, and Adams properties. Harquahala Mountains, Bonanza mine (Bancroft, 1911; Wilson, 1933; Wilson et al., 1951). Plomosa Mountains, in barite veins with manganese oxides in layered volcanic rocks (Hewett, 1964).

## *GEARKSUTITE

Calcium aluminum hydroxide fluoride hydrate, $CaAl(OH)F_4 \cdot H_2O$. An uncommon mineral, formed in pegmatites, hydrothermal veins, sedimentary rocks, and as a result of hot spring activity.

*Graham County:* Aravaipa district, Grand Reef mine, as porcelaneous masses and powder, associated with barite, linarite, and other minerals in quartz lined vugs. (Richard L. Jones, pers. comm.).

## GEDRITE

Magnesium iron aluminum silicate hydroxide $(Mg,Fe,Al)_7(Al,Si)_8O_{22}(OH)_2$. Similar to the species anthophyllite but containing aluminum (Deere et al., 1962). Fleischer (1975), following the work of Rabbitt (1948), considers the mineral to be a variety of anthophyllite. Typically restricted in occurrence to metamorphic rocks.

*Santa Cruz County:* Patagonia Mountains, Duquesne-Washington Camp area, in contact metamorphosed limestone at the Westinghouse property (Schrader and Hill, 1915).

## GERHARDTITE

Copper nitrate hydroxide, $Cu_2(NO_3)(OH)_3$. A rare secondary mineral formed under oxidizing conditions in copper deposits in arid and semiarid regions. Occurs with such minerals as atacamite, brochantite, malachite, and azurite. Originally described by Wells and Penfield (1885) from the United Verde mine.

*Greenlee County:* On cliffs of granite porphyry in Chase Creek Canyon, as a bright green coating of small, rough, mammillary forms (Lindgren and Hillebrand, 1904; Lindgren, 1905; Guild, 1910).

*Pima County:* Pima district, Daisy shaft, Mineral Hill mine, as thin seams of granular material in cuprite which is coated by aurichalcite (Williams, 1961).

*Yavapai County:* United Verde mine, as small crystals on fractures in massive cuprite (Wells and Penfield, 1885). Analysis by R. C. Wells is as follows:

| CuO 66.26 | $N_2O_5$ 22.25 | $H_2O$ 11.49 |
|---|---|---|
| | | TOTAL 100.00% |

## GIBBSITE

Aluminum hydroxide, $Al(OH)_3$. A secondary mineral derived from the alteration of aluminous minerals; locally the chief constituent of bauxite deposits formed by weathering of aluminous rocks. Also as a low-temperature hydrothermal mineral in veins or cavities in igneous rocks.

*Cochise County:* Warren district, Sacramento pit, as massive, pale blue botryoidal material, formed as an alteration product of chalcoalumite (which see) (Williams and Khin, 1971).

## *GISMONDINE

Calcium aluminum silicate hydrate, $CaAl_2Si_2O_8 \cdot 4H_2O$. A member of the zeolite group; typically associated with lavas, especially basalts.

*Yuma County:* Found near Salome, with dioptase (William Panczner, pers. comm., 1972).

**Gismondine.**
Horseshoe Dam, Maricopa County. Julius Weber.

## GLAUBERITE

Sodium calcium sulfate, $Na_2Ca(SO_4)_2$. A constituent of sedimentary salt deposits; also as isolated crystals in sedimentary rocks and associated with fumerolic activity.

*Mohave County:* Detrital Valley, as crystals associated with halite and anhydrite in fine-grained sediments in the subsurface of Secs. 12 and 13, T.29N, R.21W. (H. Wesley Peirce, pers. comm., 1973).

*Yavapai County:* About 2 miles southwest of Camp Verde and in nearby Copper and Lucky Canyons, abundant as tabular crystals up to 2 to 3 inches in length, occurring loose in silts and clays and also in masses of thenardite in the Verde Formation. Commonly replaced by gypsum and by calcite and aragonite to form pseudomorphs (Snyder, 1971; see also Blake, 1890; Guild, 1910).

## GLAUCONITE

Potassium sodium aluminum iron magnesium silicate hydroxide, $(K,Na)(Al,Fe^{3+},Mg)_2(Al,-Si)_4O_{10}(OH)_2$. A member of the mica group. Formed during marine diagenesis under a rather restricted set of conditions in arenaceous sedimentary rocks; the usual coloring agent in "greensands." Believed to have formed by the alteration of other silicate minerals.

*Cochise County:* Present in certain beds in the Bolsa Quartzite of Cambrian age, north of the Swisshelm Mountains.

*Greenlee County:* Clifton-Morenci district, in shale of the Morenci Formation and in green shales above the Coronado Quartzite (Lindgren, 1905).

*Mohave County:* Valley of the Big Sandy, east of Wikieup, as extensive sand beds of glauconite-coated analcime grains (Wilson, 1944; Robert O'Haire, pers. comm., 1973).

## *GMELINITE

Sodium calcium aluminum silicate hydrate, $(Na_2,Ca)Al_2Si_4O_{12}\cdot6H_2O$. A member of the zeolite group, usually found in cavities in basaltic and related igneous rocks.

*Cochise County:* Reported as occurring near Bowie (Ted H. Eyde, pers. comm.).

## GOETHITE

Iron oxide hydroxide, $FeO(OH)$. Trimorphous with lepidocrocite and akaganeite. After hematite, the most common and abundant of the iron oxides. Formed under a wide range of oxidizing conditions but most typically as a weathering product of iron-bearing minerals. The abundant component of limonites. So abundant in the vicinity of mineralized areas in Arizona as to be practically ubiquitous.

*Cochise County:* Warren district, as thick botryoidal crusts having fibrous structure at the Shattuck and Copper Queen mines; may be mixed with cuprite, native copper, or tenorite (Frondel, 1941).

*Greenlee County:* With hematite, the most abundant oxidation product of the Morenci open pit mine, Clifton-Morenci district (Moolick and Durek, 1966).

*Mohave County:* Cerbat Mountains, Wallapai district, in widespread limonites in the oxidized zones of vein sulfide deposits; may be associated with hematite, jarosite, and plumbojarosite (Thomas, 1949).

*Pima County:* Princess claim, Cerro Colorado district (UA 7472). Cababi district, Mildren and Steppe claims (Williams, 1963). Ajo district, as fine, earthy powder at the New Cornelia mine (Gilluly, 1937). Pima district, abundant in the oxidized capping at the Twin Buttes open pit mine (Stanley B. Keith, pers. comm., 1973).

*Pinal County:* Mineral Creek deposit, northeast of Ray Hill in the Mineral Creek district, present in copper-rich Holocene gravels, associated with hematite, azurite, malachite, cuprite, and tenorite (Phillips et al., 1971). Mammoth district, San Manuel mine, abundant in surface rocks which have been stained red by hematite (Schwartz, 1953).

*Yuma County:* Near Parker, as well-crystallized sheaves and fanlike aggregates with malachite and hematite (Robert O'Haire, pers. comm., 1972).

## GOLD

Gold, Au. Widely distributed, but usually in very small amounts, as the native element. Seldom combines with other elements to form minerals, with the important exception of the gold tellurides and selenides. Most typically occurs in hydrothermal quartz veins and, because of its superior chemical inertness and mechanical resistivity, in placer deposits derived from them. Also associated with copper sulfide deposits where it may most readily be observed in the oxidized zones.

Placer deposits had early been worked by the Indians and by the Spanish explorers of the southwest, but attention became focused on the mineral wealth of Arizona during the middle and latter half of the nineteenth century with the modern re-discovery and exploitation of placer gold deposits, especially those of Mohave, Yuma, and Cochise Counties. With the subsequent rapid depletion of these limited placer and lode deposits, and the ascendancy of modern copper mining, especially of the huge, low-grade sulfide ore bodies, more and more gold came into production as a by-product of the smelting of base metal ores. The bulk of the placer production of the state had been made by the year 1885 (Moore, 1969), and the total output probably did not exceed much more than 400,000 ounces of gold (Elsing and Heineman, 1936). Since the late 1930s, substantially all gold production in the state has been from secondary sources. Shortly before World War II, gold production in Arizona reached an all-time high annual rate of about 300,000 ounces. Arizona ranks high among the states of the Union as a producer of gold.

The number of lode and placer gold mines and prospects of the state is large, and only a few can be listed here. The reader is referred to Arizona Bureau of Mines Bulletins 137 (lode deposits), 168 (placer deposits), and 180 (general) for more detailed information.

## Lode Deposits

*Cochise County:* Warren district, with ores of copper, lead, and silver (UA 9525) on bornite; Shattuck mine, as rich, spongy, gold matte in conglomerate (S 95730) (Ransome, 1904; Bain, 1952). Turquoise district, with copper, silver, and lead ores. Tombstone district, with lead-silver ores (Butler et al., 1938). Pearce district, with silver ores; in broad splotches and leaf form in the Pearce mine (Endlich, 1897). Dos Cabezas Mountains, Dos Cabezas and Teviston districts. Also from the Huachuca and Swisshelm Mountains.

*Gila County:* Globe-Miami district, with ores of copper and silver (Ransome, 1903; Peterson, 1938), as fine specimens from the Old Dominion mine. Dripping Spring Mountains, Barnes district, with copper, lead, and silver ores. Payson district. In a prospect ⅛ mile east of the Cowboy mine, about 8 miles northwest of Christmas (Stanley B. Keith, pers. comm., 1973).

*Graham County:* Galiuro Mountains, Rattlesnake district (Blake, 1902). Pinaleño Mountains, Aravaipa and Stanley districts. Gila Mountains. Santa Teresa Mountains.

*Greenlee County:* Clifton-Morenci district, with copper and silver ores (Lindgren, 1905).

*Maricopa County:* Vulture district (Metzger, 1938), with lead and silver ores (with wulfenite) (H 100312). Cave Creek district. Phoenix Mountains. Big Horn Mountains.

*Mohave County:* Black Mountains, San Francisco (Silliman, 1866; Ransome, 1923) and Katherine (Gardner, 1936) districts; massive in quartz in the Ruth vein, one quarter mile south of the Moss mine, San Francisco district (Lausen, 1931) (H 81355). Cerbat Mountains, in the Chloride, Cerbat, Stockton Hill, Mineral Park, and Wallapai districts (Thomas, 1949). Hualpai Mountains, Maynard district, with silver ores. Gold Basin district (Schrader, 1907). Williams Fork of the Colorado River (Blake, 1865).

*Pima County:* Ajo district, with copper and silver ores (Gilluly, 1937); sometimes found cementing fractured bornite in the New Cornelia pit. South Comobabi Mountains, Cababi district, Mildren and Steppe claims (Williams, 1963). Cahuabi mine, as flecks in quartz, with limonite, galena, and malachite (Pumpelly collection, H 81380). Santa Catalina Mountains, Molino Basin, in quartz stained with chrysocolla and shattuckite. Tucson Mountains, on limonite which is pseudomorphous after magnetite (UA 9724). Sungold mine, Arivaca district. Santa Rita Mountains, Greaterville area, Golden Gate mine (UA 344); rusty wire gold (UA 6734) (Schrader, 1917).

*Pinal County:* Superior and Ray districts, with copper, silver, and lead ores. Mammoth district, Mammoth–St. Anthony mine, as flecks in andesite(?) porphyry (H 99916) (Peterson, 1938); near Oracle, from John's ranch, as hackly gold in quartz (S R221); also in rusty pockets in quartz, with brochantite (H 108579). Casa Grande district, with copper and silver ores. Goldfield Mountains, Goldfield district. Pinto Creek, as stout, one-inch wires (H 100386).

*Santa Cruz County:* Santa Rita and Patagonia Mountains (Schrader, 1917). Oro Blanco Mountains, with ores of lead, silver, and copper; "white gold" in crystalline flakes was reported from the district; the color was said to have been due to small amounts of gallium (*Engr. Mining Jour.*, 1900). Nogales district, Little Annie mine (UA 2180).

*Yavapai County:* Verde district, with copper and silver ores. Bradshaw Mountains, Big Bug, Peck, Walker, and Tiger districts with ores of copper, silver, and lead (Anderson and Creasey,

1958); Hassayampa and Black Canyon (Guiteras, 1936) districts, with lead-silver ores; Pine Grove and Agua Fria districts, with copper-silver ores; Groom Creek, Turkey Creek, Bradshaw, and Tip Top districts, with silver ores. Santa Maria Mountains, Eureka district, with copper-silver-lead ores. Date Creek Mountains, Martinez district (Metzger, 1938). Wickenburg Mountains, Black Rock district (Lindgren, 1926). Copper Basin district (Johnston and Lowell, 1961). Red Bank, Weaver district, where coarse, gold-bearing quartz veins yielded a nugget shaped like a human molar which measured 53 x 47 mm and weighed 270.90 grams (Heineman, 1931). A nugget weighing 4.81 ounces was produced from placers near Congress Junction (UA 3470).

*Yuma County:* Kofa (Jones, 1915) and Gila Mountains, with silver ores. Harquahala Mountains, with ores of copper and lead, Penniehatche-pet, rich in quartz (H 100665). Williams Mountains, Cienega district, with copper ores. Castle Dome Mountains, with lead-silver ores. Also found in the Plomosa, Sheep Tank, Dome Rock, Laguna, and Trigo Mountains. Gila Bend Mountains, Fortuna mine (Blake, 1897). Dutchman mine, Bouse area, as flecks on massive, red, earthy hematite (S R15089).

## Placer Deposits

*Cochise County:* Small production was made from the Dos Cabezas, Teviston, Huachuca, Gleeson, Pearce, Gold Gulch (Bisbee) and Hartford (Huachuca) placers. Also 10 miles south of Hereford in Ash Canyon, as coarse placer gold (Robert O'Haire, pers. comm., 1972).

*Coconino County:* White Mesa district, near Lee's Ferry, Paria Creek, in the shales of the Chinle Formation as flakes, associated with small amounts of mercury (Lausen, 1936).

*Gila County:* Small production was made from Green Valley, Dripping Spring, Barbarossa, Globe-Miami, and Payson placers.

*Greenlee County:* Clifton-Morenci placers, along San Francisco and Chase Creeks.

*Maricopa County:* Cave Creek, Agua Fria, Pikes Peak, Big Horn, Vulture, San Domingo, and Hassayampa (Carter, 1911) placers. Near Phoenix, a water-worn nugget weighing 3½ ounces was found in a gravel pit near the Salt River (Scott Williams collection, H 106179).

*Mohave County:* Gold Basin, Chemehuevis, Lost Basin, Lewis, Wright Creek, Lookout, and Silver Creek placers.

**Gold in quartz.**
Rich Hill, near Congress, Yavapai County.
Jelks collection. Jeff Kurtzeman.

*Pima County:* Greaterville, Quijotoa, Horseshoe Basin, Arivaca, and Papago placers. A nugget valued at $228 was found at Greaterville in 1924. Of less commercial importance are the Las Guijas, Old Baldy, Old Hat, Baboquivari, Armagosa, and Alder Creek placers. Papago Indian Reservation, Golden Green mine, where William Coplen discovered a placer nugget which weighed a little over 8 ounces (*Arizona Daily Star,* March, 1967). Cañada del Oro placers, on the northwest flank of the Santa Catalina Mountains.

*Santa Cruz County:* In many small placer deposits including the Oro Blanco, Mowry, Harshaw, Patagonia, Tyndall, Nogales, and Palmetto placers.

*Yavapai County:* Weaver Creek, Rich Hill, Lynx Creek, Big Bug, Minnehaha, Hassayampa, Groom Creek, Copper Basin, Placerita, and Black Canyon (Guiteras, 1936) areas are the most important of the many placer deposits in the region. A nugget weighing 271 grams was found on Weaver Creek in 1930, and in 1932–33 several nuggets weighing 3 ounces or more were recovered from that area.

*Yuma County:* Most of the placer production has come from the La Paz, Dome, Plomosa Laguna, Castle Dome, Trigo, Muggins, and Colorado River districts. Of less importance are the Kofa, Fortuna, and Harquahala placers. A water-worn nugget, produced in July of 1863 by dry washing from the Myers claim on the Colorado River near Yuma, was ½ inch across (S 5549).

## GOSLARITE

Zinc sulfate hydrate, $ZnSO_4 \cdot 7H_2O$. A water-soluble secondary mineral formed through the alteration of sphalerite; commonly seen as efflorescences on the walls of old mine openings.

*Gila County:* Globe-Miami district, as efflorescences at the Continental and Old Dominion mines; also at the Castle Dome mine (Peterson et al., 1951; Peterson, 1962).

*Greenlee County:* Clifton-Morenci district, as efflorescences; Arizona Central mine (Lindgren, 1905; Guild, 1910).

*Mohave County:* Cerbat Range, Chloride district, at the de la Fontaine property.

*Pima County:* As cuprogoslarite in old mine workings of the Silver Bell district. South Comobabi Mountains, Mildren and Steppe claims (Williams, 1963).

## *GRAEMITE

Copper tellurite hydrate, $CuTeO_3 \cdot H_2O$. A rare secondary mineral known only from the two Arizona localities noted here; the Cole mine is the type locality.

*Cochise County:* Warren district, in a single, isolated specimen from the 1200 level of the Cole mine; the graemite is attached to the surface of and replacing teineite crystals embedded in malachite associated with a loose spongy aggregate of cuprite crystals (Williams and Matter, 1975). Average of several analyses gave:

CuO 31.0        TeO$_2$ 61.2        H$_2$O 8.2

                                    TOTAL 100.4%

Density: 4.13.

*Yuma County:* In a small prospect in the Dome Rock Mountains, associated with goethite, gypsum, and teineite in small cavities in djurleite in quartz-tourmaline gangue; as at Bisbee, the graemite appears to replace teineite (Williams and Matter, 1975).

## GRAPHITE

Carbon, C. Formed by the reduction of carbon-containing compounds during metamorphism or hydrothermal activity; possibly also formed as a primary constituent in igneous rocks.

*Cochise County:* Dos Cabezas Mountains, in thin veins or streaks in gold quartz veins. Graphitic clay is reported from near Benson (Guild, 1910). Warren district (Richard Graeme, pers. comm., 1973).

*Coconino County:* As small nodules in the Canyon Diablo and Elden meteorites (Ksanda and Henderson, 1939).

*Gila County:* Uraniferous graphite occurs in the northwestern part of the county in the Dripping Spring Quartzite at the Rainbow deposit (Granger and Raup, 1969).

*Mohave County:* Disseminated in Precambrian schist at Canyon Station Wash in the Cerbat Mountains (Schrader, 1907).

## GREENOCKITE

Cadmium sulfide, the hexagonal dimorph of CdS. A rare mineral most commonly found as coatings on sphalerite; rarely in cavities in mafic igneous rocks. It is probable that much of what has been called greenockite is hawleyite, the cubic modification of CdS.

*Cochise County:* Reported as a yellow coating on sphalerite from the Warren district.

*Coconino County:* Cameron district, Huskon mine, with pitchblende, bravoite, marcasite, pyrite, calcite and siliceous gangue, replacing wood structures as well as forming cement in sandstones of the Chinle Formation (Bollin and Kerr, 1958; Maucher and Rehwald, 1961).

## GROSSULAR

Calcium aluminum silicate, $Ca_3Al_2(SiO_4)_3$. A member of the garnet group. Typically a product of thermal and contact metamorphism of calcareous and aluminous rocks.

*Cochise County:* Warren district, as rounded crystals in unoxidized pyritic ores. Tombstone district, of cinnamon brown color in contact metamorphic zones; forms massive beds at Comstock Hill (Butler et al., 1938). Little Dragoon Mountains, as a gangue mineral with copper ores in the Johnson area (Kellogg, 1906; Cooper and Huff, 1951; Cooper, 1957; Cooper and Silver, 1964).

*Pima County:* Pima district, Pima mine, abundant in limestone hornfels with diopside and tremolite (Journeay, 1959; Himes, 1972); found in the skarns of the Twin Buttes mine (Stanley B. Keith, pers. comm., 1973). Silver Bell district, Atlas mine area, abundant in tactites as a product of pyrometasomatic alteration of limestone (Agenbroad, 1962). Helvetia-

Rosemont district, Peach-Elgin copper deposit, with garnet and diopside, as bedded replacements (Heyman, 1958).

*Santa Cruz County:* Santa Rita and Patagonia Mountains, in the latter range at Duquesne and Washington Camp, in limestone in contact metamorphic deposits associated with other contact silicate minerals (Schrader, 1917).

*Yavapai County:* With fluorite, at Bagdad (UA 8039).

## GROUTITE

Acid manganese oxide, $HMnO_2$. Originally described from the iron ranges of Minnesota where it occurs in the iron ores with manganite in vuggy cavities.

*Cochise County:* Warren district, Campbell mine, as tiny dark-brownish-black crystals coating sooty manganese oxides.

## GRUNERITE

Iron magnesium silicate hydroxide, $(Fe,Mg)_7Si_8O_{22}(OH)_2$. Characteristically in iron-rich, thermally metamorphosed, rocks; also found in the contact metamorphic environment with fayalite, hedenbergite, and almandine.

*Graham County:* Santa Teresa Mountains, in contact metamorphosed limestone.

*Yavapai County:* Tip Top district, north of the Foy mine.

## GUILDITE

Copper iron aluminum sulfate hydroxide hydrate, $CuFe(SO_4)_2(OH)\cdot4H_2O$. A very rare mineral formed during a mine fire in the United Verde mine at Jerome which is the type and only locality.

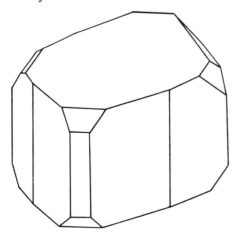

**Guildite.**
United Verde mine (Lausen, 1928).

*Yavapai County:* Black Hills, United Verde mine, formed under fumerolic conditions as a result of burning pyritic ores; relatively rare at the locality. Crystals up to 5 mm in size (Lausen, 1928; Laughon, 1970).

## GUMMITE

Hydrated uranium oxides. A generic term for colorful oxides of uranium whose true identity is not known; usually contains Pb and Th, with relatively large amounts of water. The term is similar in usage to wad and limonite. Gummites are usually derived through the alteration of uraninites.

*Apache County:* Monument Valley, Monument No. 2 mine, with other oxidized uranium and vanadium minerals (Mitcham and Evensen, 1955).

*Pima County:* Linda Lee claims, with torbernite and hematite in a vein cutting arkose (Robinson, 1955).

## GYPSUM

Calcium sulfate hydrate, $CaSO_4\cdot2H_2O$. A common mineral of widespread and abundant occurrence. Formed during the evaporation of the waters of inland seas and salt lakes and by the hydration of anhydrite. A common constituent of the oxidized zones of sulfide ores.

*Apache and Navajo Counties:* Monument Valley, in the copper-uranium-vanadium ores in channels at the base of the Shinarump conglomerate (Mitcham and Evensen, 1955).

*Cochise County:* Tombstone district, widespread as small crystals and scales lining small fissures in shale and as coatings in stopes in the mines (Butler et al., 1938). Near Douglas, where it was quarried (Santmyers, 1929). Sulphur Springs Valley, as beds in recent lake sediments. San Pedro Valley, deposits reported from both north and south of Benson (Guild, 1910). Warren district, as excellent examples of the curved "rams horn" variety (UA 294); Gleeson district, Shannon mine (UA 6790).

*Coconino County:* Along the eastern bank of the Little Colorado River, between Cameron and Leupp, as beds of varying thickness alternating with mudstones, in the Moenkopi Formation (Anthony et al., 1955; Baldwin, 1971).

*Gila County:* Globe-Miami district, as large crystals in the lower levels of the Copper Cities mine (Peterson, 1962).

*Greenlee County:* Clifton-Morenci district, common in oxidized deposits in limestone (Lindgren, 1905; Guild, 1910).

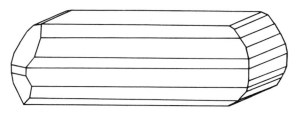

**Gypsum.**
Little Mary mine (Williams, 1962).

*Maricopa County:* 15 miles south of Gila Bend, as beds in sandstone and conglomerate, with celestite. Reported from the Superstition Mountains.

*Mohave County:* Abundant in the Virgin Valley badlands, as thick beds in South Mountain and Quail Canyon. Williams River in beds northeast of the Planet mine. As satin spar, Mammoth claim, 60 miles southeast of Kingman.

*Navajo County:* Mined near Winslow. As large plates of selenite in the upper member of the Permian Supai Formation (Peirce and Gerrard, 1966; Moore, 1968).

*Pima County:* Empire and Santa Rita Mountains, as beds up to 50 feet thick, of Permian age (Schrader and Hill, 1915). Santa Catalina Mountains foothills, north of Tucson. In recent sediments near Vail (Guild, 1910). Pima district, Minnie mine (UA 7162), and from near the San Xavier mine, as clear crystals to 3 to 4 inches. Mission mine, enclosing copper and, rarely, chalcotrichite (UA 9198). South Comobabi Mountains, Cababi district, Mildren and Steppe claims (Williams, 1963).

*Pinal County:* As thick beds in lake deposits on the east side of the San Pedro River, near Feldman (Hardas, 1966). Galiuro Mountains, Copper Creek district, as layers in Gila Conglomerate.

*Santa Cruz County:* Santa Rita Mountains, Montosa Canyon area, as thick beds of Permian age (Anthony, 1951); Cottonwood Canyon, as a secondary mineral in the Glove mine (Olson, 1966).

*Yavapai County:* In salt deposits of the Verde Valley (Guild, 1910); as pseudomorphs after glauberite (UA 9872). Abundant in decomposed dikes at the United Verde mine (Lindgren, 1926). Eureka district, Hillside mine, in the oxidized portion of the vein, associated with secondary uranium minerals (Axelrod et al., 1951).

*Yuma County:* Plomosa Mountains, at Mud-

ersbach Camp, as a bed several feet thick. As beds at the eastern base of the Harquahala Mountains. Massive and crystalline in the upper portions of veins in the Castle Dome, Silver, and Neversweet districts (Wilson, 1933); as gangue in veins in andesite, with fluorite, calcite, and barite at the Red Cloud mine (Foshag, 1919).

# H

## HALITE

Sodium chloride, NaCl. An important source of common salt, found in sedimentary beds formed by the evaporation of inland seas and salt lakes. Commonly associated with gypsum, anhydrite, and other salts of sodium and potassium.

*Apache and Navajo Counties:* In the subsurface of the southern portions of the counties in the Supai Salt basin which embraces about 2300 square miles; with anhydrite and dolomite, and clastic red beds (Peirce and Gerrard, 1966; Peirce, 1969).

*Cochise County:* San Simon Basin, 7 miles northeast of Bowie, in bedded lake deposits, associated with thenardite and a variety of zeolite minerals (Regis and Sand, 1967).

*Gila and Maricopa Counties:* Occurs throughout the Salt River Valley as incrustations derived from evaporation of saline springs (Guild, 1910; Wilson, 1944).

*Maricopa County:* A well drilled at Sec. 19, T.2N, R.1W, encountered substantial thicknesses of halite at a depth of 2350 feet in Cenozoic rocks. A large deposit of halite is suggested (Peirce, 1969).

*Mohave County:* With gypsum in the badlands of the Virgin River Valley near the Nevada border (Wilson, 1944). In the subsurface of Detrital and Hualpai Valleys with anhydrite and glauberite at Secs. 12 and 13, T.29N, R.21W; the deposits south of Red Lake are at least 1200 feet thick (Peirce, 1969, pers. comm., 1973). Deposits northwest of Red Lake playa are 500–700 feet thick (Pierce and Rich, 1962).

*Pima County:* Reported from the Papago Indian Reservation.

*Pinal County:* A drill hole sunk by the Humble Oil and Refining Company in 1972 penetrated 80 feet of halite and about 6000 feet of anhydrite, just west of the Picacho Mountains, Sec. 5, T.8S, R.8E (H. Wesley Peirce, pers. comm., 1973).

*Yavapai County:* In the salt deposits of the Verde Valley, associated with glauberite, gypsum, mirabilite, and thenardite (Blake, 1890; Guild, 1910).

## HALLOYSITE

Aluminum silicate hydroxide hydrate, $Al_2Si_2O_5(OH)_4 \cdot 2H_2O$. A hydrated member of the kaolinite group; sometimes formed by the action of sulfate-containing waters on kaolinite, but also formed independently of it. Observed as a hydrothermally-formed alteration mineral associated with porphyry copper deposits. The distinction has generally not been made between halloysite and the more hydrated mineral, hydrohalloysite for Arizona halloysites; hydrohalloysite is regarded by Fleischer (1975) as a synonym of endellite.

*Cochise County:* Warren district, Southwest mine, as a product of hydrothermal alteration (Schwartz, 1956).

*Gila County:* Globe-Miami district, a product of hydrothermal alteration in granite porphyry, in the vicinity of ore bodies in schist (Schwartz, 1947); Castle Dome mine, present in small amounts in capping and in the chalcocite zone in quartz monzonite (Peterson et al., 1946).

*Graham County:* Lone Star district, Gila Mountains.

*Greenlee County:* Morenci district, with allophane as a product of intense hydrothermal activity (Schwartz, 1947, 1958).

*Pima County:* Silver Bell Mountains, Silver Bell mine.

*Pinal County:* Mammoth district, San Manuel mine, in small veinlets in alunite-kaolinite rock (Lovering et al., 1950; Schwartz, 1953). Mineral Creek district, Ray ore body, in the hydrothermally altered porphyry stock, associated with sericite, kaolinite, and hydromuscovite; as veinlets and coating fracture surfaces; often exhibits colloform texture and may be intergrown with chrysocolla (Schwartz, 1934, 1947; Stephens and Metz, 1967).

## HALOTRICHITE

Iron aluminum sulfate hydrate, $FeAl_2(SO_4)_4 \cdot 22H_2O$. Water soluble, formed from the weathering of pyritic and aluminous rocks; observed in old mine workings. Commonly associated with gypsum and other secondary sulfate minerals.

*Cochise County:* Tombstone district.

*Coconino County:* Cameron district, as crossfiber veins in carbonaceous fossil wood and in the surrounding sediments, some stained inky blue by ilsemannite(?) (Austin, 1964).

*Gila County:* Banner district, 79 mine, as an abundant secondary mineral throughout the oxidized portions of the mine, often as arching whiskers up to 12 inches long growing from the walls of the workings and twisting in all directions. Growth rates of up to 3 inches per year have been measured for the currently forming mineral (Keith, 1972).

*Santa Cruz County:* Patagonia Mountains (UA 4992); Harshaw district, at the Chief mine (UA 9592).

## *HARMOTOME

Barium potassium aluminum silicate hydrate, $(Ba,K)(Al,Si)_2Si_6O_{16} \cdot 6H_2O$. A member of the zeolite group. Occurs principally in cavities and veins in igneous rocks.

*Greenlee County:* From an unspecified locality near Duncan (UA 5991).

## *HAUSMANNITE

Manganese oxide, $Mn_3O_4$ or $Mn_2^{2+}Mn^{4+}O_4$. Typically, hausmannite occurs in high temperature veins; it is also a contact metamorphic mineral and forms as a recrystallization product of pre-existing manganese minerals under conditions of metamorphism.

*Cochise County:* Replaces limestone (with braunite) at the White-Tailed Deer mine, Warren district.

*Pima County:* Arivaca district, COD mine, in veinlets cutting a matrix of braunite (Hewett, 1972).

*Santa Cruz County:* Tyndall district, Cottonwood Canyon, Glove mine, as a constituent of fine-grained brownish manganese oxides (Bideaux et al., 1960; Olson, 1966). Harshaw district, at the Hardshell mine, as a constituent of the manganese ores (Davis, 1975).

## HAWLEYITE (see GREENOCKITE)

## HEDENBERGITE

Calcium iron silicate, $CaFeSi_2O_6$. A member of the pyroxene group. Typically occurs in contact metamorphosed limestones, associated with other calcium silicates.

*Graham County:* Stanley Butte (UA 6987).

*Pima County:* Silver Bell Mountains. Sierrita Mountains, Pima district, common as gangue in contact metamorphosed limestones (Eckel,

1930; Irvin, 1959); Atlas mine area (Agenbroad, 1962).

*Santa Cruz County:* Patagonia Mountains, Westinghouse property, with diopside and other contact silicates (Schrader and Hill, 1915; Schrader, 1917).

## *HELVINE

Manganese iron zinc beryllium silicate sulfide, $(Mn,Fe,Zn)_4Be_3(SiO_4)_3S$. Helvine occurs in granites and granite pegmatites and in contact metamorphic rocks where it may be associated with other beryllium-bearing minerals.

*Cochise County:* Dragoon Mountains, Abril and Gordon (San Juan) mines, with berylliumbearing epidote (?) in tactite in a limestone replacement deposit (Warner et al., 1959; Meeves, 1966).

## HEMATITE

Iron oxide, $Fe_2O_3$. A mineral having great diversity of origin, found in igneous, metamorphic, and sedimentary rocks; in the latter as bedded deposits of commercial significance. Also in hydrothermal veins and in the gossans or leached cappings of base metal deposits. Only a few typical occurrences can be listed here.

*Cochise County:* Near Willcox, as finegrained, massive specularite. Tombstone district, near the Lucky Cuss mine, as radiating crystal aggregates in a vein in granodiorite (Butler et al., 1938). Warren district, Lavender pit (UA 7827). Dragoon Mountains, Black Diamond claim (UA 7357).

*Gila County:* Globe-Miami district, as massive bodies of specularite along the veins of the Old Dominion fault system as replacements of limestone or diabase; also in the Buckeye, Black Oxide, Big Johnnie, Stonewall, and other veins (Sanders, 1911; Peterson, 1962). Fort Apache Indian Reservation, Bear Spring Canyon, as extensive deposits with chert beds in the Mescal Limestone (Burchard, 1930, 1943). Aztec Peak area, reniform (Sinkankas, 1964).

*Gila and Navajo Counties:* Canyon Creek area, extensive occurrences in the middle member of the Mescal Limestone of younger Precambrian age, interbedded with ferruginous cherts, sandstones, and shales (Moore, 1968).

*Graham County:* Grand Reef area, as specularite, very abundant in the Lead King and Cobre Grande mines (Simons, 1964).

*Greenlee County:* Clifton-Morenci district, Manganese Blue mine (Lindgren, 1905).

*Maricopa County:* Pikes Peak area, northeast of Beardsley, as replacements in schist (Farnham and Havens, 1957). Painted Rock Mountains, Rowley mine, near Theba, where the powdery mineral is responsible for the ubiquitous red coloration (Wilson and Miller, 1974).

*Mohave County:* Cerbat and Aquarius Ranges, widely distributed, chiefly as the specular variety (Blake, 1865); as red ocher, at Chloride (UA 1836).

*Navajo County:* Sierra Ancha Mountains, Canyon Creek, as a large deposit of siliceous hematite, estimated at 10 million tons, ranging from soft, pulverulent, bright red material to hard, dark, blue oxide (Stewart, 1947).

*Pima County:* Santa Rita Mountains, Cuprite mine area. Empire Mountains, as red ocher at the Hilton mines (UA 7388). Pima district, at the Copper Glance mine (UA 7389); Twin Buttes mine (Stanley B. Keith, pers. comm., 1973). Ajo district, New Cornelia mine, as specularite in brilliant adamantine crystals as much as 2 inches across, of hypogene origin; also as the earthy variety, of supergene origin (Gilluly, 1937).

*Pinal County:* Near Winkelman, as an iridescent variety. Mammoth district, Mammoth–St. Anthony mine, as glistening black masses in the lower levels of the Collins vein (Peterson, 1938).

*Santa Cruz County:* Santa Rita Mountains, Tyndall district, Montosa Canyon, Isabella mine, as massive material of the specular variety with cerussite ("sand carbonate"), replacing Permian-age limestones (Schrader and Hill, 1915; Anthony, 1951). Patagonia Mountains, widely distributed throughout the range in mines and prospects; Patagonia district at Duquesne and Washington Camp, in limestone in contact metamorphic deposits, Mowry mine; also as a primary mineral in numerous siliceous veins filling fault fissures (Schrader, 1917).

*Yavapai County:* On the McBride claims, 17 miles south of Seligman, where large deposits of earthy material form irregular lenses in limestone near the contact with diorite. A high concentration of hematite and magnetite is noted in schist near Townsend Butte and the Howard Copper property. From near Camp Wood. Verde district, United Verde mine, the specular variety, as late-stage veinlets cutting massive sulfides; also common in Precambrian rocks south of Jerome (Anderson and Creasey, 1958).

*Yuma County:* Buckskin Mountains, Planet mine, as extensive replacement deposits with carbonate and silicate copper ores in limestone (Cummings, 1946b). As high quality crystals to over 2 inches from the vicinity of Bouse; these crystals, sometimes mirror bright and twinned, with quartz crystals, are perhaps the finest United States occurrence (UA 8640). Trigo Mountains, southeast of Ehrenberg (Bancroft, 1911). Harquahala Mountains, Harquahala mine, as tiny, lustrous black crystals with dioptase and malachite on brownish "cherty-appearing" rock (John S. White, Jr., pers. comm., 1972).

## *HEMIHEDRITE

Zinc lead chromate silicate fluoride, $ZnF_2$-$[Pb_5(CrO_4)_3(SiO_4)]_2$. A rare secondary mineral formed in the oxidized portions of galena-sphalerite veins.

**Hemihedrite penetration twin.**
Florence Lead-Silver mine, Tortilla Mountains,
Pinal County. J. W. Anthony.

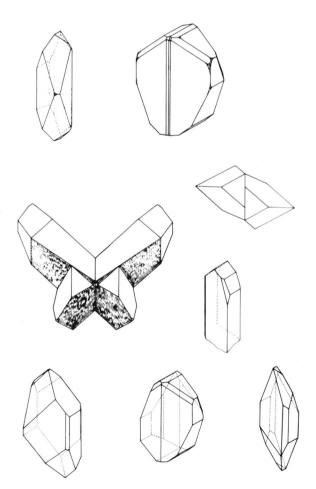

**Hemihedrite.**
Florence Lead-Silver mine
(Williams & Anthony, 1970).

*Maricopa County:* Wickenburg district, at several localities including the Pack Rat claim, Moon Anchor mine, and the Potter-Cramer property, in the oxidized portion of galena-bearing veins where it forms by alteration of galena in quartz veins which cut andesite agglomerate; associated minerals include phoenicochroite, vauquelinite, willemite, and mimetite. Crystals are orange to nearly black in color (Williams and Anthony, 1970; Williams et al., 1970; McLean and Anthony, 1970).

*Pinal County:* Tortilla Mountains, Florence Lead-Silver mine, in the oxide zone of lead-bearing veins cutting strongly brecciated Precambrian limestone intruded by an altered latite porphyry, and Precambrian quartzite containing a diabase dike. Associated primary minerals are galena, sphalerite, pyrite, and tennantite; oxidation products include cerussite, phoenicochroite, vauquelinite, wulfenite, and willemite. Hemihedrite forms contemporaneously with wulfenite, following the formation of cerussite, and may be replaced by wulfenite. Crystals are bright orange to henna brown to nearly black in color (Williams and Anthony, 1970). This is the type locality.

## HEMIMORPHITE

Zinc silicate hydroxide hydrate, $Zn_4Si_2O_7$-$(OH)_2 \cdot H_2O$. A secondary mineral which occurs in the oxidized portion of zinc deposits; commonly associated with smithsonite, cerussite, anglesite, galena, and sphalerite.

*Cochise County:* Tombstone district, sparingly, as radiating aggregates in oxidized ore at the Empire and Toughnut mines (Butler et al., 1938). Warren district, with aurichalcite (UA 9343) and rosasite (UA 8685). Turquoise district, as incrustations and druses at the Mystery and Silver Bill mines. Little Dragoon Mountains, Cochise district, Johnson Camp (Cooper and Huff, 1951). Gunnison Hills, as small colorless crystals; the most abundant oxidized ore mineral in the Texas-Arizona mine (Cooper, 1957; Cooper and Silver, 1964).

*Gila County:* Globe-Miami district, where it is common in the oxidized zones of veins at the Irene, Albert Lea, and Defiance deposits (Peterson, 1962); Castle Dome mine, as tiny rounded grains in clay (Peterson et al., 1951). Cherry Creek area, Horseshoe deposit, where it occurs sparingly coating fracture surfaces (Granger and Raup, 1969). Banner district, 79 mine, as particularly fine specimen material in a variety of habits, associated with rosasite, chrysocolla, and cerussite (UA 1164, 8307), and, altering to chrysocolla, as pseudomorphs on wad (UA 7587); as curious blue, but sometimes white, hollow eggshell-like balls on matrix (Keith, 1972).

*Greenlee County:* Clifton-Morenci district, as small transparent crystals in decomposed garnet rock at the Shannon mine (Lindgren and Hillebrand, 1904; Lindgren, 1905).

*Maricopa County:* White Picacho district, as a supergene mineral after sphalerite; also associated with hydrozincite (Jahns, 1952).

*Mohave County:* McCracken mine, near Signal, with fluorite (UA 9271).

*Pima County:* Empire Mountains, as small colorless crystals in smithsonite, Hilton mines; Total Wreck mine, with plattnerite on limonite (Bideaux et al., 1960). Pima district, as light green-blue mammillary crusts on mine dumps (T. Murchison, pers. comm., 1972); San Xavier West mine (Arnold, 1964); also found at the Queen mine (UA 5634). Waterman Mountains, Silver Hill mine (UA 6005). Tucson Mountains, as botryoidal crusts found at the north base of Amole Peak (Robert O'Haire, pers. comm., 1972).

*Pinal County:* Mammoth district, Mammoth–St. Anthony mine, as porous to compact masses and as slender needles bristling from quartz crystals on the walls of open cavities (Peterson, 1938; Palache, 1941b; Fahey, 1955).

*Santa Cruz County:* Patagonia district, Flux mine, on limonite with aurichalcite (Tom Trebisky, pers. comm., 1972).

*Yavapai County:* Eureka district, Hillside mine, in the oxidation zone of a sulfide-bearing vein in mica schist, associated with silver, anglesite, cerussite, chlorargyrite, and smithsonite (Axelrod et al., 1951).

## *HERCYNITE

Iron magnesium aluminum oxide, $(Fe,Mg)Al_2O_4$. A member of the spinel group; found most commonly in metamorphosed argillaceous sediments.

*Gila County:* In olivine bombs from an unspecified locality near Rice (UA 3701).

## *HERSCHELITE

Sodium calcium potassium aluminum silicate hydrate, $(Na,Ca,K)AlSi_2O_6 \cdot 3H_2O$. The uncommon sodium end-member of the chabazite-herschelite series; a member of the zeolite group.

*Cochise County:* San Simon Basin, 7 miles northeast of Bowie, in bedded lake deposits with analcime, chabazite, erionite, clinoptilolite, halite, and thenardite. Composition of the herschelite varies laterally with its position within the

**Herschelite.**
Horseshoe Dam, Maricopa County. Julius Weber.

deposit. It occurs as minute spherules and crystal aggregations formed by alteration of volcanic pyroclastic material in the bedded lake deposits. Regis and Sand (1967) give as the composition: $(0.91Na_2,0.05K_2,0.04Ca)O \cdot Al_2O_3 \cdot 6.4SiO_2 \cdot 7.6H_2O$.

## HESSITE

Silver telluride, $Ag_2Te$. Occurs in veins with other tellurides, gold, and native tellurium.

*Cochise County:* Tombstone district (Romslo and Ravitz, 1947). West Side mine, as bands and disseminations in quartz with chlorargyrite and gold (Genth, 1887); Flora Morrison mine, altering to native silver (Butler et al., 1938; Rasor, 1938).

## HETAEROLITE

Zinc manganese oxide, $ZnMn_2O_4$. A rare secondary mineral found in association with other oxidized manganese and zinc minerals.

*Cochise County:* Tombstone district, Lucky Cuss mine, as tiny veinlets in manganite (Butler et al., 1938; Rasor, 1939; Hewett and Fleischer, 1960). Warren district, as splendent botryoidal and stalactitic masses and coatings from the Campbell (UA 7717) and Junction mines; 1300-foot level Cole shaft (Richard Graeme, pers. comm., 1969).

*Gila County:* Banner district, 79 mine, as massive encrustations on hemimorphite in the stope system extending above the 470 level (Keith, 1972).

*Pinal County:* Pioneer district, with psilomelane and pyrolusite at the Domeroy property, 4 miles north of Superior (Dean et al., 1952).

## *HEULANDITE

Sodium calcium aluminum silicate hydrate, $(Na,Ca)_{4-6}Al_6(Al,Si)_4Si_{26}O_{72} \cdot 24H_2O$. A member of the zeolite group. Common in vesicles in basaltic rocks and can form as a result of low temperature alteration.

*Gila County:* Globe-Miami district, Black Copper portion of the Inspiration ore body, interlayered with chrysocolla in some of the schist outcroppings in the area (Throop and Buseck, 1971).

**Heulandite and mesolite.**
Safford, Graham County. Julius Weber.

*Greenlee County:* Five miles south of Hannegan Meadow along road cuts on U.S. Highway 666, in vesicles in basalt with phillipsite crystals.

*Mohave County:* With thomsonite in andesite from an unspecified locality near Oatman (UA 8476).

*Pinal County:* Mineral Creek district, Ray area, mixed with green and black chrysocolla; the former presence of volcanic ash is indicated (Throop and Buseck, 1971). In a railway cut in Malpais Hill north of Mammoth, as crystals with calcite in vesicles (Bideaux et al., 1960). Mammoth district, Mammoth–St. Anthony mine, rarely as twinned microcrystals with wulfenite.

## HEWETTITE

Calcium vanadium oxide hydrate, $CaV_6O_{16} \cdot 9H_2O$. A rare secondary mineral found in the oxidized portions of vanadium or uranium-vanadium deposits.

*Apache County:* Monument Valley, where it is moderately abundant in the Monument No. 2 mine, as individual crystals up to 15 mm long and about 0.5 mm thick, as cross-fibrous seams and in crusts up to 10 cm thick (Weeks et al., 1955; Witkind and Thaden, 1963). Also found in the Carrizo Mountains as hairlike crystals and as fibrous incrustations in sandstone, with carnotite. In the Salt Wash Member of the Morrison Formation, Lukachukai Mountains (Chenoweth, 1967).

*Navajo County:* Monument Valley, Monument No. 1 and Mitten No. 2 mines (Evensen and Gray, 1958; Holland et al., 1958; Witkind, 1961; Witkind and Thaden, 1963).

## *HEXAHYDRITE

Magnesium sulfate hydrate, $MgSO_4 \cdot 6H_2O$. An uncommon secondary mineral found sparingly in mine workings with epsomite from which it may form by dehydration.

*Pinal County:* Mammoth district, San Manuel mine, where it occurs on the 2015 and 2700 levels with epsomite, starkeyite, and an unknown selenium-bearing aluminum sulfate hydrate mineral (material collected by Joseph Urban) (Anthony and McLean, 1975).

## *HIDALGOITE

Lead aluminum sulfate arsenate hydroxide, $PbAl_3(SO_4)(AsO_4)(OH)_6$. A rare member of the beudantite group; of secondary origin.

*Yavapai County:* Silver Crown mine, in the oxidized zone of massive sulfide ores where it occurs alone as a white crust on silicified schist. Unit cell constants: $a = 7.012$ Å, $c = 17.272$ Å.

## HILLEBRANDITE

Calcium silicate hydroxide, $Ca_2SiO_3(OH)_2$. A rare mineral formed during contact metamorphism of limestone.

*Cochise County:* Tombstone district, Lucky Cuss mine, with monticellite and vesuvianite (Butler et al., 1938).

## *HINSDALITE

Lead aluminum phosphate sulfate hydroxide, hexagonal $PbAl_3(PO_4)(SO_4)(OH)_6$. A rare primary or early supergene mineral, found in sulfide veins.

*Mohave County:* Ithaca Peak mine, where it is found in abundance in a vein of pyrite, sphalerite, bornite, and minor galena with abundant supergene chalcocite. The hinsdalite occurs in voids on corroded, sooty sulfides as spherical clusters of pearly white, scaly tablets.

## *HISINGERITE

Iron silicate hydroxide hydrate, $Fe^{3+}Si_2O_5$-$(OH)_4 \cdot 2H_2O$. A secondary mineral formed as a result of alteration in mineral deposits, or as a result of weathering in iron-bearing silicates in rocks.

*Gila County:* Globe-Miami district, Castle Dome mine, a supergene mineral associated with malachite, limonite, jarosite, and wulfenite (Peterson, 1947; Peterson et al., 1951).

*Pima County:* Cababi district, South Comobabi Mountains, Mildren and Steppe claims, with a variety of secondary minerals which are products of the oxidation of sulfide ores in quartz veins cutting andesite (Williams, 1963).

*Pinal County:* Mammoth district, Mammoth–St. Anthony mine, where it occurs in the central parts of radiating fibrous creaseyite spherules.

## *HOLLANDITE

Barium manganese oxide, $Ba(Mn^{2+},Mn^{4+})_8$-$O_{16}$. An associate of several other manganese oxide minerals which occur both in veins and as bedded deposits.

*Coconino County:* Peach Springs district, fine-grained, massive, intimately associated with cryptomelane; contains 0.23 percent thallium and 5.5 percent BaO (Crittenden et al., 1962).

*Gila County:* Sierra Ancha district, Apache mine, one of several manganese oxides concentrated in breccia zones of a sandstone conglomerate; botryoidal, intimately associated with cryptomelane; contains 0.34 percent thallium, about 9 percent BaO, and about 2 percent $K_2O$ (Crittenden et al., 1962); Hewett et al., 1963). Banner district, a constituent of wad in the 79 mine (Keith, 1972).

*Mohave County:* Artillery Mountains, as one of the most abundant minerals in the Black Jack and Priceless deposits where it occurs as very fine anhedral grains or as needles; associated with cryptomelane, psilomelane, pyrolusite, coronadite, ramsdellite, and lithiophorite; also in the Price vein and the Maggie Canyon bedded deposit (Mouat, 1962).

*Pinal County:* Mammoth district, Mammoth-St. Anthony mine, where it is the most widespread manganese oxide mineral.

## HORNBLENDE

Calcium sodium magnesium iron aluminum silicate hydroxide, $(Ca,Na)_{2-3}(Mg,Fe^{2+},Fe^{3+},-Al)_5(Al,Si)_8O_{22}(OH)_2$. The commonest member of the amphibole group of rock-forming minerals. Found in a wide variety of igneous and metamorphic rocks. Some hornblende in igneous rocks is a product of alteration of primary pyroxenes.

*Cochise County:* Tombstone district, as long, prismatic crystals in the Schieffelin Granodiorite (Butler et al., 1938).

*Gila and Pinal Counties:* The principal constituent of the greenschist facies of the Pinal Schist. In intrusive bodies in the vicinity of Picket Post Mountain, near Superior.

*Greenlee County:* Abundant in diorite porphyry, Clifton-Morenci district.

*Pima County:* At Ajo, as bodies of hornblendite, the largest of which is about 2000 by 1000 feet in planar dimensions, in the Cardigan Gneiss (Gilluly, 1937). Empire Mountains, as phenocrysts in a diorite porphyry dike at the Prince mine. Near Vail (UA 5556). Quinlan Mountains, 40 miles southwest of Tucson (UA 5459).

*Yavapai County:* Lenticular bodies composed largely of hornblende are found at many places in the Yavapai Schist.

*Yuma County:* Harcuvar Mountains, as crystals more than an inch long near dikes in Precambrian granite.

## HUEBNERITE (see WOLFRAMITE)

## HUMMERITE

Potassium magnesium vanadium oxide hydrate, $KMgV_5O_{14} \cdot 8H_2O$. A rare water-soluble secondary mineral which occurs as veins and efflorescences in sedimentary rocks.

*Apache County:* Mesa No. 1 mine, Lukachukai Mountains.

## HUREAULITE

Acid manganese phosphate hydrate, $H_2Mn_5(PO_4)_4 \cdot 4H_2O$. Formed by alteration of primary phosphate minerals such as triphylite, and associated with other secondary phosphates in granite pegmatites.

*Yavapai County:* White Picacho district, occurs locally between crystals of lithiophilite and triphylite and as a coating on sicklerite, also as crystalline aggregates and in fractures in these crystals; amber- to flesh-colored (Jahns, 1952). Eureka district, on the 7U7 Ranch west of Hillside, associated with bermanite in triplite seams, and with metastrengite and leucophosphite (Leavens, 1967).

## *HYDROBASALUMINITE

Aluminum sulfate hydroxide hydrate, $Al_4(SO_4)(OH)_{10} \cdot 36H_2O(?)$. A rare secondary mineral associated with basaluminite to which it alters on dehydration.

*Cochise County:* Warren district, as pearly white flakes with basaluminite and other aluminum sulfates encrusting silicified limestones in the Holbrook pit.

## Hydrobiotite

Potassium iron aluminum silicate hydroxide, a mixed-layer mica composed of interstratified sequences of vermiculite and biotite. Although the material termed hydrobiotite may be questioned as a distinct mineral species, its abundance and prevalence in certain mineral deposits warrants its inclusion here.

*Cochise County:* Warren district, as an alteration product associated with the intrusion of a quartz monzonite.
*Pinal County:* Mammoth district, San Manuel mine, a minor constituent of hydrothermal alteration of monzonite and quartz monzonite porphyries in the marginal, less intense alteration areas (Schwartz, 1958; Creasey, 1959; Lowell, 1968). Cottonwood Canyon district southeast of Apache Junction; abundant in selvages of ore veins carrying lead and copper.

## HYDROCERUSSITE

Lead carbonate hydroxide, $Pb_3(CO_3)_2(OH)_2$. A rare mineral of secondary origin usually found with other products of the alteration of galena.

*Maricopa County:* As thin, milky films on cerussite at the Potter-Cramer mine, Wickenberg district.
*Pinal County:* Mammoth district, Mammoth-St. Anthony mine, in the Collins vein, as snow-white hexagonal pyramidal crystals accompanied by diaboleite and leadhillite. These steep pyramidal crystals to nearly an inch in length are the finest occurrence of the mineral (Fahey et al., 1950).

Hydromuscovite (see ILLITE)

## *HYDRONIUM JAROSITE

Hydronium iron sulfate hydroxide, $(H_3O)Fe_3(SO_4)_2(OH)_6$. At an occurrence in Poland, the origin of the mineral is ascribed to low alkali content of mine waters resulting from more rapid breakdown of sulfide than of rock-forming minerals. The mineral previously termed carphosiderite was shown by Moss (1957) and by Van Tassel (1958) to be hydronium jarosite (Brophy and Sheridan, 1965).

*Pinal County:* Cherry Creek area, Black Brush deposit (Granger and Raup, 1969).

## HYDROZINCITE

Zinc carbonate hydroxide, $Zn_5(CO_3)_2(OH)_6$. A secondary mineral formed in the oxidized portions of mineral deposits by the alteration of sphalerite; commonly associated with smithsonite, hemimorphite, and cerussite.

*Cochise County:* Tombstone district, in a small seam with aurichalcite and hemimorphite on the west side of the Quarry "roll" (Butler et al., 1938). Gunnison Hills, in metamorphosed limestone as white, chalky masses up to 6 by 24 inches in size, Texas-Arizona mine (Cooper, 1957; Cooper and Silver, 1964).
*Coconino County:* Havasu Canyon, with other secondary lead and zinc minerals.
*Maricopa County:* As a supergene mineral in the pegmatites of the White Picacho district, from the alteration of sphalerite and associated hemimorphite (Jahns, 1952).
*Pinal County:* Pioneer district, as a white film on sphalerite from the 1600-foot level, Magma mine (Short et al., 1943). Also from the Hancock property near Superior. Slate Mountains, Jackrabbit mine, intimately associated with pul-

verulent yellow limonite and as fracture fillings (Hammer, 1961).

*Yuma County:* Castle Dome Mountains, with gypsum, in fissures, Senora claims; in vugs and channels associated with smithsonite, wulfenite, vanadinite, and mimetite, Castle Dome district (Wilson, 1933).

## HYPERSTHENE

Magnesium iron silicate, $(Mg,Fe)SiO_3$. An orthorhombic pyroxene intermediate in composition between enstatite and orthoferrosilite; associated with many ultrabasic rocks, especially norites; as a metamorphic mineral in charnockites.

*Apache County:* Monument Valley, Garnet Ridge, as sparse crystals in ejection boulders of garnet gneiss (Gavasci and Kerr, 1968).

*Navajo County:* As dikes of hypersthenite in shale, 20 miles west of Dilkon.

*Yavapai County:* In tholeiitic basalts near the summit of Mingus Mountain (McKee and Anderson, 1971).

## I

## Idocrase (VESUVIANITE)

## *ILLITE

Potassium aluminum silicate hydroxide, $K_{1-1.5}Al_4(Si_{7-6.5}Al_{1-1.5}O_{20})(OH)_4$, having generally higher OH and lower K or K plus Al content than muscovite which it resembles structurally. The name is widely used to signify clay minerals having both the muscovite and biotite structure types. A widespread and common mica component of clays. Illites are especially abundant in the Colorado Plateaus where they constitute, for example, a prominent constituent of the Salt Wash member of the Morrison Formation (Keller, 1962); it is the dominant clay mineral component of most of the Moenkopi Formation, often associated with kaolinite, chlorite, or montmorillonite, in places forming mixed-layer assemblages of illite-montmorillonite, and chlorite-illite (Schultz, 1963). Illite is a common constituent of many clay-bearing sedimentary rocks throughout the state. Also important as a widespread product of hydrothermal alteration of some mineral deposits, especially certain of the porphyry copper ore bodies. Only a few examples are mentioned.

*Cochise County:* Warren district, Sacramento Hill, in a hydrothermally altered granite porphyry stock with sericite, kaolinite, allophane, and alunite as an alteration of feldspar; also in dikes cutting limestones peripheral to the stock (Schwartz, 1947).

*Gila County:* Globe-Miami district, Miami ore body, associated with kaolinite, alloysite, and sericite as a product of hydrothermal alteration of granite porphyry and schist (Schwartz, 1947).

*Greenlee County:* Morenci district, as pseudomorphs after plagioclase, with sericite, in the intensely altered porphyry ore body with a large suite of hydrothermal alteration products including kaolinite, leucoxene, and allophane (Schwartz, 1947, 1958).

*Pinal County:* Ray area, in a hydrothermally altered stock within the ore body, as pseudomorphs after plagioclase phenocrysts, associated with sericite, allophane, and kaolinite; also as pseudomorphs after biotite (Schwartz, 1947, 1952). Mammoth district, San Manuel mine, in hydrothermally altered monzonite and quartz monzonite porphyries as an abundant minor constituent (Schwartz, 1947, 1958; Creasey, 1959). Kalamazoo ore body (Lowell, 1968).

## ILMENITE

Iron titanium oxide, $FeTiO_3$. Occurs in close association with mafic igneous rocks as veins, disseminated deposits, and as dikes; common as an accessory mineral in certain igneous rocks, and, locally, as an important detrital constituent of black sands.

*Apache County:* Monument Valley, Garnet Ridge, as fragments in a breccia dike piercing sediments; as needles, the most common inclusions in garnets (Gavasci and Kerr, 1968).

*Gila and Pinal Counties:* A constituent of the Pinal Schist and diabase (Peterson, 1962). Also as tabular pieces in quartz, near Castle Dome (Peterson et al., 1946, 1951). Sierra Ancha Mountains, intimately associated with hematite in Precambrian quartzite (Peterson, 1966).

*Graham County:* Galiuro Mountains, in disseminated form in the northern part of the range.

*Maricopa County:* White Tank Mountains, as grains and intergrowths with magnetite in Precambrian schist and in pegmatites (Harrer, 1964). Big Horn district, with titaniferous magnetite constituting about 3 to 7 percent of extensive placer deposits (Harrer, 1964).

*Pima County:* Ajo district, formed both as a primary constituent of the Cornelia Quartz Monzonite and by alteration of sphene (Gilluly, 1937).

*Pinal County:* Red Rock-Florence Junction area, with titaniferous magnetite in widespread alluvial deposits (Harrer, 1964).

*Santa Cruz County:* Patagonia Mountains, especially at Duquesne and Washington Camp, in a number of contact metamorphic deposits (Schrader, 1917).

*Yavapai County:* Eureka district, with magnetite, as dikes and irregular bodies in gabbro (Ball and Broderick, 1919). A low-grade deposit of large size is reported not far from the Bagdad mine. Bradshaw Mountains, in granite pegmatite near Cleator.

## *ILSEMANNITE

Molybdenum oxide hydrate, $Mo_3O_8 \cdot nH_2O(?)$. A secondary mineral formed by oxidation of molybdenite and other molybdenum minerals, sometimes of post-mine origin.

*Apache County:* Monument Valley, Monument No. 2 mine as dispersed, powdery, fine-grained material associated with corvusite, navajoite, hewettite, uraninite, and gypsum (Witkind and Thaden, 1963).

*Cochise County:* From a prospect pit near Warren (Warren district) as blue stains on intensely silicified, brecciated limestone, with fluorite and scheelite.

*Coconino County:* Cameron area, Huskon No. 11 mine, with marcasite in sandstone; as inky blue masses and stains; Huskon No. 10 mine, staining halotrichite a deep inky blue. (AEC, Grand Junction mineral collection.) East of Jacob Lake, Sun Valley mine, abundant in uranium deposits in a paleo-stream channel filled with Shinarump Conglomerate where it forms on the walls of older mine workings (Petersen et al., 1959).

## *ILVAITE

Calcium iron silicate hydroxide, $CaFe_2^{2+}Fe^{3+}$-$Si_2O_8(OH)$. Found in contact metamorphic deposits, typically in limestones and dolomites, with other calcium silicate minerals.

*Cochise County:* Abundant in hedenbergite hornfels from several mines in Middlemarch Pass, Dragoon Mountains. Crystals up to ½ inch long were found but are poorly developed. Other associated species include sphalerite and fluorite.

*Graham County:* Aravaipa district, Iron Cap mine, as vitreous, fine-grained material associated with fluorite, sphalerite, and quartz (Don Burt, pers. comm., 1971).

## IODARGYRITE (Iodyrite)

Silver iodide, AgI. A secondary mineral found in the oxidized zone of silver deposits; alters to native silver.

*Cochise County:* Pearce Hills, with chlorargyrite, bromargyrite, embolite, and acanthite at the Commonwealth mine (Endlich, 1897; Guild, 1910).

*Mohave County:* Silliman (1866) reported the occurrence of "iodyrite" in quartz veins in "ash-colored feldspathic porphyry," associated with "fluorspar, green carbonate of copper, free gold, and abundant iron gossan in cellular quartz," from Mohave County. San Francisco district, in several lodes in the Caledonia, Dayton, Quackenbush, and Knickerbocker properties. Cerbat Mountains, Wallapai district, as the chief near-surface ore mineral in oxidized sulfide vein deposits of the district (Thomas, 1949).

*Pima County:* Cerro Colorado Mountains, Cerro Colorado mine (Guild, 1910). Cababi district, South Comobabi Mountains, Mildren and Steppe claims (Williams, 1960).

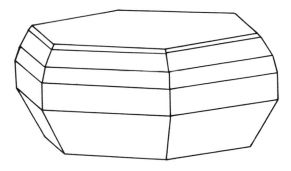

**Iodargyrite.**
Silver-Lead claim, Mildren and Steppe area
(Williams, 1962).

*Pinal County:* Mammoth district, Mammoth-St. Anthony mine, as pale green droplets with caledonite and boleite.

*Santa Cruz County:* Patagonia district, Worlds Fair mine, as small, isolated crystals, globules, and specks in "contour boxwork" limonite deposits, the limonite having been derived from tetrahedrite (Blanchard and Boswell, 1930).

## IODOBROMITE

Silver bromide chloride iodide, $Ag(Br,Cl,I)$. A secondary mineral found in oxidized silver deposits.

*Gila County:* Reported by W. P. Blake (1905) from an unspecified locality near Globe; also in thin seams and crusts in a quartz vein at the Hechman mine.

Iodyrite (IODARGYRITE)

Iolite (a variety of CORDIERITE)

IRON (see KAMACITE and TAENITE)

## J

## *JADEITE

Sodium aluminum iron silicate, $Na(Al,Fe)Si_2O_6$. A member of the pyroxene group.

*Apache County:* Monument Valley area, Garnet Ridge, as diopsidic jadeite with an appreciable proportion of acmite, coexisting with pyrope-almandine garnet in eclogite inclusions from kimberlite pipes (Watson and Morton, 1969).

## JAMESONITE

Lead iron antimony sulfide, $Pb_4FeSb_6S_{14}$. Occurs in hydrothermal veins formed under low to moderate temperatures; typically associated with galena, sphalerite, stibnite, tetrahedrite, and other sulfosalts.

*Yavapai County:* Reported in some of the ores of the Bradshaw Mountains, with free gold.

## JAROSITE

Potassium iron sulfate hydroxide, $KFe_3(SO_4)_2(OH)_6$. A member of the alunite group. A secondary mineral of widespread occurrence in the southwestern United States; found in the oxidized cappings of base metal deposits where it may be a major component of the "limonites" of gossans.

*Apache County:* Monument Valley, Monument No. 2 mine, in Shinarump conglomerate (Mitcham and Evensen, 1955; Witkind and Thaden, 1963).

*Cochise County:* Tombstone district, abun-dant at the Toughnut and Empire mines (Butler et al., 1938). Turquoise district, as small, flaky bunches. Commonwealth mine at Pearce, with chlorargyrite and native silver. Warren district, Sacramento Hill, in the hydrothermally altered granite porphyry stock (Schwartz, 1947).

*Coconino County:* Cameron area, where it is a very abundant mineral in the uranium deposits (Austin, 1964).

*Gila County:* Globe-Miami district, as an abundant supergene mineral at the Castle Dome and Copper Cities mines (Peterson, 1947, 1962; Peterson et al., 1951). Sierra Ancha Mountains, Little Joe deposit; Cherry Creek, at the Shepp 2 deposit (Granger and Raup, 1969).

*Graham County:* Lone Star district, Gila Mountains, Safford area, with pseudomalachite, malachite, brochantite, antlerite, and other secondary oxidized minerals (Hutton, 1959b); Safford porphyry copper deposit, in fractures and veins as a replacement of primary minerals, fault gouge, and clay; it is intimately associated with turquoise and alunite (Robinson and Cook, 1966).

*Greenlee County:* Clifton-Morenci district, Morenci open pit mine, where it is "widespread but most dramatic as an oxidation product of pyritic veinlets in areas of weak (copper) mineralization" (Moolick and Durek, 1966).

*Maricopa County:* Vulture Mountains, Vulture mine, as "fine, transparent, yellow and dark brown rhombic crystals" filling cavities formed from the oxidation of pyrite; associated with gold particles (Silliman, 1879). An analysis by S. F. Penfield (1881) gave:

| | | |
|---|---|---|
| $SO_3$   30.42 | $K_2O$   8.53 | $Na_2O$   0.28 |
| $Fe_2O_3$   48.27 | | $H_2O$   12.91 |
| | | TOTAL   100.41% |

Specific gravity: 3.09.

Also from the Black Rock mine, Vulture district.

*Mohave County:* As a secondary mineral in the Wallapai district, Cerbat Mountains (Thomas, 1949; Field, 1966). Jarosite from the Sunset claim, Kingman, has the composition $(K_{.33}Na_{.66})_3Fe_9(SO_4)_6(OH)_{18}$ (Brophy and Sheridan, 1965).

*Navajo County:* Monument Valley, Monument No. 1 mine, in the largely-oxidized uranium-vanadium ores, associated with torbernite, carnotite, corvusite, tyuyamunite (Holland et al., 1958).

*Pima County:* Empire Mountains, Total Wreck mine, with wulfenite, vanadinite, chlorargyrite,

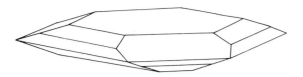

**Jarosite.**
Beacon claim (Williams, 1962).

cerussite, and smithsonite, and in the Jerome No. 2 mine in lead-zinc veins, Hilton group of claims (Schrader and Hill, 1915). With alunite at the Silver Bell mine, Silver Bell Mountains (Kerr, 1951). Sierrita Mountains, Pima district, Mineral Hill area (UA 7857). South Comobabi Mountains, Cababi district, Mildren and Steppe claims (Williams, 1963) as lovely color-zoned tablets in vugs in milky quartz up to 5 mm in size. Ajo district, where it is largely confined to weathered capping of areas of notably pyritic ore (Gilluly, 1937). Helvetia-Rosemont district (Schrader, 1917). Pima district, at the Twin Buttes open pit mine (Stanley B. Keith, pers. comm., 1973).

*Pinal County:* Mineral Creek deposit near Ray, where it is one of the oxidized iron minerals in a copper-rich Holocene gravel deposit leached from a nearby secondary copper sulfide blanket; may be associated with goethite, hematite, malachite, azurite, cuprite, and tenorite (Phillips et al., 1971). Copper Creek region, where it is a minor oxidation product of pyrite (Simons, 1964). Mammoth district. San Manuel deposit, in small amounts, formed from the breakdown of pyrite (Schwartz, 1953).

*Santa Cruz County:* Red Mountain, near Patagonia, as small (less than 0.5 mm) honey-brown, transparent crystals coating breccia and altered porphyritic rock (Kenneth W. Bladh, pers. comm., 1974); Patagonia Mountains, Mowry mine; Flux mine, Harshaw district.

*Yavapai County:* Copper Basin district, found in the oxidized zone of mineralized breccia pipes with hematite, limonite, chrysocolla, and other secondary minerals (Johnston and Lowell, 1961). Eureka district, Bagdad, in leached cappings of sulfide deposits, with goethite (Anderson, 1950). Jerome, United Verde mine, in minor amounts dusted over yavapaiite and other secondary sulfides formed from the burning of pyritic ores (Hutton, 1959).

*Yuma County:* Engesser prospect, Copper Mountains (Wilson, 1933).

## *Jeromite (?)

Arsenic selenide sulfide, near $As(S,Se)_2(?)$. An amorphous substance of uncertain nature formed as a result of a mine fire at Jerome. Its validity as a mineral species has been questioned.

*Yavapai County:* Jerome, United Verde mine, as "a coating on fragments of rock beneath iron hoods placed over vents from which sulfur dioxide gases are issuing," a product of burning sulfide ores; occurs as black globular opaque masses, translucent on thin edges. This is the type and only known locality (Lausen, 1928). A chemical analysis by Buehrer gave:

| | | | | | |
|---|---|---|---|---|---|
| S | 40.8 | As | 46.8 | Insol. | |
| Se | 7.5 | Sb | tr. | (by diff.) | 4.9 |
| Te | tr. | | | TOTAL | 100.00% |

## JOHANNITE

Copper uranyl sulfate hydroxide hydrate, $Cu(UO_2)_2(SO_4)_2(OH)_2 \cdot 6H_2O$. A secondary mineral usually formed by the alteration of uraninite in sulfide veins.

*Yavapai County:* Hillside mine, with several other uranium minerals (Axelrod et al., 1951).

## *JOHANNSENITE

Calcium manganese iron silicate, $Ca(Mn,Fe)Si_2O_6$. A member of the pyroxene group, near diopside-hedenbergite in properties and paragenesis, but much less common in occurrence.

*Graham County:* Aravaipa district, Black Hole prospect, NW ¼, Sec. 36, T.5S, R.19E, in tabular bodies and irregular masses replacing limestone, as radiating or spherulitic aggregates of prisms or needles a few centimeters across and larger masses a few feet thick and several tens of feet long; associated with neotocite, chalcopyrite, galena, and sphalerite (Simons and Munson, 1963).

## *JORDISITE

Amorphous molybdenum sulfide, $MoS_2$. A poorly characterized black powdery amorphous material which may alter to ilsemannite.

*Coconino County:* East of Jacob Lake, Sun Valley mine, found in a uranium deposit located in a bend in a paleostream channel filled with Shinarump Conglomerate, associated with uraninite, ilsemannite, pyrite, sphalerite, and hematite (Petersen, 1960).

## *JUNITOITE

Calcium zinc silicate hydrate, $CaZn_2Si_2O_7 \cdot H_2O$. Probably of late-stage hydrothermal origin; known only from its type locality at Christmas.

*Pinal County:* Occurs in the open pit mine at Christmas, Banner district, in tactite ores with kinoite and apophyllite. As clear, platy crystals up to 5 mm across, typically associated with pink smectite (clay) in rings around nuggets of unaltered sphalerite.

## *JURBANITE

Basic aluminum sulfate hydrate, $AlSO_4OH \cdot 5H_2O$. A very sparse secondary mineral formed at the San Manuel mine under conditions of high humidity; associated with other secondary water-soluble sulfates including epsomite and pickeringite. The San Manuel mine occurrence is the type and only known locality.

*Pinal County:* Mammoth district, where it occurs sparsely on the 2075 level of the San Manuel mine as minute, clear, colorless crystals intimately associated with epsomite, hexahydrite, pickeringite, and starkeyite deposited on lagging and overhead pipes (Anthony and McLean, 1975).

# K

## Kaersutite
### (titanian HORNBLENDE)

Calcium sodium potassium magnesium iron titanium aluminum silicate hydroxide fluoride, $Ca_2(Na,K)(Mg,Fe^{2+},Fe^{3+})_4Ti(Si_6Al_2O_{22})-(O,OH,F)_2$. May be regarded as a titanian oxy-hornblende; a member of the amphibole group. Occurs in volcanic and alkalic plutonic rocks.

*Gila County:* Near San Carlos on the San Carlos Indian Reservation. A thorough study of an occurrence of the variety kaersutite, which occurs in xenocrysts from a basalt flow, was made by Mason (1968). The mineral is associated with spinel and augite. An analysis by J. Nelen gave the following result:

| | | | | | |
|-------|-------|-----|------|---------|------|
| $SiO_2$ | 39.63 | FeO | 10.36 | $Na_2O$ | 2.96 |
| $TiO_2$ | 4.93 | MnO | 0.14 | $K_2O$ | 1.60 |
| $Al_2O_3$ | 14.36 | MgO | 10.90 | $H_2O(+)$ | 0.72 |
| $Fe_2O_3$ | 3.20 | CaO | 10.93 | $H_2O(-)$ | 0.10 |

TOTAL (incl. 1.0% ilmenite) 99.83%

*Mohave County:* As large phenocrysts in a camptonite dike, about 8 miles south of Hoover Dam on U.S. Highway 93, in the Minnesota district (Campbell and Schenk, 1950). At the southern end of the Uinkaret Plateau, as large poikilitic crystals in ultramafic inclusions in geologically young flows and associated cinder cones of basanitic composition (Best, 1970).

## KAMACITE

Nickel iron, $\alpha$-(Fe,Ni) (body-centered cubic structure) containing about 5.5% nickel. Occurs as a major constituent of iron meteorites, commonly associated with taenite. A constituent of Arizona iron meteorites.

## KAOLINITE

Aluminum silicate hydroxide, $Al_2Si_2O_5(OH)_4$. The most important member of the kaolinite group of clay minerals which also includes dickite, nacrite, anauxite, halloysite, and meta-halloysite. Members of the group are formed by hydrothermal processes during alteration accompanying mineral deposit formation; they also are found extensively in sedimentary formations where they have been introduced either by transportation from previously weathered rocks or by weathering *in situ* of feldspathic rocks. Only a few Arizona occurrences can be listed.

*Colorado Plateau:* Prevalent in certain sedimentary formations of the region, including the sandstone of the Shinarump Member of the Chinle Formation (Schultz, 1963) and the Salt Wash Member of the Morrison Formation (Keller, 1962).

*Apache County:* Associated with the uranium-vanadium ores in the Monument No. 2 mine (Witkind and Thaden, 1963).

*Cochise County:* Warren district, as nearly pure kaolinite in white, waxy masses from the second level of the Copper Queen mine; Sacramento Hill, in hydrothermally altered granite porphyry stock with sericite, hydromuscovite, allophane, and alunite, as an alteration product of feldspar (Schwartz, 1947, 1958). Tombstone district, Toughnut mine (Butler et al., 1938). Turquoise district, Silver Bill mine. Very sparingly present in the clays near the surface of Willcox Playa (Pipkin, 1968).

*Gila County:* Globe-Miami district, with chalcocite in the oxidized zones, Old Dominion mine; Castle Dome mine, associated with halloysite and endellite (hydrohalloysite) as small masses filling open fractures in quartz monzonite (Schwartz, 1947); Miami and Inspiration ore bodies where it is a product of hydrothermal alteration (Schwartz, 1947).

*Greenlee County:* Clifton-Morenci district, where it is widespread as a product of pervasive hydrothermal alteration (Schwartz, 1947, 1958; Moolick and Durek, 1966); in large masses at the Longfellow mine, and in snow-white mammillary masses with azurite and malachite at the Copper Mountain and Mammoth mines (Lindgren, 1905); also at the Humboldt, Ryerson, and other properties.

*Maricopa County:* Vulture district, on the west side of the Hassayampa River, associated with alunite, making up a major phase within a hydrothermally altered rhyolite (Sheridan and Royse, 1970).

*Navajo County:* As cement or matrix of sandstones in the Westwater Canyon Member of the Morrison Formation and in the Cow Springs, and Mesa Verde Formations, along the eastern and southeastern edge of Black Mesa; also found near Coal Mine Canyon and the Hopi Mesas (Kiersch, 1955).

*Pima County:* Silver Bell Mountains, as large masses in wall rock on the upper levels of the El Tiro mine (Kerr, 1951). Pima district, Twin Buttes mine, as a product of hydrothermal alteration (Stanley B. Keith, pers. comm., 1973).

*Pinal County:* Mammoth district, San Manuel mine, as a common product of hydrothermal alteration of igneous rocks; with alunite, it makes up most of the rock in the intensely altered zones (Schwartz, 1947, 1953, 1958; Creasey, 1959). Ray area, with sericite, hydromuscovite, and alunite, in hydrothermally altered granitic rocks (Ransome, 1919; Schwartz, 1947, 1952).

## KASOLITE

Lead uranyl silicate hydrate, $Pb(UO_2)SiO_4 \cdot H_2O$. A secondary mineral, probably formed by the reaction of silica-bearing meteoric waters with earlier-formed secondary uranium minerals.

*Maricopa County:* South of Buckeye, in a pegmatite, associated with polycrase.

*Navajo County:* Seven Mile Canyon, Shinarump 1B mine, as yellow specks on gray sandstone (Atomic Energy Commission Grand Junction office mineral collection).

*Santa Cruz County:* Santa Rita Mountains, Kinsley property, east of Amado (UA 2704); Duranium claims, disseminated in arkosic sandstone with uranophane and autunite (Robinson, 1954; Shawe, 1966). Nogales district, as good crystalline material at the White Oak property, with uranophane, dumontite, autunite, associated with oxidized lead ore in a shear zone in rhyolite (Granger and Raup, 1962). Walnut Canyon, as incrustations along fractures in felsite porphyry (Robert O'Haire, pers. comm., 1972).

## *KINOITE

Calcium copper silicate hydrate, $Ca_2Cu_2Si_3O_{10} \cdot 2H_2O$. A rare mineral found in skarn; formed late in the paragenetic sequence with apophyllite and copper. The Pima County occurrence is the type locality.

*Gila County:* Banner district, present in the Christmas open pit mine where it is almost always coated by apophyllite.

*Pima County:* Northern Santa Rita Mountains, between Helvetia and Rosemont on claims controlled by The Anaconda Company and American Metals Climax, found in a core from diamond drill holes which penetrated contact metamorphosed Paleozoic limestones and dolomites; with apophyllite, copper, djurleite, bornite, and chalcopyrite, as small tabular euhedral crystals and in veinlets (Anthony and Laughon, 1970; Laughon, 1971). A chemical analysis by H. M. Ochs gave:

| CuO 31.10 | MgO 0.15 | SiO$_2$ 35.90 |
| CaO 23.55 | | H$_2$O 8.16 |
| | | TOTAL 98.86% |

Specific gravity: 3.193.

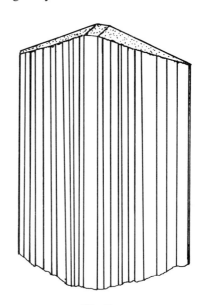

**Kinoite.**
Rosemont area (Anthony & Laughon, 1970).

## KORNELITE

Iron sulfate hydrate, $Fe_2(SO_4)_3 \cdot 7H_2O$. An uncommon secondary mineral formed by the oxidation of pyrite; seen as efflorescences in old mine workings.

*Cochise County:* Warren district, Copper Queen mine, as irregular porous crusts (Merwin and Posnjak, 1937).

## *KRINOVITE

Sodium magnesium chromium silicate, $NaMg_2CrSi_3O_{10}$. Identified in iron meteorites.

*Coconino County:* In the Canyon Diablo octahedrite meteorite, as minute, deep emerald-green, subhedral grains disseminated within graphite nodules, associated with roedderite, high albite, richterite, ureyite, and chromite (Olsen and Fuchs, 1968). This is the type locality.

## *KTENASITE

Basic copper zinc sulfate hydrate, $(Cu,Zn)_3$-$(SO_4)(OH)_4 \cdot 2H_2O$. A rare secondary mineral, previously known only from the Kamaresa mine, Laurium, Greece, where it is associated with serpierite and glaucocerinite on smithsonite.

*Gila County:* Banner district, 79 mine, where it is abundant as blue crusts in the 31 stope area (Thomas Trebisky, pers. comm., 1975).

## KYANITE

Aluminum silicate, $Al_2SiO_5$. Occurs typically in schists and gneisses as a product of thermal metamorphism of aluminous rocks; commonly associated with andalusite or sillimanite.

*Maricopa County:* Phoenix area, Squaw Peak, where 38 tons of kyanite are reported to have been mined (Wilson and Roseveare, 1949).

*Mohave County:* Specimens have been received from the county by the Arizona Bureau of Mines, but the locality is not known.

*Yuma County:* Gila Mountains, about 8 miles west of Wellton, on the east side of the range. Near Clip, as long-bladed crystals in quartzose schist found in Colorado River terrace gravels; associated with dumortierite (Wilson, 1933). An analysis of Clip material by W. F. Hillebrand (Schaller, 1905) gave:

| | | | |
|---|---|---|---|
| $SiO_2$ | 36.30 | FeO | undet. |
| $Al_2O_3(+TiO_2)$ | 62.51 | CuO | tr. |
| $Fe_2O_3$ | 0.70 | Loss on Ign. | 0.40 |
| | | TOTAL | 99.91% |

Specific gravity at 18.5° C: 3.656.

# L

## *LANGITE

Copper sulfate hydroxide hydrate, $Cu_4(SO_4)$-$(OH)_6 \cdot 2H_2O$. A secondary mineral formed as a result of the oxidation of copper sulfides, commonly associated with gypsum.

*Cochise County:* Warren district, as thin sky-blue crusts of small tabular crystals on fractures in or near chalcopyrite. Occurs with greenish films of brochantite. An uncommon mineral but widely distributed in the deeper workings of mines of the district.

*Coconino County:* Horseshoe Mesa, Grand Canyon National Park, Grandview (Last Chance) mine, as silky greenish crusts lining cavities (Leicht, 1971).

## LAUMONTITE

Calcium aluminum silicate hydrate, $CaAl_2$-$Si_4O_{12} \cdot 4H_2O$. A member of the zeolite group. Found in veins and cavities in igneous rocks where it is of hydrothermal origin; also in skarns and as a product of incipient metamorphism in sedimentary rocks.

*Cochise County:* Variety leonhardite, as pinkish crystals up to 1 inch with calcite crystals, Huachuca Mountains, 1½ miles east of Sunnyside (H 107581). Warren district, where it is a product of hydrothermal alteration.

*Gila County:* Christmas mine, variety leonhardite, in vugs and veinlets in diorite and replacing garnet in skarn (Perry, 1969).

*Maricopa County:* Wickenburg district, at several localities including the Moon Anchor mine and Potter-Cramer property (Williams, 1968).

*Navajo County:* Hopi Buttes volcanic field, Seth-La-Kai diatreme, in volcanic sandstone and tuffs, with limonite, gypsum, celadonite, and montmorillonite, with weak uranium mineralization (Lowell, 1956).

*Pinal County:* Azurite mine, Tortilla Mountains, with stilbite in quartz veins with copper silicates (Bideaux et al., 1960).

## LAUSENITE

Iron sulfate hydrate, $Fe_2(SO_4)_3 \cdot 6H_2O$. A rare mineral originally named rogersite; found only at Jerome where it formed as a result of a mine fire. This is the type and only locality.

*Yavapai County:* United Verde mine, formed as a result of the burning of a pyritic ore body (Lausen, 1928; Butler, 1928). The chemical analysis made on "rogersite" by T. F. Buehrer gave:

| | | |
|---|---|---|
| $H_2O$ 20.64 | $Fe_2O_3$ 28.07 | $K_2O$ 0.06 |
| $SO_3$ 47.90 | $Al_2O_3$ 1.40 | $Na_2O$ 1.23 |

TOTAL 99.30%

## LAWRENCITE

Ferrous chloride, $FeCl_2$. A rare mineral found in meteoric and terrestrial iron and associated with volcanic fumerolic action. Reported from some of the meteoric irons of Arizona.

## LAZULITE

Magnesium iron aluminum phosphate hydroxide, $(Mg,Fe)Al_2(PO_4)_2(OH)_2$. A member of the lazulite-scorzalite $[(Fe,Mg)Al_2(PO_4)_2(OH)_2]$ series. Occurs in granite pegmatites, quartz veins, and in aluminous, high-grade thermally metamorphosed rocks.

*Maricopa County:* Reported from a locality on the Phoenix-Cave Creek road, 1.2 miles north of Hyatt's Camp.

*Pima County:* Santa Rita Mountains, on a ridge between Stone Cabin and Madera Canyons (Robert O'Haire, pers. comm., 1972).

*Yuma County:* Near Quartzsite, in quartzite, with kyanite, andalusite, pyrophyllite, and dumortierite (Bideaux et al., 1960).

## LEAD

Lead, Pb. Of uncommon occurrence as the native metal in the oxidized zone of lead-bearing vein deposits. All occurrences noted with the exception of that at Tubac require confirmation.

*Maricopa County:* Reported from the benches of Oxbow Creek, Old Woman Gulch, and Little San Domingo Creek, in red sands with magnetite.

*Santa Cruz County:* At Tubac, replacing tree roots (UA 8577).

*Yavapai County:* Reported in fist-sized masses in Gold Crater, 15 miles west of Congress; and at La Paz, in red-stained quartz.

## LEADHILLITE

Lead sulfate carbonate hydroxide, $Pb_4(SO_4)(CO_3)_2(OH)_2$. A rare secondary mineral formed in oxidized lead deposits.

*Cochise County:* Warren district, with malachite and native silver on chalcocite (H 94731, 94733), Campbell mine, 1800-foot level (Harvard study collection), and Cole mine (Harvard study collection). Tombstone district, a fine specimen was found in 1942 on the dump of the old Manila mine on the Tombstone-Charleston road, adjoining the Gallagher properties, about 1 mile from the Charleston railroad crossing.

*Maricopa County:* Painted Rock Mountains, at the Rowley mine, near Theba, sparingly as fine, water-clear, highly modified plate-like crystals up to 6 mm; associated with caledonite and anglesite (Wilson and Miller, 1974). Some crystals are sky blue, some yellow, while others are pale green; some are probably paramorphous after susannite.

*Pima County:* Cababi district, South Comobabi Mountains, Mildren and Steppe claims, with wide variety of other oxidized minerals (Williams, 1963).

*Pinal County:* Mammoth district, Mammoth–St. Anthony mine, with other rare oxide zone minerals, 400-foot level of the Collins vein; as crystals up to ½ inch with basal cleavage having brilliant luster. Some of the crystals are prismatic, composed of sectors of monoclinic symmetry; others are pseudo-rhombohedral or tabular, composed of 2, 3, or 6 individuals, twinned after the Artini law (Palache, 1941b; Fahey et al., 1950) (H 104513).

**Leadhillite.**
Mammoth–St. Anthony mine, Pinal County.
Rock Currier.

## LECHATELIERITE (fused silica)

Fused silica (glass). Formed by the application of intense heat (as by lightning strokes) from rock and soil containing quartz. Also formed as a result of the intense heat generated by meteoric impact into quartz sandstone. Not uncommonly noted beneath power lines where it has formed by electrical discharge between the earth and the conductors during electrical storms.

*Coconino County:* At Meteor Crater, west of Winslow, where it was formed by fusion of fine-grained Coconino Sandstone (Rogers, 1930). A number of occurrences of fulgurites, as black to green obsidian-like coatings and tubes, have been observed on the summit of Mt. Humphreys and other peaks of the San Francisco Mountains near Flagstaff. The composition of these fulgurites has not been established (Davis and Breed, 1968).

*Pima County:* Numerous occurrences of fulgurites have been noted under power lines along Ajo Road, near the road to Tucson Mountain Park, formed by lightning activity. Such electrical discharges have, in some instances, followed the root portions of desert plants downward into the soil, carbonizing the plant material, and forming a sheaf of fused soil or rock (which may include lechatelierite) in and adjacent to the root.

Leonhardtite (see STARKEYITE)

## LEPIDOCROCITE

Iron oxide hydroxide, $\gamma$-FeO(OH). Formed under essentially the same conditions as goethite with which it is frequently associated. Probably far more abundant in Arizona than indicated by the few occurrences noted.

*Cochise County:* Warren district, with goethite (UA 8969).
*Gila County:* Banner district, 79 mine, where it is a constituent of the limonites formed in the oxidized portions of the deposit (Keith, 1972).
*Graham County:* Lone Star district, near Safford, as a minor constituent of metasomatized volcanic rocks (Hutton, 1959b).
*Pima County:* Cababi district, South Comobabi Mountains, Mildren and Steppe claims (Williams, 1963).
*Pinal County:* Reported but unconfirmed at the Mammoth–St. Anthony mine, Mammoth district.

## LEPIDOLITE

Potassium lithium aluminum silicate fluoride hydroxide, $K(Li,Al)_3(Si,Al)_4O_{10}(F,OH)_2$. A member of the mica group. A typical constituent of lithium-bearing granite pegmatites where it is associated with spodumene and amblygonite.

*Maricopa County:* White Picacho district, as light pinkish-gray to deep lilac or lavender compact aggregates and small books, in lithium-bearing pegmatites (Jahns, 1952). Southwest of Wickenburg, on the Boyd-Fortner claims. West of Morristown, in the Garcia Mountains. Vulture Mountains, at the eastern end. In the Harquahala Mountains, in masses which are nearly identical in appearance to muscovite.
*Yavapai County:* Eureka district, in pegmatites of the Bagdad area.

## *LEUCITE

Potassium aluminum silicate, $KAlSi_2O_6$. Leucite occurs most typically in potassium-rich, silica-poor, volcanic flow rocks as well as in chemically equivalent hypabyssal rocks.

*Apache County:* Buell Park, in a cocite ring dike which curves around the south margin of the diatreme (Bideaux et al., 1960), and reported by Williams (1936) to occur in certain olivine-rich dike rocks on the Navajo Indian Reservation.

## *LEUCOPHOSPHITE

Potassium iron phosphate hydroxide hydrate, $KFe_2(PO_4)_2(OH)\cdot2H_2O$. An uncommon secondary mineral found in pegmatites; also of sedimentary origin.

*Yavapai County:* Eureka district, 7U7 Ranch, with bermanite in seams in triplite, also associated with phosphosiderite (metastrengite) and hureaulite (Leavens, 1967).

## LIBETHENITE

Copper phosphate hydroxide, $Cu_2(PO_4)(OH)$. A rare secondary mineral formed in the oxidized zones of copper deposits.

*Cochise County:* From an unspecified locality in the Little Dragoon Mountains (Robert O'Haire, pers. comm., 1972).
*Gila County:* Globe-Miami district, Castle Dome mine, where it forms crusts composed of small emerald-green prismatic crystals or as drusy mats of acicular crystals along open fractures (Peterson et al., 1946, 1951). (Note —

**Libethenite.**
Ray, Pinal County. Julius Weber.

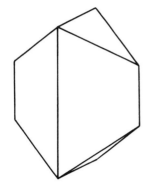

**Libethenite.**
Morenci
(Lindgren & Hillebrand, 1904).

Jewell J. Glass, of the United States Geological Survey, observed that the mineral from the Castle Dome mine exhibited inclined extinctions, as with dihydrite, and should properly be termed "clinolibethenite" [*in* Peterson, 1962, p. 75].) Banner district, 79 mine (Robert O'Haire, pers. comm., 1970).

*Greenlee County:* Morenci district, as small crystals in cavities at the Coronado mine (identified by S. F. Penfield). This discovery was the first of the mineral in the United States (Lindgren and Hillebrand, 1904).

*Pima County:* Southern side of Saginaw Hill, about 7 miles southwest of Tucson, associated with other oxidized minerals including cornetite, pseudomalachite, malachite, atacamite, and chrysocolla, coating fractures in chert. As pale green, greasy masses surrounding cornetite crystals, sometimes apparently formed from partially corroded, fairly large cornetite crystalline masses; occasionally found as clusters of yellow-ish-green prismatic crystals (Khin, 1970). East end of the South Comobabi Mountains, on fractures in quartz monzonite, with pseudomalachite. Santa Rita Mountains, near Rosemont (Bideaux et al., 1960). Silver Bell mine (Robert O'Haire, pers. comm., 1972).

*Pinal County:* Copper Creek district, Old Reliable mine (UA 9609, 529). Ray area (UA 9611).

## LIEBIGITE

Calcium uranium carbonate hydrate, $Ca_2U(CO_3)_4 \cdot 10H_2O$. An uncommon secondary mineral formed from alkaline carbonate solutions; may be associated with calcite, schroeckeringite, bayleyite, and gypsum.

*Colorado Plateau:* Reported from a diatreme of the Hopi-Navajo county, but the exact location is unknown.

## Limonite

A general term for mixtures of cryptocrystalline minerals, predominantly goethite and hematite. Most of the brownish material seen in the oxidized outcroppings of copper and other base metal deposits in Arizona is limonite. Note, however, that jarosite, alunite, and other secondary minerals may be common constituents of the so-called "limonites" which occur in oxidized cappings (gossans). Abundant in the southwestern United States and found in all mineral deposits in the region containing iron-bearing minerals which have been subjected to oxidation. Deposited in bogs and marshes forming low-grade iron deposits.

## LINARITE

Lead copper sulfate hydroxide, $PbCu(SO_4)(OH)_2$. A sparse but widely distributed secondary mineral found in the oxidized zone of copper-lead deposits. Easily mistaken for azurite.

*Cochise County:* At the Tranquillity mine (H 101593).

*Graham County:* Aravaipa district, at the Ten Strike group of claims (Ross, 1925a); Grand Reef mine, as brilliant druses and splendent groups of crystals to one inch, associated with cerussite, anglesite, and leadhillite in quartz-lined cavities (Richard L. Jones, pers. comm., 1969).

*Maricopa County:* Painted Rock Mountains, at the Rowley mine near Theba, associated with atacamite and diaboleite (Wilson and Miller, 1974).

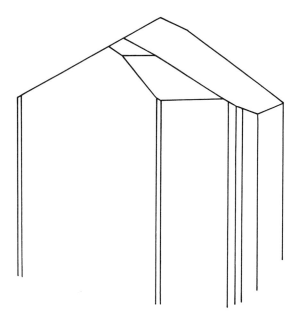

**Linarite.**
Mildren mine (Williams, 1962).

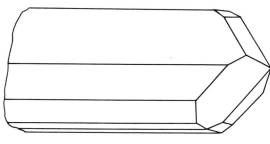

**Linarite.**
Mammoth–St. Anthony mine (Guild, 1911).

*Pima County:* Cababi district, South Como-babi Mountains, Mildren mine, in anglesite-cerussite aggregates with paratacamite, chlorargyrite, leadhillite, and matlockite, as excellent crystals whose forms include: {100}, {001}, {101}, {110}, {012}, {011}, {$\bar{1}$01}, {$\bar{2}$03}, {409}, {2.0.15}, and {$\bar{7}$04}. The crystals are flattened on {$\bar{1}$01} and elongate on [010] (Williams, 1962).

*Pinal County:* Mammoth district, Mammoth–St. Anthony mine, in places as excellent crystals, as thin films filling crevices in brecciated rock, and as druses of small to large, euhedral crystals reported to have been up to 4 inches in length; invariably associated with brochantite (Guild, 1911). Forms observed by Palache

(1941b) include: {110}, {101}, {$\bar{1}$01}, {001}, {111}, {201}, {$\bar{2}$12} (see also Sinkankas, 1964); reported as being associated with malachite and brochantite (Omori and Kerr, 1963).

## LINDGRENITE

Copper molybdate hydroxide, $Cu_3(MoO_4)_2(OH)_2$. An uncommon secondary mineral found in the oxidized portions of copper deposits.

*Gila County:* Globe-Miami district, Inspiration mine, Live Oak pit, as platy aggregates in hydrothermally altered schist, also in seams with

**Lindgrenite.**
Live Oak pit, Miami, Gila County. R. W. Thomssen collection. R. A. Bideaux.

molybdenite and, rarely, associated with powellite (F. Pough, pers. comm.; Peterson, 1962).

*Maricopa County:* Cave Creek district, with cuprotungstite (Schaller, 1932).

*Pima County:* Very sparsely at the Esperanza mine, Pima district (UA 6445).

*Pinal County:* Childs-Aldwinkle mine, Copper Creek district (Richard Thomssen, pers. comm.) (UA 488). On the Hull claims, south of Ray (H 108666). Superior (H 105628).

## *LITHARGE

Lead oxide, $\alpha$-PbO. One of four lead oxides, an uncommon mineral formed under very alkaline and oxidizing conditions in deposits containing lead sulfide. Litharge is probably a more common mineral in oxidized environments containing lead than is generally supposed.

*Pima County:* Cababi district, South Como-babi Mountains, Mildren and Steppe claims as an alteration product of wulfenite; occurs with a large variety of secondary minerals formed during the oxidation of sulfide-bearing veins cutting andesite (Williams, 1962, 1963).

## LITHIOPHILITE

Lithium manganese iron phosphate, $Li(Mn^{2+},Fe^{2+})PO_4$. A primary mineral which occurs in granite pegmatites. A complete substitutional series probably extends between lithiophilite and triphylite ($LiFePO_4$).

*Mohave County:* From Wikieup (UA 6437, 7889).

*Yavapai County:* White Picacho district, with triphylite, at the Midnight Owl and other pegmatites (Jahns, 1952) (UA 5880).

## *LITHIOPHORITE

Aluminum lithium manganese oxide hydroxide, $(Al,Li)MnO_2(OH)_2$. Found in vein and bedded deposits with other, more common, manganese oxide minerals.

*Mohave County:* Artillery Mountains, Priceless vein, where it is associated with cryptomelane, hollandite, pyrolusite, psilomelane, and coronadite; at the Plancha bedded deposits it is superimposed on hard silvery manganese ore, and locally it may replace ramsdellite-pyrolusite grains (Mouat, 1962).

*Yavapai County:* 5 miles north-northeast of Ashfork, as blue-black coatings in cavities in limestones of the Kaibab Formation (UA 9152) and replacing calcite cement in red beds of the Toroweap Formation (UA 9153) (Mullens, 1967).

## LOELLINGITE

Iron disulfide, $FeS_2$. The composition departs from this ideal stoichiometric form. An uncommon mineral found in mesothermal veins with iron and copper sulfides in calcite gangue.

*Maricopa County:* Disseminated in pegmatites throughout the White Picacho district (Jahns, 1952).

*Mohave County:* At the Copper World mine, near Yucca, with sphalerite, chalcopyrite, and pyrrhotite (Rasor, 1946) (H 104101). An analysis by Claude E. McLean gave:

| | | | | | |
|----|-------|----|------|--------|------|
| Fe | 25.76 | S  | 2.73 | $Al_2O_3$ | 2.02 |
| Co | 0.10  | Cu | 0.40 | $SiO_2$   | 1.30 |
| Ni | 1.55  | Zn | 0.15 | MgO    | 0.14 |
| As | 65.57 | Pb | 0.05 | CaO    | 0.11 |

TOTAL 99.88%

## *LONSDALEITE

Carbon, C. The hexagonal wurtzite-like dimorph of diamond, lonsdaleite is known to occur only in the Canyon Diablo meteorite.

*Coconino County:* Canyon Diablo area, in the Canyon Diablo meteorite, as black cubes and cubo-octahedra coated with graphite, up to 0.7 mm in size. This is the type locality (Frondel and Marvin, 1967).

## *LUDWIGITE

Magnesium iron boron oxide, $(Mg,Fe^{2+})_2$-$Fe^{3+}BO_5$. An uncommon mineral whose occurrence apparently is restricted to high-temperature contact metamorphic environments.

*Pima County:* Santa Catalina Mountains, Leatherwood mine group, in the contact zone near and west of the Control mine (James Post, pers. comm.).

## *LUETHEITE

Basic copper aluminum arsenate hydrate, $Cu_2$-$Al_2(AsO_4)_2(OH)_4 \cdot H_2O$. The analogue of chenevixite, an oxide zone mineral presently known only from its Arizona type locality.

*Santa Cruz County:* Patagonia district, in the vicinity of the Humboldt mine. Found in veinlets and vugs in rhyolite breccia as small, tabular crystals of distinctive Indian blue color. Associated with chenevixite, cornubite, and alunite (Williams, 1976). A chemical analysis by M. Duggan (Phelps Dodge Corp.) gave:

| | | | |
|------|------|--------|------|
| CuO  | 28.9 | $As_2O_5$ | 40.5 | $H_2O$ | 9.3 |
| $Al_2O_3$ | 18.4 | | | TOTAL 97.1% |

# M

## *MACKAYITE

Iron tellurium oxide hydroxide, $FeTe_2O_5$-(OH) (?). A rare secondary mineral found in the oxidized portions of tellurium-bearing base metal deposits with emmonsite and other secondary minerals.

*Cochise County:* Tombstone district, Toughnut-Empire mine, with native tellurium and emmonsite (Bideaux et al., 1960).

## *MAGHEMITE

Iron oxide, $\gamma$-$Fe_2O_3$. Maghemite possesses the magnetite structure but is deficient in iron. Formed from magnetite or lepidocrocite by slow oxidation at low temperatures; also from the oxidation of meteoric iron.

*Coconino County:* Meteor Crater area, in elongate isotropic grains alternating with a goethite-like iron oxide as a minor component of metallic spheroids formed as a result of impact of a large meteoric body (Mead et al., 1965).

*Pima County:* Tucson Mountains, as a thin, brownish surface alteration product on magnetite pebbles and boulders.

## MAGNESITE

Magnesium carbonate, $MgCO_3$. A member of the calcite group. Magnesite forms as a product of the metamorphism of magnesian rocks, through the alteration of calcite by magnesium-bearing waters, as sedimentary deposits, and, rarely, as a hydrothermal gangue mineral.

*Greenlee County:* Sparingly in beds of the Longfellow limestone.

*Mohave County:* Oatman district, in veins with brucite and serpentine (Wilson, 1944).

## MAGNETITE

Iron oxide, $Fe_3O_4$. The most abundant and widespread member of the spinel group. An accessory mineral in many igneous rocks, abundant in metamorphic rocks, and widely distributed in sedimentary rocks and in sands as a detrital mineral. Only a few representative occurrences can be noted.

*Cochise County:* Dragoon Mountains, Black Diamond mine, as granular masses with chalcopyrite and ilvaite.

*Coconino County:* Grand Canyon, as octahedral crystals up to an inch in diameter, in pegmatite.

*Gila County:* Globe-Miami district, as sharp, dodecahedral crystals up to ½ inch in diameter, in calcite with tremolite and serpentine at Asbestos Peak, near Globe (Sinkankas, 1964). Banner district, Christmas mine, a very abundant mineral in the lower Martin Formation ore body; locally abundant in the skarn rocks (David Perry, pers. comm., 1967; Peterson and Swanson, 1956). As large crystals in volcanic bombs of olivine from near Peridot.

*Greenlee County:* Clifton-Morenci district, abundant in metamorphosed limestone with garnet, amphibole, pyroxene, and sulfides; mined as flux at the Manganese Blue and Arizona Central mines (Lindgren, 1905). With

pyrolusite and psilomelane, from the Pyrolusite claims 12 miles southeast of Morenci (Potter et al., 1946).

*Maricopa County:* Big Horn district, with ilmenite in extensive placer deposits up to 100 feet thick (Harrer, 1964).

*Pima County:* Santa Rita Mountains, abundant in contact metamorphic copper ores at Rosemont (Schrader and Hill, 1915). Sierrita Mountains, in contact metamorphic ore bodies, Pima district (Guild, 1934). In large amounts on the surface, 5 miles from Tule Wells, near pegmatite bodies containing copper sulfides (Gilluly, 1937). Tucson Mountains, found as rounded, transported blocks, frequently pitted in such a manner as to resemble meteorites (Guild, 1910). Santa Catalina Mountains, Pontotoc mine, in an ore deposit in gneiss associated with limestone (Guild, 1934).

*Pinal County:* Red Rock-Florence Junction area, with ilmenite in widespread alluvial deposits (Harrer, 1964).

*Santa Cruz County:* Patagonia Mountains, as lodestone, in considerable quantities at the Line Boy mine, near Duquesne (Schrader, 1917).

*Yavapai County:* Bradshaw Mountains, Big Bug Creek, as large (up to 18 inches) streamworn masses in a stream bed (David Shannon, pers. comm., 1971); with hematite in schist near Townsend Butte and the Howard property; as large crystals with apatite and sphene in granodiorite at the Springfield group, Pine Grove district; as large pieces of lodestone near Stoddard (Lindgren, 1926). Eureka district, as titaniferous magnetite in dikes and irregular bodies in gabbro (Harrer, 1964). A partial analysis (Ball and Broderick, 1919) gave:

Fe 60.35          Ti 8.40          Mn trace.

McBride claims, 17 miles south of Seligman, as segregations of titaniferous magnetite in gabbro.

## MALACHITE

Copper carbonate hydroxide, $Cu_2CO_3(OH)_2$. Occurs commonly as an alteration product in oxidized copper deposits. Generally associated with other secondary copper minerals, particularly azurite, cuprite, and tenorite, and commonly with limonite. Only a few localities can be listed.

*Apache County:* Found in vein fillings and in cement in the Navajo Sandstone at Garnet Ridge where it is associated with chrysocolla, calciovolborthite, tyuyamunite, limonite, chalcopyrite, and pyrite (Gavasci and Kerr, 1968).

*Apache and Navajo Counties:* Monument Valley, associated with uranium-vanadium ores in the Moenkopi Formation, the Shinarump Member of the Chinle Formation, and the DeChelly Member of the Cutler Formation (Mitcham and Evensen, 1955; Evensen and Gray, 1958).

*Cochise County:* The Warren district has produced some of the world's most remarkable malachite, and specimens are to be found in places of honor in all of the major collections. Note was made early of the large mass of malachite at the Copper Queen mine by Kunz in 1885 (see also Douglas 1899; Ransome, 1903; Lindgren, 1904; Palache and Lewis, 1927; Frondel, 1941); Campbell mine (Schwartz and Park, 1932); Cole mine (Trischka et al., 1929; Schwartz, 1934; Hutton, 1957); Junction mine (Mitchell, 1920). Turquoise (Courtland-Gleeson) district, in large masses at the Maid of Sunshine mine, in some places as small but superb crystals. Tombstone district, where it is widespread but not abundant (Butler et al., 1938). Sulphur Springs Valley, Pat Hills, in numerous calcite veins cutting porphyritic andesite flow rocks, associated with chrysocolla (Lausen, 1927). Cochise district, Johnson Camp area, as an abundant mineral in tabular bodies with sulfides in contact metamorphosed limestone (Kellogg, 1906; Cooper and Huff, 1951).

**Malachite pseudomorph of azurite crystals.**
Bisbee, Cochise County. Arizona-Sonora Desert Museum collection. Robert H. Perrill.

*Coconino County:* Kaibab Plateau, with azurite and chrysocolla as impregnations in chert beds of wide extent; Apex copper property (Tainter, 1947a). Grand Canyon National Park, Horseshoe Mesa, Grand View (Last Chance) mine, as botryoidal crusts coating limestone and in fissures in clay (Leicht, 1971); also found in the Orphan mine with azurite. White Mesa district, with chrysocolla cementing Navajo Sandstone (Hill, 1912; Mayo, 1955). Apex mine, Jacob Lake, with azurite and minor chalcopyrite and chalcocite in the Kaibab Formation (Tainter, 1947a). Cameron area, in mineralized silica plugs (Barrington and Kerr, 1963).

**Fine-grained malachite.**
Bisbee, Cochise County. University of Arizona collection. Jeff Kurtzeman.

*Gila County:* Globe-Miami district, where it constitutes a considerable part of the ores at the Buffalo, Big Johnnie, Buckeye, and other mines, but nowhere found in large masses (Ransome, 1903); Old Dominion mine (Schwartz, 1921, 1934); Castle Dome and Copper Cities deposits (Peterson, 1947); Inspiration mine, as aggregates of malachite, chrysocolla, and chalcedony of great beauty (Ransome, 1919; Peterson, 1962; Sun, 1963). Payson district, as stout, prismatic crystals in porous quartz, at the Silver Butte mine (Lausen and Wilson, 1927). Banner district, Christmas mine, as a common supergene mineral in skarn developed in the Naco Formation (David Perry, pers. comm., 1967); 79 mine, where it is scattered throughout the supergene zone as veinlets and small spheres to ¼ inch (Keith, 1972).

*Graham County:* Klondyke area, where it is a widespread but sparse mineral (Simons, 1964).

*Greenlee County:* Clifton-Morenci district, where it is common in irregular deposits in limestone; mined at the Detroit, Manganese Blue, and Longfellow mines, and as fine radiating groups of crystals at the Standard mine, near Metcalf (Lindgren, 1903, 1904, 1905; Reber, 1916; Schwartz, 1934; Grout, 1946).

*Maricopa and Gila Counties:* Mazatzal Mountains, where it occurs in small quantities associated with mercury deposits, in schist (Lausen, 1926).

*Mohave County:* Noted at Bill Williams Fork (Genth, 1868). Cerbat Mountains, Wallapai district, widely distributed as an oxidation product of copper sulfide vein deposits; also locally as cement in alluvium, with cuprite and native copper (Thomas, 1949); Mineral Park district, in granite gneiss. Bentley district, Grand Gulch, Bronze L, and Copper King mines (Hill, 1914).

*Navajo County:* Monument Valley, where it is an associate of the uranium-vanadium deposits in sandstone (Holland et al., 1958). White Mesa district, as irregular masses in sandstone.

*Pima County:* Santa Rita Mountains, as globular masses and in veinlets in the Rosemont district; also widely distributed throughout the range in small showings. Sierrita Mountains, abundant in the Pima district, and the most important ore mineral at Mineral Hill (Schrader and Hill, 1915; Schrader, 1917); present in the oxidized portion of the Twin Buttes open pit mine (Stanley B. Keith, pers. comm., 1973). Ajo district, the most common product of the weathering of copper-bearing minerals and the dominant mineral in the leaching ores (Jora-

lemon, 1914; Gilluly, 1937); malachite occurs in the New Cornelia open pit mine in concentric structures with azurite and quartz in vugs, and covering the outside of porous nodules of cuprite which, in turn, surround native copper (Schwartz, 1934); Copper Giant deposits, Secs. 10, 11, and 15, T.13S, R.6W, in fanglomerate and volcanic rocks (Romslo and Robinson, 1952); Waterman Mountains, Silver Hill mine, with rosasite (UA 6709). Silver Bell Mountains, Silver Bell mine, with cuprite and azurite in limestone replacements (*Engr. Min. Jour.,* 1904; Stewart, 1912).

*Pinal County:* Mineral Creek district, Ray area, with chrysocolla and tenorite cementing part of the White Tail Conglomerate south of Ray (Clarke, 1953); also cementing Holocene gravels with other oxidized copper and iron minerals in the Pearl Handle open pit at Ray. This mineralization was shown to be not older than about 7,000 years on the basis of a radiocarbon date obtained from a fossil log incorporated in the mineralized area (Phillips et al., 1971). Pioneer district, with chrysocolla as the principal oxidized copper mineral at the Magma mine (Short et al., 1943). Mammoth district, San Manuel mine, where it is present in the oxidized zone of the ore body but accounts for little of the copper values; associated with chrysocolla, cuprite, and azurite (Schwartz, 1949); Mammoth–St. Anthony mine, common as powdery masses and microcrystals and as pseudo-

**Malachite.**
Chicago mine (Williams, 1962).

morphs after azurite crystals, sometimes of large size.

*Santa Cruz County:* Santa Rita and Patagonia Mountains, widespread as a product of the oxidation of sulfide copper minerals at numerous mines and prospects, especially in the Tyndall, Old Baldy, and Patagonia districts (Schrader and Hill, 1915; Schrader, 1917).

*Yavapai County:* Black Hills, as fine specimen material with crystallized azurite at the Yeager mine; also common in the oxidized portions of all of the copper deposits of the region (Schwartz, 1934, 1938; Anderson and Creasey, 1958). Walnut Grove district, Zonia Copper mine (Kumke, 1947).

*Yuma County:* Buckeye Mountains, Planet mine, where it is associated with azurite and chrysocolla (Bancroft, 1911). From near Bouse (UA 3966).

## *MANANDONITE

Lithium aluminum boron silicate hydroxide, $LiAl_5BSi_2O_{10}(OH)_8$. A member of the chlorite group.

*Yuma County:* Near Quartzsite, as minute tablets in quartzite with kyanite, andalusite, and dumortierite (Bideaux et al., 1960).

## *MANGANAXINITE

Calcium manganese iron aluminum borosilicate hydroxide, $Ca_2(Mn,Fe)Al_2BSi_4O_{15}(OH)$. A member of the axinite group. Most common occurrence is in contact metamorphic aureoles formed where intrusive rocks have invaded sediments, especially limestones; commonly associated with other calcium silicate minerals.

*Pima County:* North end of the South Comobabi Mountains north of the Ko Vaya Hills, as yellow plates being replaced by withamite (a variety of clinozoisite); also in vesicles in andesite flow boulders (Bideaux et al., 1960).

## MANGANITE

Manganese oxide hydroxide, $MnO(OH)$. Usually associated with other manganese and iron oxide minerals in low-temperature veins and in deposits of secondary origin.

*Cochise County:* Tombstone district, as needle-like crystals in parallel groups and as soft fibers lining cavities (Butler et al., 1938; Rasor, 1939; Romslo and Ravitz, 1947; Needham and Storms, 1956).

*Gila County:* Globe-Miami district, where, with pyrolusite and psilomelane, it forms the bulk of the gangue in the manganese-zinc-lead-silver deposits of the district (Peterson, 1962). As irregular masses with oxidized copper ore at the 79 mine (Kiersch, 1949).

*Maricopa County:* Bighorn Mountains, Aguila district, with pyrolusite and wad. White Picacho district, as crusts coating lithium phosphate· minerals, in pegmatites (Jahns, 1952).

*Mohave County:* Rawhide Mountains, associated with a variety of manganese oxide minerals in the Artillery Peak deposit (Head, 1941).

*Pinal County:* Silver King mine, as crystals in barite (UA 7437).

*Santa Cruz County:* Patagonia Mountains, Harshaw district, Mowry mine (Havens et al., 1954); Tyndall district, Cottonwood Canyon, Glove mine (Olson, 1966).

*Yavapai County:* With pyrolusite and wad in the northern part of the Aguila district. As shiny black crystals on the property of the Mohave Mining and Milling Company, east of Wickenburg. White Picacho district, as thick crusts coating lithium phosphate minerals, in pegmatites (Jahns, 1952). Burmeister mine, 12 miles southeast of Mayer, with pyrolusite and psilomelane (Long et al., 1948).

*Yuma County:* Planet district; Santa Maria district, Keiserdoom claims (Hewett et al., 1963). War Eagle claims, 26 miles north of Bouse (Havens et al., 1947).

## MARCASITE

Iron sulfide, $FeS_2$. A low-temperature mineral formed under near-surface or surface conditions. Less common and less stable than pyrite with which it is dimorphous. Generally in replacement deposits or as concretions in sedimentary rocks.

*Apache County:* From the vicinity of Sander (UA 7285).

*Cochise County:* Cochise district, Johnson Camp area, in narrow seams and as crystals along faults near Johnson (Cooper and Silver, 1964).

*Coconino County:* Cameron area, as cyclically-twinned inclusions in amethystine quartz crystals in petrified wood; Alyce Tolino mine, enclosed by small cubes of cobalt-rich pyrite, associated with umahoite in sooty masses and in carbonaceous trash replacements (Hamilton and Kerr, 1959); Huskon mines, with pitchblende, bravoite, greenockite, pyrite, calcite, and siliceous gangue replacing wood of a fossil Triassic conifer in the Chinle Formation (Maucher and Rehwald, 1961); Huskon No. 11 mine, in sandstone with inky blue masses and stains of ilsemannite (AEC).

*Gila County:* Common in veins in uranium deposits in the Dripping Spring Quartzite, northwestern Gila County (Granger and Raup, 1969).

*Mohave County:* Cerbat Mountains, Wallapai district, as a primary mineral deposited in a belt of sulfide-bearing fissure veins, associated with arsenopyrite (Thomas, 1949). Black Mountains, as thin plates in quartz of the Moss mine, Oatman district.

*Pima County:* Santa Rita Mountains, Old Baldy district (Schrader, 1917). Pima district, as an alteration product of pyrrhotite at the Copper Glance and Queen mines (Webber, 1929); also as thin coatings and veinlets in the Mineral Hill and San Xavier areas.

*Santa Cruz County:* Tyndall district (Schrader, 1917).

*Yavapai County:* Bradshaw Mountains, as small, colloform masses in partly oxidized ore, Iron Queen mine, Big Bug district (Lindgren, 1926). In the pegmatites of the White Picacho district (Jahns, 1952). Martinez district, Congress mine, in gold-bearing quartz veins with pyrite, associated with intrusive dikes in granite (*Engr. Min. Jour.*, 1904).

Mariposite (chromian MUSCOVITE)

## MASSICOT

Lead oxide, PbO. An uncommon mineral of secondary origin, dimorphous with litharge, formed by the alteration of galena or secondary lead-bearing minerals.

*Gila County:* Globe-Miami district, as a yellow, powdery deposit on cerussite in the Albert Lea mine (Peterson, 1962). Payson district, Silver Butte mine, as an earthy yellow powder associated with anglesite and galena (Lausen and Wilson, 1927).

*Maricopa County:* Bighorn Mountains, Tonopah-Belmont mine, with minium. South of Wickenburg, at the Moon Anchor mine, Potter-Cramer property, and the Rat Tail claim (Williams, 1968).

*Pima County:* Cababi district, South Comobabi Mountains, as an earthy, yellow, alteration product of galena, at the Little Mary mine and Silver-Lead claim (Williams, 1962).

*Santa Cruz County:* Patagonia Mountains, Flux mine, with cerussite (Schrader and Hill, 1915; Wilson, 1933).

*Yuma County:* Trigo Mountains, Silver district, as an earthy powder associated with cerussite and smithsonite.

## MATLOCKITE

Lead fluoride chloride, PbFCl. A rare secondary mineral known from few localities in the world.

*Gila County:* Apache mine, near Globe, with cerussite, brochantite, and boleite (Bideaux et al., 1960).

*Pima County:* Cababi district, South Comobabi Mountains, Mildren mine, as a single crys-

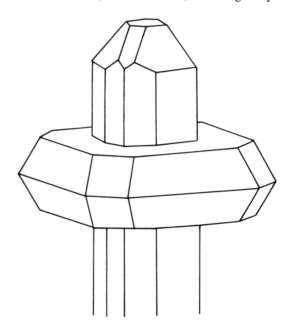

**Matlockite in parallel orientation on anglesite.**
Mildren mine (Williams, 1962).

tal in parallel growth with a prismatic anglesite crystal (Williams, 1962).

*Pinal County:* Mammoth district, Mammoth–St. Anthony mine, as minute crystals on boleite from the 400-foot level of the Collins vein, and as a large nodule coated with cerussite from the 500-foot level (Fahey et al., 1950).

Meerschaum
(a variety of SEPIOLITE)

## MELANOTEKITE

Lead iron silicate, $Pb_2Fe_2Si_2O_9$. A very rare lead silicate mineral.

*Pinal County:* Mammoth district, Mammoth–St. Anthony mine, as minute brownish spherules on diaboleite.

## MELANOVANADITE

Calcium vanadium oxide hydrate, $Ca_2V_4^{4+}$-$V_6^{5+}O_{25} \cdot nH_2O$. A rare mineral which may be of primary origin.

*Apache County:* Lukachukai Mountains, at the Mesa No. 1 and No. 5 mines; also in the Kerr-McGee 4-1 mine.

## MELANTERITE

Ferrous sulfate hydrate, $FeSO_4 \cdot 7H_2O$. Water soluble; forms commonly in old mine workings as a product of the oxidation of pyritic ores. Probably more common in abandoned mine workings than the occurrences mentioned below would suggest.

*Cochise County:* Warren district, where it is a widely distributed post-mine mineral in stalactites and coatings of walls or as large (up to 18 mm) pseudo-octahedra in stagnant waters. Usually in pyrite-rich areas.

*Gila County:* Banner district, 79 mine, sparingly as white encrustations and efflorescences in the mine workings, altering from pyrite; melanterite blankets formed on the floor of the fifth level contain short, acicular, prismatic, hairlike crystals (Keith, 1972).

*Greenlee County:* Clifton-Morenci district, sparingly in the upper levels of mines in the district (Lindgren, 1905), and locally present in the Morenci open pit mine formed by the oxidation of sulfide minerals (Moolick and Durek, 1966).

*Pima County:* Sierrita Mountains, as efflorescences on the walls of old mine workings in the Pima district; associated with chalcanthite at the San Xavier West mine (Arnold, 1964).

## MERCURY

Mercury, Hg. Comparatively rare; of secondary origin formed by the alteration of cinnabar with which it is usually associated.

*Coconino County:* Near Lee's Ferry, in minute quantities in the Chinle Formation associated with gold (Lauson, 1913; Lausen, 1936).

*Gila County:* Mazatzal Mountains, Pine Mountain mine, Sunflower district (Lausen, 1926); small amounts in the Slate Creek deposits at the Ord mine (Faick, 1958).

*Maricopa County:* Mazatzal Mountains (Beckman and Kerns, 1965), Sunflower district, Pine Mountain mine.

*Mohave County:* Maynard district, Hualpai Mountains, with cerussite.

*Navajo County:* Snowflake, in schist with cinnabar (UA 6502).

*Pinal County:* As fine globules in schist, 9 miles east of Apache Junction (Beckman and Kerns, 1965).

*Yavapai County:* In appreciable quantity in the Kirkland placers, lower Copper Basin Wash, probably derived from the low-grade cinnabar deposits in the vicinity. Wagoner district, Walnut Grove, associated with cinnabar (*Engr. Min. Jour.,* 1897b).

## *MESOLITE

Sodium calcium aluminum silicate hydrate, $Na_2Ca_2(Al_6Si_9)O_{30} \cdot 8H_2O$. In cavities in volcanic rocks associated with other zeolites.

*Gila County:* Banner district, Christmas mine, as fine hairlike crystals in hydrothermally altered andesite (David Perry, pers. comm., 1967).

*Santa Cruz County:* As silky white fibers in veinlets, with calcite, cutting fresh granodiorite. The samples came from core in a hole drilled southwest of Patagonia.

## *META-AUTUNITE

Calcium uranyl phosphate hydrate, $Ca(UO_2)_2$-$(PO_4)_2 \cdot 2$-$6H_2O$. Closely associated with the higher hydrate, autunite; a secondary uranium mineral formed by the oxidation of other uranium minerals, notably uraninite.

*Cochise County:* As scaly yellow crusts on pyrite ores, Courtland-Gleeson district.

*Coconino County:* Cameron district, Black Point-Murphy mine; Jackpot No. 24 mine (Austin, 1964).

*Gila County:* Northwestern Gila County in weathered deposits in the Dripping Spring Quartzite at the Sue, Red Bluff, and Little Joe deposits (Granger and Raup, 1962, 1969).

## METACINNABAR

Mercury sulfide, HgS. The dimorph of cinnabar. An uncommon mineral of secondary origin found in the upper portion of cinnabar deposits.

*Gila and Maricopa Counties:* Mazatzal Mountains, Alder and Slate Creeks area, in small quantities with cinnabar (Lausen, 1926).

*Yuma County:* Dome Rock Mountains, as thin coatings on cinnabar at the Colonial property.

## *METAHALLOYSITE

Basic aluminum silicate, $Al_4(OH)_8Si_4O_{10}$. A clay mineral found in weathered soils or as a supergene mineral in maturely weathered gossans.

*Cochise County:* Warren district, Holbrook pit where it occurs as greenish gray crusts and pods resembling impure chrysocolla in silicified shale.

## METAHEWETTITE

Calcium vanadium oxide hydrate, $CaV_6O_{16} \cdot 9H_2O$. A secondary mineral found as impregnations in sandstone.

*Apache County:* Monument Valley, Monument No. 2 mine, where it is one of the principal ore minerals which occur in a channel in the DeChelly Sandstone filled by Shinarump Conglomerate, as impregnation of sandstone and replacement of fossil plant matter; associated with tyuyamunite, carnotite, becquerelite, corvusite, hewettite, rauvite, navajoite, and uraninite (Finnell, 1957).

## *METANOVACEKITE

Magnesium uranyl arsenate hydrate, $Mg(UO_2)_2(AsO_2) \cdot 4\text{-}8H_2O$. A lower hydrate of novacekite. A rare secondary mineral of which this may be only the second known occurrence.

*Gila County:* Cherry Creek area, in a Dripping Spring Quartzite at the Sue mine where it occurs as individual flakes and as alteration rims 0.2–0.5 mm wide on bassetite and saleeite (Granger and Raup, 1969).

## METAROSSITE

Calcium vanadium oxide hydrate, $CaV_2O_6 \cdot 2H_2O$. A rare mineral found in sandstone uranium-vanadium deposits of the Colorado Plateau; from alteration of rossite with which it is intimately associated.

*Apache County:* An alteration product of rossite, found in the Salt Wash Member of the Morrison Formation, Lukachukai Mountains (Chenoweth, 1967). Also reported as occurring in the Monument No. 2 mine (Evensen and Gray, 1958).

## *METASIDERONITRITE

Sodium iron sulfate hydroxide hydrate, $Na_4Fe_2(SO_4)_4(OH)_2 \cdot 3H_2O$. A rare secondary mineral associated with other secondary sulfates.

*Coconino County:* Cameron area, as a deep yellow, cleavable, fibrous material from the Yazzie No. 101 mine which reportedly constitutes the first North American occurrence (Austin, 1957, 1964).

## Metastrengite
## (see PHOSPHOSIDERITE)

## METATORBERNITE

Copper uranyl phosphate hydrate, $Cu(UO_2)_2(PO_4)_2 \cdot 8H_2O$. A secondary mineral which may in part crystallize directly from solution; much like torbernite in occurrence and association.

*Apache and Navajo Counties:* Monument Valley district, in the oxidized "yellow ore" of uranium-vanadium deposits in the Shinarump Member and DeChelly Formation, associated with conglomerate-filled paleo stream channels; associated with carnotite, tyuyamunite, metarossite, calciovolborthite, and hewettite (Evensen and Gray, 1958).

*Coconino County:* Cameron area, present in uranium deposits in the Shinarump Member of the Chinle Formation, associated with uraninite and meta-autunite (Holland et al., 1958), Huskon No. 7 mine and Riverside No. 1 claims (Austin, 1964), Arrow Head claim; with uraninite in the River View collapse feature, southeast of Cameron (Barrington and Kerr, 1963).

*Gila County:* Associated with nearly all of the uranium deposits in the Dripping Spring Quartzite, northwestern Gila County (Granger and Raup, 1969); Workman Creek, Sierra Ancha Mountains, in the upper Dripping Spring Quartzite, with crocoite (Robert O'Haire, pers. comm., 1972). Wilson Creek area, in Dripping Spring Quartzite (Wells and Rambosek, 1954). Globe-Miami district, Copper Cities deposit, as small amounts in disseminated copper ore in quartz monzonite, as tiny rosettes on the walls of minute fractures along the Coronado fault (Peterson, 1954, 1962); Melinda mine; Castle Dome mine, where it usually occurs on wavellite crusts (Peterson, 1947).

*Mohave County:* Hacks Canyon mine (Granger and Raup, 1962).

*Navajo County:* Monument Valley, Monument No. 1 and Mitten No. 2 mines (Witkind,

1961; Witkind and Thaden, 1963). Also in the Ruth group of claims, near Holbrook.

## METATYUYAMUNITE

Calcium uranyl vanadate hydrate, $Ca(UO_2)_2(VO_4)_2 \cdot 3\text{-}5H_2O$. A secondary mineral commonly associated with tyuyamunite in the oxidized portions of uranium-vanadium deposits; more likely to form than carnotite in limestones or calcareous sandstones.

*Apache County:* Carrizo Mountains, at Cove Mesa and in the Sycamore group of claims and in the King Tut mine. Monument Valley, Monument No. 2 mine, where it is associated with tyuyamunite forming a thick zone around the dark resinous material which is the matrix for small grains of uraninite (Rosenzweig et al., 1954).

*Coconino County:* Cameron area, in Shinarump Conglomerate at the Montezuma group; also at the Huskon No. 12 mine (Austin, 1964).

*Navajo County:* Ruth group of claims, Holbrook district; also near Tuba City. Monument Valley, Monument No. 1 and Mitten No. 2 mines (Witkind, 1961; Witkind and Thaden, 1963).

## *META-URANOCIRCITE

Barium uranyl phosphate hydrate, $Ba(UO_2)_2(PO_4)_2 \cdot 6\text{-}2H_2O$. An uncommon member of the meta-autunite group; of secondary origin.

*Coconino County:* Cameron area, an abundant ore mineral which occurs as fine-grained yellow masses in association with fossil logs in the Petrified Forest member of the Chinle Formation (Austin, 1964).

*Gila County:* Reported from uranium deposits in the Dripping Spring Quartzite (Granger, 1955).

## *METAVOLTINE

Potassium sodium iron sulfate hydroxide hydrate, $(K,Na,Fe)_5Fe_3(SO_4)_6(OH)_2 \cdot 9H_2O(?)$. A rare mineral found in arid climates, often as a result of the oxidation of pyritic ores.

*Cochise County:* Warren district, 2100 level of the Campbell mine, as vermicular stacks of minute hexagonal platelets of greenish yellow color. Associated species are copiapite, coquimbite, voltaite, and roemerite. The indices of refraction suggest that the Na:K ratio is about 4:1 (Fabien Cesbron, pers. comm., 1975).

## METAZEUNERITE

Copper uranyl arsenate hydrate, $Cu(UO_2)_2(AsO_4)_2 \cdot 8H_2O$. A secondary mineral formed in the oxidized portions of uranium deposits.

*Apache County:* Monument Valley, Monument No. 2 mine (Witkind and Thaden, 1963).

*Coconino County:* Cameron area, present in minor amounts in uranium ore in the Shinarump Conglomerate, in paleo stream channels and in Chinle Formation in the sandy portions of mounds, associated with uraninite, metatorbernite, and meta-autunite (Holland et al., 1958). Grand Canyon National Park, Horseshoe Mesa, Grandview (Last Chance) mine, as transparent, emerald-green to leek-green, tabular crystals, associated with scorodite and olivenite (x-ray data suggest that the leek-green variety is close to metazeunerite, the emerald-green variety to zeunerite) (Leicht, 1971).

*Gila County:* Sierra Ancha Mountains, Easy deposit, where it coats limonite on fracture surfaces and is locally coated by hyalite (Granger and Raup, 1969).

*Navajo County:* At the Ruth group of claims, near Holbrook.

## MIARGYRITE

Silver antimony sulfide, $AgSbS_2$. Occurs in low temperature hydrothermal veins with galena and other silver minerals.

*Mohave County:* Cerbat Mountains, Wallapai district, as a primary mineral in minor amounts with pyrargyrite and polybasite; accompanies proustite in veinlets cutting galena and chalcopyrite (Thomas, 1949).

## MICROCLINE

Potassium aluminum silicate, $KAlSi_3O_8$. Triclinic. A widespread rock-forming feldspar which forms under nearly the same conditions as orthoclase. Much of the potash feldspar commonly classed as orthoclase is probably microcline. The mineral is far more prevalent in the rocks of Arizona than the few localities listed would suggest.

*Cochise County:* Reported from the Warren district.

*Maricopa and Yavapai Counties:* White Picacho district, as perthite, the most abundant mineral constituting the pegmatites of the district; as crystals up to 13 feet in maximum dimension (Jahns, 1952).

*Mohave County:* Cerbat Range, Kingman

Feldspar mine, commercially mined from pegmatites since 1924 (Heinrich, 1960).

*Pima County:* Ajo district, as coarse crystals in the pegmatite masses in the south-central part of the New Cornelia ore body (Gilluly, 1937); all has been removed by mining.

## MICROLITE (see PYROCHLORE)

### *MIERSITE

Silver copper iodide, (Ag,Cu)I. A rare secondary mineral found in oxidized zones of base metal deposits with malachite and cerussite.

*Pima County:* Cababi district, South Comobabi Mountains, Mildren and Steppe claims, as overgrowths on iodargyrite; a product of the oxidation of sulfide ores in quartz veins cutting andesite (Williams, 1962, 1963).

### *MILLERITE

Nickel sulfide, NiS. Usually a low temperature mineral found in cavities in veins, where it is frequently associated with carbonates; also found in geodes in limestone.

*Coconino County:* Orphan mine, near Grand Canyon, in cavities in barite.

### MIMETITE

Lead arsenate chloride, $Pb_5(AsO_4)_3Cl$. Structurally, a member of the apatite group; an end-member of the pyromorphite-mimetite series. A secondary mineral found in the oxidized portions of lead deposits; typically associated with cerussite, plattnerite, wulfenite, smithsonite, hemimorphite, anglesite, limonite, and other oxide zone minerals.

*Cochise County:* Gallagher Vanadium property, near Charleston. Warren district, with wulfenite, etc. in the Campbell mine, and with malachite in a fault zone on the 600-foot level, Cole mine; also found with cerussite in the Shattuck mine.

*Gila County:* Banner district, 79 mine, as fine specimens, notably as brilliant orange, orange-yellow, and bright yellow non-crystallized crusts and reniform masses some of which are stalactitic. Some of the finest specimen material is associated with thick, clear orange wulfenite crystals many of which are ½ inch on an edge; occasionally found as small canary-yellow crystals (Keith, 1972).

*Maricopa County:* Near Gila Bend and also east of the Alaska mine, southwest of Aguila.

Painted Rock Mountains, Rowley mine, near Theba, as microcrystals of a great variety of habits: as minute, perfect hexagonal crystals less than 1 mm in length and as nearly perfect spherical aggregates which have been interpreted by Wilson and Miller (1974) as being the extreme case of "wheat-sheaf"-like growth of mimetite crystals in which the extremities of the crystal aggregates have folded outward and back upon themselves and met, to produce the spherical shape. Colors range from red to yellow; some crystals are color zoned, the centers being orange and the ends yellow. Vulture district, at the Domingo mine.

*Mohave County:* Cerbat Mountains, reported from the Wallapai district.

*Pinal County:* Mammoth district, Mammoth–St. Anthony mine, as bright orange and canary-yellow crusts and as coatings of tiny prismatic to tabular crystals with wulfenite.

*Yavapai County:* Bradshaw Mountains, with pyromorphite and mottramite as botryoidal crusts at a prospect on the Slate Creek property of Kalium Chemicals Ltd. (W. C. Berridge, pers. comm., 1973).

### MINIUM

Lead oxide, $Pb_3O_4$. A secondary mineral formed in lead deposits under extreme oxidizing conditions.

*Maricopa County:* Big Horn Mountains, Tonopah-Belmont mine, with massicot. South of Wickenburg, at the Moon Anchor mine, Potter-Cramer property, and Rat Tail claim, with a variety of exotic secondary minerals formed by the oxidation of galena-sphalerite ores (Williams, 1968).

*Mohave County:* Black Mountains, as pulverulent material in cavities in the Big Jim vein, Oatman district.

*Pima County:* South Comobabi Mountains, Cababi district, as coatings on wulfenite and other secondary minerals, at the Silver-Lead and Beacon claims, and the Mildren mine (Williams, 1962).

*Pinal County:* Tortilla Mountains, Florence Lead-Silver mine (Williams and Anthony, 1970). Casa Grande area (UA 7324). Mammoth–St. Anthony mine, Mammoth district, rarely as a powdery coating on wulfenite crystals.

*Santa Cruz County:* Patagonia Mountains, Flux mine, with cerussite (Schrader and Hill, 1915).

*Yavapai County:* From near Salome (UA 373—specimen of questionable natural origin).

*Yuma County:* Castle Dome Mountains, Castle Dome district, with cerussite.

## MIRABILITE

Sodium sulfate hydrate, $Na_2SO_4 \cdot 10H_2O$. A water soluble mineral found in desert playas, saline lakes, and in clay soils of the desert environment; also found in old mine workings.

*Yavapai County:* In salt deposits of the Verde Valley, associated with halite, glauberite, and thenardite (Silliman, 1881; Blake, 1890; Guild, 1910; Peirce, 1969).

## *MIXITE

Basic bismuth copper arsenate hydrate, $Bi_2Cu_{12}(AsO_4)_6(OH)_{12} \cdot 6H_2O$. A rare secondary mineral found at some base metal deposits.

*Gila County:* Banner district, at the Christmas mine, as spherules of delicate twisted or matted fibers in cavities in gangue; associated with fibrous malachite. The peculiar yellow green color strongly resembles that of creaseyite.

## MOISSANITE

Silicon carbide, SiC. Observed for the first time naturally in the Canyon Diablo meteorite. Some authorities, however, have questioned the occurrence and suggest that the material is synthetic SiC (Carborundum) used in sample sawing. Moissan, who first observed SiC in the Canyon Diablo iron, used no silicon carbide in sample preparation (Kunz, 1905).

## MOLYBDENITE

Molybdenum sulfide, $MoS_2$. The only common mineral of molybdenum. Widely distributed, but in small amounts. A primary sulfide in granitic rocks or in quartz-orthoclase veins with chalcopyrite or with tin and tungsten ores. Also in contact metamorphic deposits. An important associate of copper mineralization in southwestern "porphyry" copper deposits.

*Cochise County:* Little Dragoon Mountains, in copper ores at Johnson Camp, Cochise district (Ransome, 1919; Cooper and Huff, 1951; Cooper, 1957; Cooper and Silver, 1964). Rare, as films on pyritic ores at Bisbee.

*Gila County:* Globe-Miami district, in small quantities in the ores of the area, particularly at the Castle Dome mine (Peterson, 1947, 1962; Peterson et al., 1946, 1951); Copper Cities deposit (Peterson, 1954). Banner district, with chalcopyrite and pyrite, locally coating the walls of veinlets in the 79 mine (Keith, 1972). Workman Creek area, where it occurs in hornfels near diabase; at the Suckerite deposit it is associated with uraninite (Granger and Raup, 1969).

*Greenlee County:* Clifton-Morenci district, in veins with pyrite, chalcopyrite, and sphalerite (Lindgren, 1905; Guild, 1910; Schwartz, 1947), and as thin films in fractures devoid of other sulfides (Moolick and Durek, 1966).

*Maricopa County:* Sparsely scattered in the pegmatites of the White Picacho district; locally abundant (Jahns, 1952).

*Mohave County:* Cerbat Mountains, O.K. mine, Gold Basin district, with galena and wolframite; Mineral Park mine, Wallapai district (Garrison, 1907; Thomas, 1949; Field, 1966); Samoa mine, Chloride district (Schrader, 1909; Blanchard and Boswell, 1930). Hualpai Mountains, Leviathan and American mines, Maynard district. Deluge Wash area, in small quantities at several properties (Frondel and Wickman, 1970). Ithaca Peak, where it is present with pyrite in the core of a quartz monzonite stock (Eidel, 1966).

*Pima County:* Santa Rita Mountains, Helvetia-Rosemont district (Frondel and Wickman, 1970), at the Leader, Ridley, and Pauline mines, and in many prospects in Madera and Providencia Canyons (Guild, 1907; Schrader and Hill, 1909; Creasey and Quick, 1955); as small masses in chalcopyrite ore, Cuprite mine (Browne, 1958). Silver Bell district, at a small prospect north of the Kurtz shaft (Guild, 1910; Stewart, 1912). Baboquivari Mountains, Gold Bullion mine, in quartz veins. Pima district, in the Mineral Hill-Twin Buttes area (Eckel, 1930; Guild, 1934); a primary mineral at the Twin Buttes open pit mine (Stanley B. Keith, pers. comm., 1973); and an important primary ore mineral at the Pima mine (Himes, 1972). At Ajo, sparingly (Gilluly, 1937; Schwartz, 1947). Near Redington, in limestone (UA 7288). South Comobabi Mountains, Cababi district, Mildren and Steppe claims (Williams, 1963).

*Pinal County:* Galiuro Mountains, Copper Creek district, the most important ore mineral at the Childs-Aldwinkle mine from which 70 million pounds were produced in the period 1933–1938 (Anderson, 1969); as fine crystallized specimens (Kuhn, 1941; Fleischer, 1959). In lesser quantities at the Copper Prince, Old

Reliable, and other properties (Simons, 1964). (The rhenium content of the molybdenite concentrates from the Childs-Aldwinkle mine contained from 320 to 580 parts per million, among the highest known.) Mineral Creek district, at Ray (UA 7129). Mammoth district, San Manuel mine (Lovering et al., 1950; Schwartz, 1953); Kalamazoo ore body (Lowell, 1968).

*Santa Cruz County:* Patagonia Mountains, Santo Niño mine, 2½ miles southwest of Duquesne, as large bodies of fine-grained massive material, and as good crystals in quartz veins with pyrite (Blanchard and Boswell, 1930; Frondel and Wickman, 1970); Bonanza mine, Duquesne, where small quantities were mined; Benton and Line properties (Schrader and Hill, 1915; Schrader, 1917).

*Yavapai County:* Sierra Prieta Range, as extensive deposits at the Copper Hill, Loma Prieta, and other properties, Copper Basin district (Johnston and Lowell, 1961). Bradshaw Mountains, Black Hawk, Blue Bird, and Squaw Peak mines. Eureka district, in thin veins at the Bagdad mine (Lindgren, 1926; Schwartz, 1947; Anderson, 1950; Dale, 1961). In the pegmatites of the White Picacho district (Jahns, 1952).

## MONAZITE

Rare earth (primarily cerium) phosphate, $(Ce,La,Y,Th,Ca)PO_4$. Contains other rare earth elements in addition to the preponderant cerium; also commonly contains lanthanum, thorium, uranium, calcium, silicon, and other vicarious elements. An important ore mineral of the rare earths and thorium. Occurs as an accessory mineral in granites and pegmatites; also found in thermally metamorphosed rocks, and as an abundant constituent in certain sands.

*Graham County:* Santa Teresa Mountains, as small crystals in pegmatite.

*Maricopa and Yavapai Counties:* White Picacho district, as a minor accessory mineral, associated with tantalum-columbium minerals in several of the pegmatites of the district (Jahns, 1952).

*Mohave County:* Chemehuevis district, about 20 miles southeast of Topock, where it occurs sparingly in stream gravels (Heineman, 1930; Overstreet, 1967). Virgin Mountains, with xenotime, in Precambrian gneiss (Young and Sims, 1961). Aquarius Range, Rare Metals mine, in pegmatites, associated with sphene, chevkinite, apatite, cronstedtite (Kauffman and Jaffe, 1946; Heinrich, 1960). Near the Nevada state line,

opposite Mesquite (Clark County, Nevada), in granite augen gneiss, with xenotime (Overstreet, 1967). From near Hoover Dam (H 102368).

*Yavapai County:* Black Canyon Creek, sparse in sands with magnetite, hematite, garnet, and gold (Day and Richards, 1906).

## MONTICELLITE

Calcium magnesium silicate, $CaMgSiO_4$. An uncommon member of the olivine group.

*Cochise County:* Tombstone district, as narrow bands in a contact metamorphic zone with calcite, thaumasite, clinozoisite, and vesuvianite, on the fourth level of the Toughnut mine (Butler et al., 1938).

## MONTMORILLONITE

Sodium calcium aluminum magnesium silicate hydroxide hydrate, $(Na,Ca)_{.33}(Al,Mg)_2Si_4O_{10}(OH)_2 \cdot nH_2O$. The most common and widespread member of the smectite group of clay minerals. The principal constituent of the bentonite clays which result from the alteration of volcanic ash and tuffs; also formed by hydrothermal activity. Montmorillonites are characterized by high ion exchange capacities and by their ability to swell markedly when wetted.

*Colorado Plateau Region:* An abundant constituent of various sedimentary rock units, notably the Chinle Formation (Schultz, 1963), and in various units of the Morrison Formation, with illite and chlorite (Keller, 1962).

*Apache County:* Santer-Defiance Plateau area, at Cheto, Allentown, Barnwater Wash, and Ganado Mesa, in linear channels and lenselike bodies underlying the upper Bidahochi Member, derived from volcanic tuff; a calcian montmorillonite (Kiersch and Keller, 1955); a chemical analysis of white material from Sanders by J. G. Fairchild (Wells, 1937) gave the following result:

| | | | | | |
|------|-------|--------|------|---------|-------|
| $SiO_2$ | 51.20 | CaO | 2.71 | $H_2O(-)$ | 16.78 |
| $Al_2O_3$ | 15.12 | $Na_2O$ | 0.50 | $H_2O(+)$ | 6.92 |
| $Fe_2O_3$ | 1.41 | $K_2O$ | 0.08 | $TiO_2$ | 0.10 |
| MgO | 5.22 | | | MnO | 0.03 |

TOTAL 100.07%

Monument Valley, Monument No. 2 mine, where it is associated with the uranium-vanadium ores (Witkind and Thaden, 1963).

*Cochise County:* Reported from an area east of Elgin. Also reported from 2 miles south of Benson. After illite, the most abundant clay mineral in the sediments of the Willcox Playa (Pipkin, 1967).

*Coconino County:* An abundant alteration product of mafic dike rocks associated with the Black Peak breccia pipe, near Cameron (Barrington and Kerr, 1961).

*Gila County:* Globe-Miami district, abundant in the host rocks of the Castle Dome and Copper Cities mines where it formed by the hydrothermal alteration of rock-forming silicate minerals (Peterson, 1962).

*Greenlee County:* Morenci district, as yellowish brown material in small amounts in the less-intensely altered rock below the supergene zone, formed as a result of hydrothermal alteration; associated with kaolinite, allophane, and beidellite (Schwartz, 1947; Moolick and Durek, 1966).

*Maricopa County:* In bentonites, reported from an occurrence 2 miles northeast of Wickenburg. Also reported from the vicinity of Phoenix. In bentonites of poor quality near Carl Pleasant Dam.

*Mohave County:* Reported as bentonite from the southern part of the county and east of the Big Sandy River. In altered tuff, near Kingman (UA 8859).

*Pima County:* An alteration mineral which replaces feldspar in dacite in the Silver Bell area (Kerr, 1951). Pima district, present in the Twin Buttes mine as product of hydrothermal alteration (Stanley B. Keith, pers. comm., 1973).

*Pinal County:* As bentonite in the vicinity of Ray and Superior.

*Yavapai County:* As bentonite, in Thompson Valley, between Kirkland and Yava, at the Lyles deposit (W.½, Sec. 12, T.12N, R.6W): the clay is characterized as being intermediate between normal montmorillonite and hectorite, and contains 0.3 to 0.5 percent $Li_2O$ (Norton, 1965). Also reported from near Wagoner.

*Yuma County:* Reported from near Wellton and Bouse, in bentonite.

## MONTROSEITE

Vanadium iron oxide hydroxide, (V,Fe)O-(OH). An essentially unoxidized vanadium mineral which occurs with uraninite and sulfide minerals and is believed to be of primary origin. Alters to paramontroseite, corvusite, and melanovanadite.

*Apache County:* Carrizo Mountains, at the Martin mine, Mesa 4½, and Cove mines. Also reported from the Lukachukai Mountains.

## *MORDENITE

Calcium sodium potassium aluminum silicate hydrate, $(Ca,Na_2,K_2)Al_2Si_{10}O_{24} \cdot 7H_2O$. A member of the zeolite group. Occurs as an alteration product of volcanic glass in tuffs; also in fissures and vesicle fillings in mafic volcanic rocks.

*Greenlee County:* About 6 miles north of Morenci in Tertiary volcanic tuffs (Sheppard, 1969).

*Mohave County:* Union Pass district, north side of Union Pass, in tuff and lapilli tuff in the Golden Deer volcanics (Sheppard, 1969).

*Pinal County:* As vesicle fillings in flow rock, Midway Station (Bideaux et al., 1960).

## MORENOSITE

Nickel sulfate hydrate, $NiSO_4 \cdot 7H_2O$. A rare secondary mineral formed through the oxidation of nickel-bearing sulfides.

*Maricopa County:* Reported from near Wickenburg.

## MOTTRAMITE

Lead copper zinc vanadate hydroxide, $Pb(Cu,Zn)(VO_4)(OH)$. An uncommon secondary mineral formed in the oxidized portions of base metal deposits, commonly associated with vanadinite and cerussite.

*Cochise County:* Warren district, as crystals from the Higgins mine (Taber and Schaller, 1930; Schaller, 1934), and as reniform masses in the Shattuck mine (Taber and Schaller, 1930). An analysis of the mineral from the Higgins mine by W. T. Schaller (Wells, 1937) yielded the following result:

| | | | | | |
|---|---|---|---|---|---|
| CuO | 19.10 | $V_2O_5$ | 21.11 | Loss | |
| ZnO | nil | MnO | 3.06 | on Ign. | 4.79 |
| PbO | 50.13 | | | TOTAL | 98.19% |

Tombstone district, as brilliant black crystals at the Lucky Cuss and Toughnut mines (Guild, 1911). An analysis by Hillebrand (1889b) on material from the Lucky Cuss mine which formed botryoidal encrustations up to ½ inch thick gave:

| | | | | | |
|---|---|---|---|---|---|
| PbO | 57.00 | $As_2O_5$ | 1.10 | CaO | 1.01 |
| CuO | 11.21 | $P_2O_5$ | 0.19 | MgO | 0.04 |
| FeO | tr. | $H_2O$ | 2.50 | $K_2O$ | 0.10 |
| ZnO | 4.19 | Cl | 0.07 | $Na_2O$ | 0.17 |
| $V_2O_5$ | 19.79 | $SiO_2$ | 0.80 | $CO_2$ | 0.82 |
| | | | | TOTAL | 98.99% |

Specific gravity at 19° C: 5.88.

Also reported from the Pat Hills. Cochise district, Texas Canyon Quartz Monzonite, in the quartz-tungsten veins of the area (Cooper and Silver, 1964). Charleston, at the Gallagher Vanadium property (UA 6061).

*Gila County:* Apache mine, as rich, black druses associated with vanadinite, rarely as free-standing, arborescent-botryoidal forms (Wilson, 1971). Banner district, 79 mine, on the fourth level as crystal druses encrusting wulfenite and mimetite in breccia of the Main fault zone (Keith, 1972).

*Pima County:* Tucson Mountains, at the Old Yuma mine (Guild, 1910, 1911). An analysis by Guild is as follows:

| | | |
|---|---|---|
| PbO 52.26 | ZnO 6.71 | $V_2O_5$ 23.02 |
| CuO 11.64 | MnO 2.16 | $H_2O$ 2.52 |
| | | TOTAL 98.31% |

Empire Mountains, at the Total Wreck mine in part as complete replacements of wulfenite crystals. Cababi district, South Comobabi Mountains, found at the Mildren and Steppe claims (Williams, 1963). Ajo district, a chromian variety which occurs in small quantities as crusts on weathered keratophyre southwest of the New Cornelia pit (Gilluly, 1937). Pima district, at the Twin Buttes open pit mine (Stanley B. Keith, pers. comm., 1973).

*Pinal County:* Mammoth district, Mammoth–St. Anthony mine, as crusts of small pointed crystals (Guild, 1911; Galbraith and Kuhn, 1940).

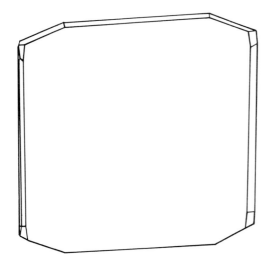

**Mottramite.**
Silver-Lead claim, Mildren and Steppe area
(Williams, 1962).

*Santa Cruz County:* "Cuprodescloizite" from an unspecified locality near Nogales, as fibrous, reddish chestnut brown layers up to ½ inch thick, enclosed in crystallized calcite which may, in turn, be covered by a layer of amber-colored indistinct crystals (Headden, 1903; Guild, 1911). An analysis provided by Headden gave:

| | | |
|---|---|---|
| $V_2O_5$ 19.014 | CuO 8.506 | MnO tr. |
| $As_2O_5$ 3.842 | ZnO 12.450 | $H_2O$ 2.650 |
| PbO 52.954 | $Fe_2O_3$ 0.200 | Insol. 0.350 |
| | | TOTAL 99.966% |

Specific gravity: 6.176.

*Yavapai County:* Bradshaw Mountains, at a prospect on the Slate Creek property of Kalium Chemicals Ltd., associated with pyromorphite and mimetite (W. C. Berridge, pers. comm., 1973).

## MURDOCHITE

Lead copper oxide, $PbCu_6O_8$. A rare secondary mineral formed in the oxidized portion of copper-lead deposits; associated with wulfenite, in which it may be embedded, and with hemimorphite, willemite, and quartz. The Mammoth–St. Anthony mine is the type locality (Fahey, 1955).

*Cochise County:* Higgins mine, Warren district (UA 7715).

*Gila County:* Banner district, 79 mine, where it is of rare occurrence; associated with plattnerite, aurichalcite, and rosasite, as tiny black crystals (up to 0.1 mm) of cubic habit (Keith, 1972).

*Pima County:* Waterman Mountains, Silver Hill mine, with plattnerite and aurichalcite (Bideaux et al., 1960).

*Pinal County:* Mammoth district, Mammoth–St. Anthony mine, as tiny black octahedrons on the surface of and embedded within plates of wulfenite, and on the surfaces of fluorite crystals, with hemimorphite and willemite (Fahey, 1955). An analysis by Fahey gave:

| | | |
|---|---|---|
| CuO 67.24 | $Fe_2O_3$ 0.17 | Insol. |
| $PbO_2$ 30.53 | $SiO_2$ 0.05 | in $HNO_3$ 1.11 |
| | | TOTAL 99.10% |

Specific gravity: 6.47.

## MUSCOVITE

Potassium aluminum silicate hydroxide, $KAl_2(AlSi_3)O_{10}(OH)_2$. An important rock-forming member of the mica group. Most abundant in schists and gneisses and in granite pegmatites where it may occur in large "books." As sericite, it is an abundant alteration product of the wall rocks of many mineral deposits. Also widely distributed in sediments and sedimentary rocks. An abundant and widely distributed mineral in Arizona, and only a few relatively unusual occurrences are listed.

*Maricopa County:* White Picacho district, as an abundant varietal mineral of the pegmatites of the district (Jahns, 1952).

*Mohave County:* In clear, transparent sheets up to 6 x 10 inches in size, at an unspecified locality north of the Colorado River (*Engr. Min. Jour.*, 1892a).

*Pima County:* Ko Vaya Hills, north of Sells, as the variety fuchsite (UA 4174). As large masses at the San Antonio Mica mine, near Ajo.

*Pinal County:* Willow Springs ranch, Oracle, as pseudomorphs after tourmaline (UA 8617).

*Yavapai County:* Weaver Mountains, near Peebles Valley, as segregations in Yavapai Schist. Bradshaw Mountains, as segregations in a pegmatite dike which extends from Middleton to Horse Thief, a distance of about 5 miles. White Picacho district, as an accessory mineral in the pegmatites of the district (Jahns, 1952).

*Yuma County:* Trigo Mountains, Eureka district, as the chromian variety, mariposite; disseminated in schist and accompanied by chromite.

## Sericite

Most, if not all, mica which is termed sericite is muscovite. Sericite is a finely-divided, shreddy variety of muscovite formed primarily as a result of hydrothermal alteration processes and a common associate of the porphyry copper deposits. Only a few representative occurrences are listed here.

*Cochise County:* Warren district, in the Sacramento Hill stock (Schwartz, 1947, 1958).

*Gila County:* Globe-Miami district, Castle Dome mine (Peterson et al., 1946; Schwartz, 1947; Creasey, 1959); Copper Cities deposit (Peterson, 1954); Miami and Inspiration ore bodies (Schwartz, 1947; Olmstead and Johnson, 1966).

*Graham County:* Lone Star district, Gila Mountains, Safford area, at and in the vicinity of the Kennecott Safford copper deposit (Robinson and Cook, 1966; Rose, 1970).

*Greenlee County:* Clifton-Morenci district (Reber, 1916; Schwartz, 1947, 1958; Creasey, 1959).

*Mohave County:* Cerbat Mountains, Wallapai district, associated with sulfide-bearing veins (Thomas, 1949).

*Pima County:* Ajo district, New Cornelia mine area (Gilluly, 1937; Schwartz, 1958; Creasey, 1959). The most widespread alteration product in the Silver Bell area (Kerr, 1951).

*Pinal County:* Mammoth district, San Manuel ore body (Schwartz, 1947; Creasey, 1959); Kalamazoo ore body (Lowell, 1968). Mineral Creek district, Ray (Schwartz, 1947, 1952, 1959; Rose, 1970).

*Yavapai County:* Copper Basin district (Johnston and Lowell, 1961); Big Bug district, Iron King mine (Creasey, 1952; Moxham et al., 1965). Bagdad area (Schwartz, 1947; Anderson, 1950; Creasey, 1959; Moxham et al., 1965). United Verde mine (Moxham et al., 1965).

# N

## *NACRITE

Basic aluminum silicate, $Al_4Si_4O_{10}(OH)_8$. A less common member of the kaolinite group.

*Greenlee County:* A widespread constituent of solfatarically-altered dacites and quartz latites in the Steeple Rock area, northwest of Duncan. Sometimes accompanied by diaspore. The nacrite, when pure, is dull chalky white.

## *NATROALUNITE

Sodium aluminum sulfate hydroxide, $NaAl_3(SO_4)_2(OH)_6$. A relatively rare member of the alunite group formed during alunitization, a process of solfataric action, commonly accompanied by kaolinization and silicification.

*Yuma County:* Sugarloaf Butte, near Quartzsite (Omori and Kerr, 1963) (UA 9348).

## NATROJAROSITE

Sodium iron sulfate hydroxide, $NaFe_3(SO_4)_2(OH)_6$. Isostructural with jarosite. Formed, with other sulfates, as an oxidation product of pyrite.

*Gila County:* From the Globe area, exact locality unknown.

*Mohave County:* At the Georgia Sunset claim, 4 miles south of Kingman, as compact to earthy, golden brown to yellow masses made up of tabular crystals. An analysis reported in Shannon and Gonyer (1927) gave:

| | | | | | |
|---|---|---|---|---|---|
| $Fe_2O_3$ | 48.23 | $K_2O$ | 2.28 | $(NH_4)_2O$ | none |
| $Al_2O_3$ | 0.09 | $Na_2O$ | 4.28 | $SO_3$ | 33.71 |
| FeO | 0.58 | $Ag_2O$ | none | $H_2O$ | 10.76 |
| CaO | 0.05 | PbO | tr. | Insol. | 0.22 |
| MgO | 0.05 | | | TOTAL | 100.25% |

## *NATROLITE

Sodium aluminum silicate hydrate, $Na_2Al_2Si_3O_{10} \cdot 2H_2O$. A member of the zeolite group.

*Mohave County:* On U.S. Highway 93-446, 7 miles southeast of Hoover Dam, in a kaersutite-camptonite dike (Bideaux et al., 1960).

**Natrolite.**
Hillside mine, Yavapai County. Julius Weber.

## NAVAJOITE

Vanadium oxide hydrate, $V_2O_5 \cdot 3H_2O$. An uncommon secondary mineral which occurs as impregnations of sandstone, associated with oxidized vanadium-uranium minerals. The Monument No. 2 mine is the type locality.

*Apache County:* Monument Valley, Monument No. 2 mine, as dark brown, silky to fibrous minute columns normal to fracture surfaces in channel fillings in Shinarump Conglomerate, with a host of other secondary uranium and vanadium minerals (Weeks et al., 1954, 1955; Finnell, 1957; Ross, 1959; Witkind and Thaden, 1963; Young, 1964). An analysis by Sherwood (Weeks et al., 1954) gave:

| | | | | | |
|---|---|---|---|---|---|
| $V_2O_5$ | 71.68 | $Fe_2O_3$ | 3.58 | CaO | 0.22 |
| $V_2O_4$ | 3.08 | $SiO_2$ | 1.20 | $H_2O$ | 20.30 |
| | | | | TOTAL | 100.06% |

Specific gravity: 2.56.

## *NEKOITE

Calcium silicate hydrate, $CaSi_2O_5 \cdot 2H_2O$. A dimorph of okenite; first described from the Crestmore Quarries, California.

*Graham County:* Aravaipa district, Landsman's Camp, in contact metamorphosed limestone as white, radiating fibers, filling fractures in quartz-epidote hornfels.

## *NEOTOCITE

Manganese magnesium iron silicate hydroxide, $(Mn,Mg,Fe)_4(SiO_2)_3(OH)_{10}(?)$. Of uncertain composition, a poorly defined species which appears to form in part through alteration of manganese-bearing silicate minerals.

*Graham County:* Aravaipa district, in drusy cavities associated with johannsenite which has replaced limestone containing galena, sphalerite, chalcopyrite, quartz, and calcite (Simons and Munson, 1963).

## *NEPHELINE

Sodium potassium aluminum silicate, $(Na,K)AlSiO_4$. A feldspathoid mineral, nepheline occurs most characteristically in alkaline igneous rocks and similar rocks thought to be of metasomatic origin.

*Coconino County:* Cameron area, as small anhedral crystals filling interstices in the augite matrix of mafic dike rocks associated with a breccia pipe at Black Peak, near Cameron (Barrington and Kerr, 1961); as anhedral crystals between blades of augite in the monchiquite Tuba dike (Barrington and Kerr, 1962).

*Pima County:* Gunsight Mountain, Sierrita Mountains, sparse in a pulaskite dike (Bideaux et al., 1960). Esperanza mine, in nepheline syenite encountered in an exploratory drill hole (Schmidt et al., 1959).

Niccolite (see NICKELINE)

## NICKELINE

Nickel arsenide, NiAs. An uncommon mineral of primary origin, found in significant quantity at only a few localities. Usually associated with cobalt and silver-arsenic minerals and with primary native silver.

*Yavapai County:* With chloanthite and native silver, Monte Cristo mine (Bastin, 1922).

## NITER

Potassium nitrate, $KNO_3$. A product of evaporation formed from guano or by bacterial action upon other animal remains. Found in caves or old mine workings in small amounts.

*Coconino County:* Walnut Canyon, as a thin white covering on limestone shelves in ancient cliff dwellings (Guild, 1910).
*Graham County:* Peloncillo Mountains, in caves.
*Pinal County:* Galiuro Mountains, Aravaipa Canyon, as thin crusts and in cracks in rock below caves.

## NITROCALCITE

Calcium nitrate hydrate, $Ca(NO_3)_2 \cdot 4H_2O$. A water-soluble mineral which occurs as efflorescences in limestone caverns; locally, the nitrate is derived from guano.

*Gila and Pinal Counties:* In fissures up to 6–8 inches wide in Mississippian limestone, along the Gila River, 2 miles above Winkelman.
*Pinal County:* Casa Grande area (UA 7837).

## NONTRONITE

Sodium iron aluminum silicate hydroxide hydrate, $Na_{.33}Fe_2^{3+}(Al,Si)_4O_{10}(OH)_2 \cdot nH_2O$. An iron-rich end member of the montmorillonite (smectite) group of clay minerals. A secondary mineral of uncertain origin, known to occur in hydrothermally altered copper deposits. *Morencite* has been shown to be identical with nontronite.

*Gila County:* Globe-Miami district, Miami ore body, where it is abundant as pseudomorphs after plagioclase phenocrysts, associated with sericite, kaolinite, and hydromuscovite (Schwartz, 1947). Workman Creek area, associated with sulfide minerals in Dripping Spring Quartzite (Granger and Raup, 1969).
*Graham County:* Stanley district, at the Friend mine.
*Greenlee County:* Clifton-Morenci district, as silky seams in limey shale, at the Arizona Central mine (Schwartz, 1947, 1958), and around the periphery of the Morenci ore body (Moolick and Durek, 1966).
*Pima County:* Santa Rita Mountains, in metamorphosed wall rock at the Pauline mine, Helvetia-Rosemont district (Schrader and Hill,

1915). Ajo district, where it is uncommon as a secondary mineral (Gilluly, 1937). Pima district, associated with supergene collapsed breccias in the Twin Buttes open pit mine (Stanley B. Keith, pers. comm., 1973).
*Pinal County:* Mineral Creek district, Ray deposit, where it is found in hydrothermally altered quartz monzonite with sericite, hydromuscovite, and quartz (Schwartz, 1952). Mammoth district, impregnating younger sediments on Copper Creek where it debouches from the box canyon (Herbert E. Hawkes, pers. comm., 1973).
*Yavapai County:* Black Hills, with limonite in gossans, at the United Verde mine.

# O

## *OFFRETITE

Potassium calcium aluminum silicate hydrate, $(K,Ca)_3(Al_5Si_{13})O_{36} \cdot 14H_2O$. An uncommon mineral of the zeolite group, found in amygduloidal cavities in basaltic flow rocks, associated with other zeolites.

*Greenlee County:* As tiny elongate, prismatic crystals in amygdules in basalt flows exposed in a road cut on the east side of the San Francisco River, about one mile north of Clifton (William Hunt and Dan Caudle, pers. comm., 1976).

**Offretite.**
Clifton, Greenlee County. Julius Weber.

## OLIVENITE

Copper arsenate hydroxide, $Cu_2AsO_4(OH)$. A secondary mineral found in the oxidized zones of base-metal mineral deposits where it is associated with other secondary lead and copper minerals.

*Cochise County:* Tombstone district (UA 6439, 8240).

*Coconino County:* Grand Canyon National Park, Horseshoe Mesa, Grandview (Last Chance) mine, as short, olive-green, prismatic crystals and lighter-colored acicular groups, on altered metazeunerite (Leicht, 1971).

*Gila County:* 79 mine, as clusters of tiny green crystals on sphalerite and galena, associated with brochantite, siderite, chalcanthite, and anglesite (Thomas Trebisky, pers. comm., 1972; Keith, 1972).

*Pinal County:* Galiuro Mountains, Copper Creek district, as small olive-green crystals, Old Reliable mine. Pioneer district, Magma mine, as small crystals with dioptase, from the outcrop at the No. 1 glory hole.

*Yavapai County:* Copper Mountain, Old Robertson claims, near Mayer, with clinoclase (Robert O'Haire, pers. comm., 1972).

## OLIVINE

Magnesium iron silicate, $(Mg,Fe)_2SiO_4$. A group term which includes the endmembers forsterite ($Mg_2SiO_4$) and fayalite ($Fe_2SiO_4$). An abundant rock-forming group of minerals characteristically formed in mafic igneous rocks. The occurrences listed here are of olivines whose chemical compositions are insufficiently well known to warrant placing them under the names forsterite or fayalite (which see), or of olivines known to fall compositionally between the end members.

*Apache County:* As clear, green to brown material of gem quality at Buell Park, 10 miles north of Fort Defiance, and at Garnet Ridge (Gregory, 1916). At Green Knobs, near Red Lake, north of Fort Defiance, with garnet crystals, in abundance on the surfaces of ant hills near volcanic necks (diatremes) (O'Hara and Mercy, 1966).

*Cochise County:* Tombstone district, in the contact zone at Comstock Hill, and at the Lucky Cuss mine, as an important constituent of basalt

(Butler et al., 1938). Reported from the Warren district.

*Coconino County:* In the basalts of the San Francisco Mountains (Gregory, 1917).

*Gila County:* In basaltic rocks and volcanic bombs, and stream gravels of the San Carlos Indian Reservation, near Peridot and Tolklai (H 89313). A cut stone from this locality weighed 25.75 carats. An analysis of material from Rice Station School at Tolklai, by S. S. Goldich (USNM 86128), is as follows:

| | | |
|---|---|---|
| $SiO_2$ 40.90 | $Fe_2O_3$ 0.59 | NiO 0.30 |
| $TiO_2$ 0.05 | FeO 8.24 | MgO 49.78 |
| $Al_2O_3$ 0.22 | MnO 0.12 | CaO 0.05 |
| $Cr_2O_3$ 0.02 | | $H_2O(-)$ 0.01 |
| | | TOTAL 100.28% |

Near Globe, as granular masses of roughly ellipsoidal shape, associated with larger crystals of magnetite, spinel, grains of albite and diopside, hornblende, and biotite; believed to be volcanic bombs (Lausen, 1927a). Abundant in a differentiated diabase sill complex in the Sierra Ancha (Smith, 1970). An analysis (micro-probe) gave:

| | | |
|---|---|---|
| $SiO_2$ 36.6 | MnO 0.39 | CaO 0.07 |
| $TiO_2$ 0.05 | MgO 35.6 | NiO 0.11 |
| FeO 26.0 | | TOTAL 98.8 % |

*Mohave County:* In Quaternary basalts of the Cerbat Mountains (Thomas, 1953).

*Pinal County:* An accessory mineral in diabase sills which intrude rocks of the Apache Group in this and adjacent counties. Dripping Spring Mountains, with augite and iddingsite in basalts of Tertiary age. Galiuro Mountains, in the contact zone between granodiorite and Cretaceous sedimentary rocks.

*Santa Cruz County:* Sparingly in gabbro, diabase, and andesites of the Santa Rita and Patagonia Mountains.

*Yavapai County:* Big Bug district, Iron King mine, as phenocrysts in porphyritic basalt flows associated with mineralized volcanic breccias, with augite (Creasey, 1952).

Onyx (see QUARTZ)

Opal (see CRISTOBALITE)

## ORPIMENT

Arsenic sulfide, $As_2S_3$. Of secondary origin resulting from the alteration of other arsenic-bearing minerals. Commonly associated with realgar, but less common in its occurrence.

*Pinal County:* In 1915 several pounds of orpiment and realgar were discovered in an unidentified locality near the junction of the Gila River and Hackberry Wash.

## ORTHOCLASE

Potassium aluminum silicate, $KAlSi_3O_8$. Monoclinic. An important, rock-forming alkali feldspar. An abundant constituent of felsic igneous rocks and pegmatites and in detrital sedimentary rocks derived from them. Most commonly found as disseminated grains and crystals and as large masses in pegmatites. Also formed by the action of hydrothermal solutions on wall rocks of mineral deposits. Widely distributed in the felsic igneous rocks of the state. Phenocrysts up to several inches long are known from some granites in Arizona.

*Cochise County:* Orthoclase phenocrysts up to several inches long occur in the Texas Canyon stock; excellent examples of twins after the Carlsbad law abound.
*Maricopa County:* Bradshaw Mountains, in pipes in granite, up to 175 feet in diameter, Cave Creek district.
*Mohave County:* Mined from the pegmatites of the Cerbat and Hualpai Mountains (Wilson, 1944). Thick veins are reported from the vicinity of Hackberry.
*Pima County:* Ajo district, as crystals up to several inches across, as part of a massive replacement of quartz monzonite in the New Cornelia ore body (Gilluly, 1937). A late stage alteration mineral in the Silver Bell area (Kerr, 1951).
*Pinal County:* Crystal Pass, near Ray, as small euhedra, often twinned after the Carlsbad and other laws. Northern Santa Catalina Mountains, as Carlsbad twins (UA 4034, 4035). Reported as twinned crystals from the vicinity of Kearny. Copper Creek district, Childs-Aldwinkle mine, as crystals up to 6 inches long in a pegmatite zone at the base of a breccia pipe (Kuhn, 1941).
*Yavapai County:* Bradshaw Mountains, as relatively pure material near the old townsite of Middletown on the Crown King road.

## P

## *PALYGORSKITE (Attapulgite)

Magnesium aluminum silicate hydroxide hydrate, $(Mg,Al)_2Si_4O_{10}(OH)\cdot4H_2O$. Usually classified as a clay mineral because of its finely divided habit in soils. Also occurs as a product of hydrothermal activity. A variety consisting of intertwined fibers is termed mountain leather.

*Colorado Plateau:* In certain zones of the Petrified Forest and Owl Rock Members of the Chinle Formation (Schultz, 1963).
*Gila County:* Globe district, along foliation in gouge near the dacite hanging wall of the Keystone fault system in the Black Copper pit, Inspiration Consolidated Copper Company, as irregular, slickensided seams up to ½ inch thick which appear to be cementing gouge fragments; white to pale lavender in color (Bladh, 1973).

## *PAPAGOITE

Calcium copper aluminum silicate hydroxide, $CaCuAlSi_2O_6(OH)_3$. A rare mineral known only from the New Cornelia mine at Ajo, and at Messina, Transvaal, South Africa. The Ajo occurrence is the type locality.

*Pima County:* Ajo district, New Cornelia mine, in narrow veinlets and as veneers on slip surfaces in altered granodiorite porphyry. Associated with aurichalcite, shattuckite, ajoite, barite, and iron oxides as elongate, somewhat flattened, cerulean-blue crystals (Hutton and Vlisidis, 1960; Guillebert and Le Bihan, 1965) (UA 168; BM 1959, 192).

## PARALAURIONITE

Lead chloride hydroxide, $PbCl(OH)$. The monoclinic dimorph of laurionite. A rare secondary mineral known from only a few world localities; occurs in the oxidized zones of lead-copper deposits.

*Pinal County:* Mammoth district, Mammoth–St. Anthony mine, as small, slender, isolated, yellowish-white crystals in cavities with cerussite, and in coarser crystal aggregates with leadhillite and diaboleite in the Collins vein. Characterized by an extremely good cleavage and a degree of flexibility, so that the crystals are often bent (Palache, 1950; Fahey et al., 1950). An analysis by F. A. Gonyer gave:

| | | |
|---|---|---|
| Pb 77.75 | Cl 12.84 | H$_2$O 3.51 |
| O 6.00 | | TOTAL 100.10% |

**Papagoite.**
Ajo, Pima County. Julius Weber.

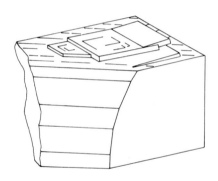

**Papagoite.**
Ajo (Hutton & Vlisidis, 1960).

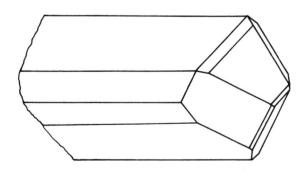

**Paralaurionite.**
Mammoth–St. Anthony mine (Palache, 1950).

## PARAMELACONITE

Copper oxide, $Cu_{1-2x}^{2+}Cu_{2x}^{1+}O_{1-x}$. A rare secondary mineral for which Bisbee is the type locality. The only other known locality is the Algomah mine, Michigan.

*Cochise County:* Warren district, Copper Queen mine, in a matrix of goethite, associated with cuprite, copper, tenorite, malachite, and connellite, as crystals of unusually large dimensions; forms present are {001}, {101}, and {100} (Koenig, 1891; Frondel, 1941).

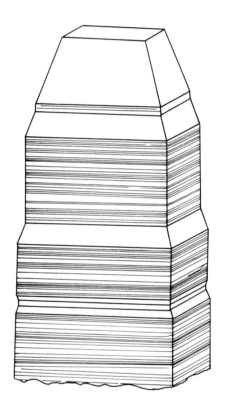

**Paramelaconite.**
Copper Queen mine, Bisbee (Frondel, 1941).

## PARAMONTROSEITE

Vanadium oxide, $VO_2$. A rare secondary mineral formed by the oxidation of montroseite.

*Apache County:* Monument Valley, Monument No. 2 mine, as shiny black crystals up to one half mm (Evensen and Gray, 1958; Witkind and Thaden, 1963; Young, 1964).

## *PARATACAMITE

Copper hydroxide chloride, $Cu_2(OH)_3Cl$. The rhombohedral dimorph of atacamite. A rare secondary mineral.

*Cochise County:* Warren district, where it is found with cuprite, malachite, and azurite in the Holbrook and Cole mines, but is rare; as crystals up to 2 mm long.

*Pima County:* Cababi district, South Comobabi Mountains, Mildren mine, where it occurs as fibrous and granular masses embaying linarite and as pseudomorphs after atacamite in quartz veins cutting andesite (Williams, 1962, 1963).

*Pinal County:* Mammoth district, Mammoth–St. Anthony mine, with cerussite, boleite, and diaboleite (UA 5154).

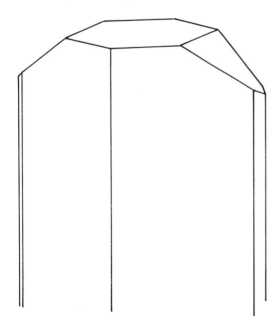

**Paratacamite (pseudomorph after atacamite).**
Mildren mine (Williams, 1962).

## *PARSONSITE

Hydrous lead uranyl phosphate, $Pb_2(UO_2)(PO_4)_2 \cdot 2H_2O$. A rare secondary mineral known only from a few localities.

*Cochise County:* As pale yellow crusts on quartz from a small lead prospect in the Huachuca Mountains.

## PASCOITE

Calcium vanadium oxide hydrate, $Ca_2V_6O_{17} \cdot 11H_2O$. A water-soluble secondary mineral formed under surficial oxidizing conditions; locally associated with carnotite.

*Apache County:* Monument Valley, Monument No. 2 mine, as water-soluble coatings formed on the mine walls (Witkind and Thaden, 1963). Lukachukai Mountains, at the Mesa No. 1, No. 5, and No. 6 mines (Chenoweth, 1967). In the Carrizo Mountains, at the Zona No. 1 claim.

## PEARCEITE

Silver arsenic sulfide, $Ag_{16}As_2S_{11}$. An uncommon mineral formed under low to moderate temperatures in veins, typically associated with lead and silver sulfosalts, galena, and acanthite.

*Mohave County:* Cerbat Range, reported in small amounts from some of the high-grade silver ores of the area; Distaff mine, Chloride district (Bastin, 1925).

*Santa Cruz County:* Trench mine, 500-foot level; Worlds Fair mine (Kartchner, 1944).

## *PENNINITE

Magnesium iron aluminum silicate hydroxide, $(Mg,Fe,Al)_6(Si,Al)_4O_{10}(OH)_8$. An important member of the chlorite group, formed during thermal metamorphism and as a result of hydrothermal processes. Extremely common despite the few occurrences noted here.

*Cochise County:* Warren district, as small scaly crystals interlayered with sericite in altered Bolsa Quartzite north of the Dividend fault; also common as a matrix of breccia dikes exposed in the upper reaches of Brewery Gulch.

*Pima County:* As large crystals pseudomorphous after hornblende in syenodiorite dikes which are abundant on the east side of the Tortolita Mountains.

## PENTLANDITE

Iron sulfide, $(Fe,Ni)_9S_8$. A primary mineral usually associated with pyrrhotite; in places thought to be a product of basic magmatic segregation. An important ore of nickel.

*Mohave County:* Reported from near Littlefield, in mafic dikes with pyrrhotite and chalcopyrite.

## *PERITE

Lead bismuth oxychloride, $PbBiO_2Cl$. A rare secondary mineral which was for some time known only from Långban, Sweden; it has since been found in several localities in the western United States.

*Cochise County:* At a locality about 8 miles northeast of Benson, it occurs as bright yellow, scaly masses in vugs in contorted quartz veins cutting epidote-plagioclase gneiss, with minor amounts of chrysocolla, pyromorphite, and vauquelinite.

### *PEROVSKITE

Calcium titanium oxide, $CaTiO_3$. Of widespread occurrence as an accessory mineral in alkaline and mafic igneous rocks, and as a mineral of deuteric origin.

*Pima County:* Gunsight Mountain, Sierrita Mountains, as polysynthetically twinned cores of sphene crystals in a pulaskite dike (Bideaux et al., 1960).

### *PHARMACOSIDERITE

Iron arsenate hydroxide hydrate, $Fe_3(AsO_4)_2$-$(OH)_3 \cdot 5H_2O$. A widespread but uncommon mineral formed from hydrothermal solutions as well as from alteration of arsenopyrite and other primary arsenic minerals.

*Cochise County:* Warren district (Richard Graeme, pers. comm., 1971) (UA 10086).
*Santa Cruz County:* Santa Rita Mountains, Tyndall district, St. Louis mine, Temporal Gulch, on barite, with azurite; resulting from the alteration of tennantite (Bideaux et al., 1960).

### *PHILLIPSITE

Potassium sodium calcium aluminum silicate hydrate, $(K_2,Na_2,Ca)(Al_2Si_4)O_{12} \cdot 4.5H_2O$. A member of the zeolite group.

*Greenlee County:* 5 miles south of Hannegan Meadow along road cuts on U.S. Highway 666, in vesicles in basalt, as limpid, clear fourlings with heulandite.
*Mohave County:* East of Big Sandy Wash, $E\frac{1}{2}$, T.16N, R.13W, in Pliocene tuff with analcime, chabazite, erionite, and clinoptilolite (Ross, 1928, 1941).

*Pinal County:* In a railway cut one mile north of Mammoth, with chabazite, heulandite, and calcite on celadonite.

### *PHLOGOPITE

Potassium iron magnesium aluminum silicate fluoride hydroxide, $K(Fe,Mg)_3(AlSi_3)O_{10}$-$(F,OH)_2$. A member of the mica group whose principal modes of occurrence are in metamorphosed limestones and dolomites and in ultramafic igneous rocks.

*Gila County:* Northwest portion of the county, associated with the uranium ores in the Dripping Spring Quartzite; Little Joe, Lucky Boy, Bix Six, and Last Chance properties (Granger and Raup, 1969). Banner district, Christmas mine, associated with talc and tremolite adjacent to anhydrite-sulfide veinlets in diopside hornfels, forming most of the footwall of the lower Martin orebody; also present in skarn rocks and in some of the hornfels (David Perry, pers. comm., 1967).
*Pinal County:* Lakeshore mine, in meta-dolomites with forsterite and chondrodite.

### PHOENICOCHROITE

Lead chromate oxide, $Pb_2O(CrO_4)$. A rare secondary mineral which occurs in association with crocoite, vauquelinite, and cerussite in oxidized galena-bearing veins.

*Maricopa County:* Collateral, Chromate, Blue Jay, and Phoenix claims, east of the present location of Trilby Wash. In 1881 Benjamin Silliman noted the presence of phoenicochroite, vauquelinite, crocoite, and the probable presence of "jossaite," volborthite, descloizite, and "chileite" from the "Vulture region." South of Wickenburg, at several localities including the Moon Anchor mine, Potter-Cramer property, and the Pack Rat claim, formed from the oxidation of lead-zinc ores, associated with wickenburgite, willemite, mimetite, hemihedrite, and vauquelinite; as dark, cochineal red cleavable and polycrystalline masses (Williams, 1968; Williams et al., 1970; Williams and Anthony, 1970). An analysis by George Roseveare gave PbO 80.88 and $CrO_3$ 18.08 percent; total, 98.96 percent. Specific gravity: 7.01.

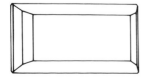

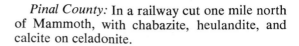

**Phoenicochroite.**
Potter-Cramer property
(Williams et al., 1970).

# PHOSGENITE

Lead carbonate chloride, $Pb_2(CO_3)Cl_2$. A rare secondary mineral formed by the oxidation of galena and other lead minerals; commonly associated with cerussite and anglesite.

*Pinal County:* Mammoth district, Mammoth–St. Anthony mine, as slender, prismatic crystals with diaboleite, paralaurionite, boleite, and cerussite, on the 400-foot level of the Collins vein (Palache, 1941b; Fahey et al., 1950) (H 104522).

# *PHOSPHOSIDERITE

Iron phosphate hydrate, $FePO_4 \cdot 2H_2O$. A dimorph of strengite. A rare mineral found in pegmatite with other phosphate minerals.

*Yavapai County:* Eureka district, on the 7U7 Ranch property, with bermanite in seams in triplite, associated also with leucophosphite and hureaulite (Leavens, 1967).

# *PHOSPHURANYLITE

Calcium uranyl phosphate hydroxide hydrate, $Ca(UO_2)_4(PO_4)_2(OH)_4 \cdot 7H_2O$. A secondary mineral commonly associated with torbernite and other uranium minerals.

*Coconino County:* Cameron district, where it replaces part of a fossilized log in the Huskon No. 17 mine; also from the Jack Daniels No. 1 mine (Austin, 1964; Finch, 1967).

# PICKERINGITE

Magnesium aluminum sulfate hydrate, $MgAl_2(SO_4)_4 \cdot 22H_2O$. Occurs as a product of weathering and is found in the gossans of oxidized pyrite-bearing bodies with other hydrated secondary sulfates.

*Cochise County:* Reported to occur in large quantities from an unspecified locality some 30 miles east of Douglas.

*Gila County:* Pueblo Canyon, at the Ancient deposit, in Mescal limestone, and near the Rock Canyon deposit, where efflorescences have been identified only as members of the pickeringite-halotrichite group (Granger and Raup, 1969).

# *PICROCHROMITE

Magnesium chromium oxide, $MgCr_2O_4$. A member of the spinel group. Occurs chiefly in peridotites and other ultramafic rocks as a primary crystallization product.

*Apache County:* Monument Valley, Garnet Ridge, where it is found as mineral fragments in a breccia dike piercing sedimentary rocks (Gavasci and Kerr, 1968).

Piedmontite (see PIEMONTITE)

# PIEMONTITE

Calcium aluminum manganese iron silicate hydroxide, $Ca_2(Al,Mn^{3+},Fe^{3+})_3(SiO_4)_3(OH)$. A member of the epidote group, found in low-grade metamorphic rocks and also formed as a product of hydrothermal activity. The variety *withamite* is a low-manganese piemontite.

*Cochise County:* Occurs typically as small rosettes of minute crystals in andesite of the Pat Hills, north of Pearce (Lausen, 1927b).

*Pima County:* In rhyolite and adjacent sandstone in the Tucson Mountain Recreational area, Tucson Mountains (Guild, 1935). Also in the Santa Rita Mountains near Madera Canyon (Guild, 1935). At the north end of the Comobabi Mountains, with withamite and manganaxinite as vesicle fillings in andesite flow boulders (Bideaux et al., 1960). Also at a locality about 6 miles from Vail (UA 2256).

*Pinal County:* Near Casa Grande (UA 2218).

*Yavapai County:* Northwest of Prescott, in the Government Canyon and Prescott granodiorites, as small veinlets and disseminated grains (Krieger, 1965).

# PIGEONITE

Magnesium iron calcium silicate, $(Mg,Fe,Ca)(Mg,Fe)Si_2O_6$. A member of the pyroxene group. A rock-forming mineral largely confined to quickly-cooled mafic flow rocks.

*Cochise County:* Tombstone district, as dark green grains with vesuvianite in the contact silicate zone at the Lucky Boy mine (Butler et al., 1938).

# PINTADOITE

Calcium vanadium oxide hydrate, $Ca_2V_2O_7 \cdot 9H_2O$. A secondary mineral which occurs as water-soluble efflorescences.

*Apache County:* In the Salt Wash sandstone at the Mesa No. 5 and No. 6 properties. Also from the Lukachukai Mountains, southwest of Cove.

# PISANITE (cuproan MELANTERITE)

Iron copper sulfate hydrate, $(Fe,Cu)SO_4 \cdot 7H_2O$. Copper substitutes into the iron site in the melanterite structure up to about 1.89:1 (18.3 percent Cu) (Palache et al., 1951). A secondary, water-soluble mineral formed by the oxidation of pyrite in cupriferous deposits; typically observed as coatings in post-mine openings.

*Gila County:* Globe district, on the 15th level of the Old Dominion mine. Banner district, 79 mine, sparingly in an area of active sulfated deposition in a crosscut on the sixth level (Keith, 1972).

*Pima County:* Silver Bell Mountains, in small amounts at the El Tiro mine.

*Santa Cruz County:* As crystal efflorescences on tunnel walls at the St. Louis mine, Temporal Gulch (Bideaux et al., 1960).

## Pitchblende (see URANINITE)

## PLAGIOCLASE

The plagioclase feldspars constitute a group of sodium calcium aluminum silicate minerals whose chemical compositions range through a substitutional series having end members $NaAlSi_3O_8$ and $CaAl_2Si_2O_8$. The sodic and calcic end members are termed albite and anorthite respectively. Throughout the series Na and Ca substitute freely for one another, electrical charge compensation in the structures being maintained by concomitant substitution of Al for Si. Individual mineral names within the series are applied to somewhat arbitrarily chosen compositional intervals. In order of increasing calcium content the specific names are: oligoclase, andesine, labradorite, and bytownite. Potassium content is usually restricted to small amounts (less than about 5 to 10%).

The plagioclase feldspars are, following quartz, the most abundant minerals in the upper portion of the earth's crust. They are characteristic of the majority of the igneous rocks and pegmatites, and are common constituents of many of the metamorphic rocks. After quartz, they are the most abundant minerals of the sandstones. Plagioclase feldspars are essential constituents of almost all of the igneous rocks of Arizona and are typically present in the gneisses and schists. It is not feasible to systematically list the many occurrences of these common rock-formers and the reader is referred to the voluminous geological literature (Moore and Wilson, 1965).

*Cochise, Graham, and Pima Counties:* Exceptional plagioclase feldspar phenocrysts (andesine-labradorite) up to 2 inches in length occur in a series of intrusive and flow porphyritic rocks locally termed "turkey-track" porphyry. Rocks containing these abundant and large phenocrysts have been observed at a number of localities within a radius of about 75 miles of Tucson. Some of the reported localities are: Twin Buttes quadrangle, in the lower part of the Helmet fanglomerate; in the Dragoon quadrangle, on the east side of the Steele Hills, and at the southern tip of the Winchester Mountains; Galiuro Mountains; Cienega Gap area, in the Pantano Formation; Graham Mountains-Fisher Hills area; Dos Cabezas Mountains; Tucson Mountains, on the southwest side of Sentinal Peak ("A" Mountain); San Xavier Indian Reservation, at Black Mountain; Mineta Ridge area, on the east flank of the Rincon Mountains (Cooper, 1961; Mielke, 1964; Percious, 1968).

*Cochise County:* Tombstone district, Lucky Cuss mine, as bytownite, with vesuvianite in metamorphosed shaly limestones.

*Pima County:* Coarse perthitic oligoclase occurs in pegmatite dikes in the Sierrita Mountains, west of Helmet Peak.

*Pinal County:* Galiuro Mountains, Copper Creek; coarse, fresh andesine crystals occur in the Copper Creek granodiorite in the Galiuro Mountains; coarse, fresh albite is also found in metasomatized granodiorite at Copper Creek.

## PLANCHEITE

Copper silicate hydroxide hydrate, $Cu_8Si_8O_{22}(OH)_4 \cdot H_2O$. A rare secondary mineral formed in the oxidized portions of copper deposits.

*Maricopa County:* Bighorn district, about 20 miles south of Aguila, as deep blue fibrous veinlets and masses in quartzite, associated with chrysocolla and "copper pitch" (Montoya, 1967).

*Pinal County:* Galiuro Mountains, Table Mountain mine near Klondyke, as blue crystal-

**Plancheite.**
Table Mountain mine, Pinal County. Julius Weber.

line grains and as masses disseminated in compact green conichalcite. Tortolita Mountains, Azurite mine, with "bisbeeite" and shattuckite (Bideaux et al., 1960). Mammoth district, San Manuel mine, as blue tablets in chrysocolla (Thomas, 1966) (UA 5390, 6461); Mammoth–St. Anthony mine, as veinlets and as coatings on chrysocolla.

## PLATINUM

Platinum, Pt. A rare mineral whose occurrence is almost always restricted to ultramafic rocks or their metamorphic equivalents, and to placers derived from them. None of the occurrences noted has been authenticated.

*Maricopa County:* Reported from the Santo Domingo placers, along the Gila River opposite the old Riverside Stage station.
*Yavapai County:* Reported in black sands near Columbia and Prescott (Guild, 1910).

## PLATTNERITE

Lead oxide, $PbO_2$. Tetragonal, an uncommon mineral found in the oxidized zone of lead deposits, formed under extreme conditions of oxidation; associated with other lead oxides, cerussite, smithsonite, pyromorphite, wulfenite, and other oxidized minerals.

*Cochise County:* Warren district, with murdochite and malachite on goethite (Bideaux et al., 1960). Turquoise district, Silver Bill, Defiance, and Tom Scott mines, with wulfenite (Bideaux et al., 1960).
*Gila County:* Banner district, 79 mine, where it occurs as tiny needle-like crystals up to 1.5 mm associated with aurichalcite, rosasite, and, rarely, murdochite (Keith, 1972).
*Pima County:* Tucson Mountains, Old Yuma mine, as tiny crystals on vanadinite (Bideaux et al., 1960). Waterman Mountains, Silver Hill mine, with murdochite, malachite, and aurichalcite (Bideaux et al., 1960). Empire Mountains, Total Wreck mine, with hemimorphite on limonite (Bideaux et al., 1960); Lone Mountain mine (UA 6364).
*Santa Cruz County:* Tyndall district, Cottonwood Canyon, Glove mine, as crystals on wulfenite.

Pleonaste (ferroan SPINEL)

## *PLUMBOGUMMITE

Lead aluminum phosphate hydroxide hydrate, $PbAl_3(PO_4)_2(OH)_5 \cdot H_2O$. A secondary mineral formed by oxidation of lead ores.

*Gila County:* Coronado fault area, Copper Cities deposit, with chalcopyrite in quartz (H 106829; specimen collected by A. R. Still).

## PLUMBOJAROSITE

Lead iron sulfate hydroxide, $PbFe_6(SO_4)_4(OH)_6$. A member of the alunite group. Less common than jarosite to which it is similar in origin and occurrence.

*Cochise County:* Tombstone district, abundant in the brown oxide ore of the Holderness "roll." Material from the Empire mine assayed 58.92 oz. in silver and from the Toughnut mine, up to 200 oz. (Butler et al., 1938).
*Gila County:* Banner district, 79 mine, as ocherous brown, massive material in the surface and near-surface workings (Keith, 1972).
*Graham County:* Aravaipa district, Dogwater mine (Simons, 1964).
*Mohave County:* Cerbat Range, Tennessee-Schuylkill mine. As a secondary mineral in the Wallapai district (Thomas, 1949).
*Pima County:* Empire Mountains, as earthy masses in the Hilton mines. Pima district, Twin Buttes mine (Stanley B. Keith, pers. comm., 1972).
*Yavapai County:* Humbug mine, Castle Hot Springs area.

## *PLUMBONACRITE (?)

Lead carbonate hydroxide oxide, $Pb_{10}(CO_3)_6(OH)_6O(?)$. An inadequately characterized mineral, long confounded with hydrocerussite and probably formed under similar conditions during oxidation of lead-rich sulfide ores.

*Pinal County:* Mammoth district, Mammoth–St. Anthony mine, identified on a specimen collected by S. C. Creasey (S-71) from the Collins vein, 1,000-foot level. As small pearly-white scales, associated with anglesite and linarite on galena; $a = 9.072$, $c = 24.55$ Å.

## POLLUCITE

Cesium sodium aluminum silicate hydrate, $(Cs,Na)_2(Al_2Si_4)O_{12} \cdot H_2O$. A member of the zeolite group. A rare mineral found in granite pegmatites.

*Maricopa County:* White Picacho district, at the Independence mine; a reported but unconfirmed occurrence.

## POLYBASITE

Silver copper antimony sulfide, $(Ag,Cu)_{16}Sb_2S_{11}$. Occurs in silver veins formed at low to moderate temperatures; associated with other silver sulfosalts.

*Mohave County:* Cerbat Mountains, common in the silver ores of a number of the mining districts (Bastin, 1925; Thomas, 1949).

*Pinal County:* Pioneer district, Silver King mine, where fine specimens were found (Guild, 1910).

*Santa Cruz County:* Patagonia Mountains, Tyndall district, near Squaw Gulch, at the Ivanhoe mine, with stromeyerite, proustite, and native silver (*Engr. Min. Jour.*, 1912).

*Yavapai County:* Bradshaw Mountains, Davis mine, Hassayampa district, with proustite (Lindgren, 1926).

## POLYCRASE

Yttrium calcium cerium uranium thorium titanium niobium tantalum oxide, $(Y,Ca,Ce,U,Th)(Ti,Nb,Ta)_2O_6$. A member of the euxenite-polycrase series. A rare mineral which occurs in granite pegmatites.

*Maricopa County:* South of Buckeye, in a pegmatite dike with kasolite (Robert O'Haire, pers. comm., 1972).

## *POLYHALITE

Potassium calcium magnesium sulfate hydrate, $K_2Ca_2Mg(SO_4)_4 \cdot 2H_2O$. Occurs primarily as a precipitate from oceanic waters, commonly associated with halite and anhydrite; of wide distribution.

*Apache and Navajo Counties:* East-central Arizona, encountered in drill holes which delineate a northeast-trending potash zone underlying an area of about 300 square miles, in Permian evaporites (Peirce, 1969). The log of a hole drilled at Sec. 24, T.18N, R.25E showed, in addition, carnallite, sylvite, halite, anhydrite and gypsum (H. Wesley Peirce, pers. comm., 1972).

## *POSNJAKITE

Copper sulfate hydroxide hydrate, $Cu_4(SO_4)(OH)_6 \cdot H_2O$. A rare secondary mineral formed under acid oxidizing conditions by alteration of copper sulfides.

*Yavapai County:* At a prospect in the Turret Peak Quadrangle, Sec. 30, T.13N, R.5E, as a powder-blue film on chalcopyrite.

*Yuma County:* A specimen in the collections of the British Museum is labeled as coming from Bouse (Peter G. Embrey, pers. comm., 1973) (BM 1972, 201).

## POWELLITE

Calcium molybdate, $CaMoO_4$. Often contains some tungsten. An uncommon secondary mineral found in tungsten ores.

*Cochise County:* Cochise district, as pseudomorphs after molybdenite, near Johnson (Cooper and Silver, 1964). Reported from the Warren district; one verified occurrence is at the Bisbee Queen shaft east of Warren.

*Gila County:* Globe-Miami district, Inspiration mine, as crusts of tiny crystals in a seam adjacent to veins containing molybdenite and lindgrenite; thought to be a product of late hydrothermal solutions which attacked molybdenite (F. E. Pough, pers. comm.).

*Maricopa County:* Near Morristown, on the upper Santo Domingo Wash, with scheelite. White Picacho district, as a rare mineral in pegmatites (Jahns, 1952). With scheelite as disseminations in granitic rocks at the Flying Saucer group, Vulture Mountains (Dale, 1959).

*Mohave County:* Reported from the Cerbat Range.

*Pima County:* Helvetia district, disseminated with scheelite in the contact zone near the Black Horse shaft (Dale et al., 1960). Pima district, Twin Buttes mine (Stanley B. Keith, pers. comm., 1973); with scheelite in quartz veins at the Senator Morgan mine (Dale et al., 1960).

*Pinal County:* With wolframite and scheelite in quartz veins in the Upshaw Tungsten mines group, near Antelope Peak (Dale, 1959).

*Yavapai County:* White Picacho district, as a rare mineral in pegmatites (Jahns, 1952).

## *PREHNITE

Calcium aluminum silicate hydroxide, $Ca_2Al_2Si_3O_{10}(OH)_2$. Prehnite occurs primarily in cavities and in veins in mafic lavas where it is commonly associated with zeolite minerals. It occurs in veins in granitic rocks and also in contact metamorphosed limestones.

*Cochise County:* One mile northwest of Portal, in vesicles in basalt with celestite, pyrite, and pumpellyite.

*Pima County:* Santa Rita Mountains, Mt. Fagan, as crystalline masses with copper, epidote, and quartz in andesite (Bideaux et al., 1960).

*Santa Cruz County:* Two miles south of the Cerro Colorado mine, on the Arivaca road (Robert O'Haire, pers. comm., 1972).

## *PROSOPITE

Calcium aluminum fluoride hydroxide, $CaAl_2(F,OH)_8$. An uncommon mineral found in greissen in tin veins, associated with other fluorine-bearing minerals; also in pegmatites.

*Yuma County:* At a copper prospect about 10 miles north of Bouse, in hematite veins in andesite with chrysocolla, malachite, and tenorite, as lovely, complex, limpid sea-green crystals in voids in hematite gangue; crystals up to 6 mm in length.

## PROUSTITE

Silver arsenic sulfide, $Ag_3AsS_3$. A late-stage mineral found in hydrothermal veins of low temperature origin with other silver sulfosalts.

*Cochise County:* Pearce Hills, Commonwealth mine, associated with tetrahedrite and silver halides.
*Maricopa County:* Wickenburg Mountains, with native silver and acanthite at the Monte Cristo mine (Bastin, 1922).
*Mohave County:* Cerbat Mountains, Minnesota-Connor, Distaff, and Merrimac mines, Chloride district (Bastin, 1925). Gold Star mine, Mineral Park district; Paymaster mine, Cerbat district (Thomas, 1949). In relatively large quantities in the Cupel mine, Stockton Hill area (Schrader, 1907).
*Pinal County:* Pioneer district, Belmont mine, as minute blebs in galena.
*Santa Cruz County:* Nogales district, Mount Benedict area, with gold in quartz monzonite (Schrader, 1917). Tyndall district, at the Ivanhoe mine, with stromeyerite, polybasite, and native silver (*Engr. Min. Jour.,* 1912).
*Yavapai County:* Bradshaw Mountains, Davis and Catoctin mines, Hassayampa district, with polybasite. With silver and chlorargyrite in a vein near the Thunderbolt mine, Black Canyon district. At the Morgan mine, Turkey Creek district. With native silver and chlorargyrite at the Tip Top mine, Tip Top district.

## *PSEUDOBOLEITE

Lead copper chloride hydroxide hydrate, $Pb_5Cu_4Cl_{10}(OH)_8 \cdot 2H_2O$. A rare secondary mineral formed in small amounts in oxidized lead-copper deposits.

*Pima County:* Daisy shaft, Banner mine, with connellite, gerhardtite, and atacamite in cuprite (Williams, 1961).

*Pinal County:* Mammoth district, Mammoth–St. Anthony mine, as overgrowths on boleite, associated with other secondary lead minerals (Palache, 1941b).

## *PSEUDOBROOKITE

Iron titanium oxide, $Fe_2TiO_5$. An uncommon mineral found in volcanic rocks as a product of pneumatolytic or fumerolic action.

*Pinal County:* Near Saddle Mountain, Winkelman area, as a few needle-like crystals, with topaz, spessartine, and bixbyite, in rhyolite (Robert L. Jones, pers. comm.).

## PSEUDOMALACHITE

Copper phosphate hydroxide hydrate, $Cu_5(PO_4)_2(OH)_4 \cdot H_2O$. A rare secondary mineral found in the oxidized zone of copper deposits with other oxidized copper minerals.

*Gila County:* Globe-Miami district, Castle Dome mine, as small, dark emerald-green crystals.
*Graham County:* Safford area, Lone Star district, as prismatic crystals, many of which are almost hairlike, in metasomatized volcanics, associated with malachite, brochantite, antlerite, carbonate-apatite, chrysocolla, jarosite, lepidocrocite, and sulfide minerals (Hutton, 1959b).
*Pima County:* East end of the South Comobabi Mountains, as lovely green crystals with slightly curved faces, up to 3 mm in length; as films on the walls of fractures in severely deformed quartz monzonite, associated with libethenite. Southern side of Saginaw Hill, about 7 miles southwest of Tucson, associated with other oxidized minerals including cornetite, brochantite, malachite, libethenite, atacamite, and chrysocolla, in fractures in chert (Khin, 1970).
*Pinal County:* Galiuro Mountains, Bunker Hill district, Copper Creek Canyon, with botryoidal malachite on quartz crystals, some of which are twinned after the Japanese law (William and Mildred Schupp, pers. comm.). Silver Bell mine (Joe Urban, pers. comm.). Sacaton Hill, found in drill core.
*Yuma County:* Harquahala Mountains, Harquahala mine, as clusters of sharp, dark green crystals in vugs, partially coated by chrysocolla (?) (John S. White, Jr., pers. comm., 1972). Cunningham Pass, near Wenden, at the Critic mine.

**Pseudomalachite, malachite.**
Harquahala mine, Harquahala Mountains,
Yuma County. Julius Weber.

## PSILOMELANE (ROMANECHITE)

As used here, psilomelane is a general term for massive, undifferentiated hard manganese oxides whose compositions approximate to barium manganese oxide hydroxide (Fleischer, 1975). Psilomelane (romanechite) is also a distinct species having the composition $BaMn^{2+}$-$Mn_8^{4+}O_{16}(OH)_4$. It is highly probable that the species is contained among the manganese oxides at many of the localities listed below. Of secondary origin, formed under surface conditions from the alteration of manganous carbonates or silicates, and associated with materials of similar origin such as pyrolusite, goethite, limonite and wad.

*Cochise County:* Tombstone district, as the most abundant manganese oxide mined in the ore deposits of the district (Rasor, 1939; Romslo and Ravitz, 1947; Hewett and Fleischer, 1960). Warren district, Higgins mine, with pyrolusite and braunite (Palache and Shannon, 1920; Taber and Schaller, 1930).

*Coconino County:* Aubrey Cliffs, with braunite at the Adams-Woody prospect, with other manganese oxides cementing rock fragments in veins which cut and in part replace Kaibab Limestone; associated with braunite, pyrolusite, and cryptomelane (Potter and Havens, 1949; Hewett et al., 1963). Long Valley district, as numerous small, irregular, disconnected lenticular masses in Kaibab Limestone, with pyrolusite (Secs. 19, 20, 29, and 30, T.14N, R.10E); also on the Denison, Shoup, Blue Ridge, and Lost Apache claims (Farnham and Stewart, 1958). Heber district, in a small deposit in parts of Secs. 17, 18, 19, and 20, T.11N, R.15E, and at the Johnson and Hayden deposit, NW¼

Sec. 2, T.26N, R.7W, in a steeply dipping fracture or brecciated zone in Kaibab Limestone (Farnham and Stewart, 1958).

*Gila County:* 6 miles north of Roosevelt Lake, at Sec. 29, T.5N, R.13E, found in Apache group rocks (Hewett and Fleischer, 1960; Hewett et al., 1963). Globe-Miami district, where, with manganite and pyrolusite, it forms the bulk of the gangue in the manganese-zinc-lead-silver deposits of the district (Peterson, 1962). Banner district, 79 mine, as a common supergene mineral in replacement deposits in limestone and rhyolite porphyry (Kiersch, 1949). Apache and Accord manganese deposits, Medicine Butte area, with wad, as fracture fillings and cement in conglomerate of Cenozoic age (Moore, 1968). Sierra Ancha Mountains, Sierra Ancha district, Armer Wash (Sunset) mine, as fine-grained mammillary material in breccia zones in quartzite of the Apache group; contains 0.02% thallium, approximately 12% BaO, and 0.2% $K_2O$ (Crittenden et al., 1962).

*Greenlee County:* Ash Peak district, forming mammillary layers on fluorite in veins in andesite porphyry at the Fourth of July mine (Hewett and Fleischer, 1960; Hewett et al., 1963; Hewett, 1964). With pyrolusite and manganite on the Pyrolusite claims, 12 miles southeast of Morenci (Potter et al., 1946).

*Maricopa County:* Black Vulture mine, 32 miles south of Wickenburg, with pyrolusite as replacements in limestone (Long et al., 1948). Big Horn district, with pyrolusite and manganite in small fissure veins in volcanic flows and breccia, Black Queen and Black Nugget mines (Sandell and Holmes, 1948; Farnham and Stewart, 1958).

*Mohave County:* Artillery Mountains, Black Jack, Price, and Priceless veins and Maggie Canyon bedded deposit, in numerous veinlets and fractures in Tertiary volcanics, associated with cryptomelane, hollandite, coronadite, pyrolusite, ramsdellite, and lithiophorite (Hewett and Fleischer, 1960; Mouat, 1962). Black Warrior mine, 6 miles northeast of Alamo Crossing, with pyrolusite (Long et al., 1948). Arizona Manganese claims, 30 miles north of Parker Dam (Havens et al., 1947).

*Navajo County:* Sonsela Buttes area, filling pore space between pebbles or fragments in sandstones and conglomerates, with pyrolusite (Mayo, 1955).

*Pima County:* Reported from near Tucson, but the exact locality is not known (Palache et al., 1944). An analysis by C. Milton (Wells, 1937) on Tucson Mountain material gave:

| Insol. | 8.35 | Na$_2$O | 0.42 | P$_2$O$_5$ | 0.05 |
|--------|------|---------|------|-----------|------|
| SiO$_2$ | 0.90 | K$_2$O | 0.11 | MnO | 6.70 |
| Al$_2$O$_3$ | 0.55 | H$_2$O(−) | 0.49 | BaO | 14.40 |
| Fe$_2$O$_3$ | 3.27 | H$_2$O(+) | 3.78 | CuO | 0.25 |
| MgO | tr. | TiO$_2$ | tr. | PbO | 0.32 |
| CaO | 0.05 | | | MnO$_2$ | 59.65 |

TOTAL 99.29%

Specific gravity: 4.21.

Northern end of the Coyote Mountains, as massive hard coatings up to 3 inches thick on porphyritic rocks (Blake, 1910).

*Pinal County:* Cochise group of claims, as fine-grained mammillary material, in veins and breccia zones in Gila (?) Conglomerate; contains less than 0.01 percent thallium and about 12 percent BaO (Crittenden et al., 1962). Copper Creek district, Blue Bird mine (Kuhn, 1951). Mammoth district, in a vein with barite and calcite on the west side of Tucson Wash (Schwartz, 1953). Riverside district, Almino and Cochise mines, cementing fragments in breccia zones (Hewett and Fleischer, 1960; Hewett et al., 1963). With manganite cementing brecciated zones of conglomerate, Geronimo claims, 8 miles west of Winkelman (Dean et al., 1952); with pyrolusite, as nodules and lenses replacing limestone in capping over an igneous dike, Benningfield property, 11½ miles south of Winkelman (Dean et al., 1952); also at the North Star group and Orson Branch claim in the vicinity of Winkelman (Dean et al., 1952). Pioneer district, with pyrolusite and hetaerolite at the Domeroy property, 4 miles north of Superior (Dean et al., 1952).

*Santa Cruz County:* Tyndall district, Cottonwood Wash, Glove mine (Olson, 1966). Patagonia district, Mowry mine, in replacement deposits in limestone, with pyrolusite and hematite (Schrader, 1917).

*Yavapai County:* Castle Creek district, Black Rock deposit, found with pyrolusite in Precambrian granite gneiss (Fleischer and Richmond, 1943; Hewett and Fleischer, 1960; Hewett et al., 1963). Burmeister mine, near Sycamore Creek above its junction with the Agua Fria River, with cryptomelane and other manganese oxides interlayered with volcanic ash, clastic sediments, and a basalt flow (Hewett and Fleischer, 1960; Hewett et al., 1963). Big Bug district, 13 miles from Mayer at the confluence of Big Bug and Sycamore creeks, as nodules and irregular veins, with minor pyrolusite (*Engr. Min. Jour.*, 1918).

*Yuma County:* Santa Maria district, Kaiserdoom claims, at Sec. 22, T.11N, R.11W, in bedded manganese oxides underlying a bed of volcanic ash in the Artillery Formation, with associated soft manganite (Hewett et al., 1963). As irregular nodular fragments cemented by clay, at the Spring mine, 40 miles west of Congress Junction (Long et al., 1948). Doyle-Smith claims, 15 miles from Artillery Peak (Ipsen and Gibbs, 1952). Sheep Tanks mine, southwest of Salome (Romslo and Ravitz, 1947).

## *PUMPELLYITE

Calcium magnesium aluminum silicate hydroxide hydrate, Ca$_2$MgAl$_2$(SiO$_4$)(Si$_2$O$_7$)(OH)$_2$·H$_2$O. Occurs with prehnite and zeolites in vesicles in volcanic rocks.

*Cochise County:* About 1 mile northwest of Portal, in vesicles in basalt with prehnite, celestite, and adularia; as amygdule fillings up to 1 inch across, in radiating fibrous masses. Also reported from the Warren district.

*Pima County:* Santa Rita Mountains, Mt. Fagan, with thomsonite, pennine, prehnite, epidote, and copper in andesite (Bideaux et al., 1960). Chicago mine, near Sells, with epidote and delessite as amygdules in altered andesite.

## PURPURITE

Manganese iron phosphate, (Mn$^{3+}$,Fe$^{3+}$)PO$_4$. A rare mineral formed in granite pegmatites as an alteration product of triphylite and lithiophilite; also associated with sicklerite.

*Yavapai County:* White Picacho district, as tiny needles and plates forming crusts and cavity fillings; associated with strengite (Jahns, 1952).

## PYRARGYRITE

Silver antimony sulfide, Ag$_3$SbS$_3$. An important ore of silver, it occurs in veins and is commonly associated with galena, tetrahedrite, pyrite, and other silver-bearing sulfosalt minerals as a product of late stage mineralization.

*Mohave County:* Cerbat Mountains, mined in the Chloride, Mineral Park, Cerbat, and Stockton Hill districts (Thomas, 1949).

*Pinal County:* Vekol mine, near Casa Grande, as massive material in quartz (AM 24969). Galiuro Mountains, Little Treasure mine, Saddle Mountain district.

*Santa Cruz County:* Patagonia Mountains, Harshaw district, at the Alta mine, with embolite and fluorite. Palmetto district, Sonoita mine (Schrader, 1917).

*Yavapai County:* Bradshaw Mountains, Tillie Starbuck and Davis mines, Hassayampa district; Thunderbolt mine, Black Canyon district; Tip Top and Tiger districts (Lindgren, 1926).

## PYRITE

Iron sulfide, $FeS_2$. The most common of all sulfide minerals, pyrite forms under a wide range of conditions. In Arizona it is typically an abundant associate of most metallic mineral deposits. The localities noted here are representative of a very widespread mineral.

*Apache and Navajo Counties:* Monument Valley uranium-vanadium deposits, found in vein fillings and as cement in sandstone (Joralemon, 1952; Rosenzweig et al., 1954; Coleman and Delavaux, 1957; Jensen, 1958).

*Cochise County:* Warren district, in large massive bodies at the Copper Queen and other mines (Ransome, 1904; Mitchell, 1920; Trischka et al., 1929; Schwartz and Park, 1932; Bain, 1952; Bryant, 1968).

*Coconino County:* South Rim of the Grand Canyon, Orphan mine, associated with copper-uranium-lead ores in Coconino Sandstone (Isachsen, 1955). Cameron area, commonly associated with the uranium mineralization in sedimentary rocks; at the Alyce Tolino mine it is cobalt bearing (Rosenzweig et al., 1954; Hamilton and Kerr, 1959).

*Gila County:* Banner district, as a massive body containing some highly modified crystals, 79 mine (Keith, 1972); Christmas mine (Ross, 1925b; Peterson and Swanson, 1956; Perry, 1969). Globe-Miami district, Copper Cities deposit (Peterson, 1954). Castle Dome mine (Peterson, 1947), Miami and Inspiration mines (Schwartz, 1947), Old Dominion mine (Ransome, 1903).

*Graham County:* Lone Star district, the primary sulfide mineral in the Safford porphyry copper deposit (Robinson and Cook, 1966).

*Greenlee County:* Clifton-Morenci district, as large crystals in the Hudson and Fairplay veins (Lindgren, 1905; Reber, 1916; Schwartz, 1947, 1958; Creasey, 1959).

*Mohave County:* Cerbat Range, Wallapai district, of widespread occurrence in a belt of sulfide-bearing fissure vein deposits (Thomas, 1949; Field, 1966). Ithaca Peak, with chalcopyrite, galena, sphalerite, and molybdenite in quartz monzonite porphyry (Eidel, 1966).

*Pima County:* Tucson Mountains, Arizona-Tucson mine, as crystals from ⅛ to ¼ inch in diameter with remarkably abundant crystal

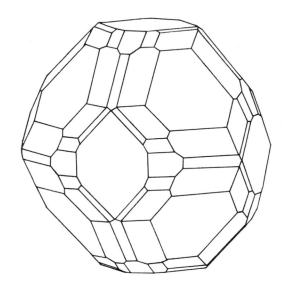

**Pyrite.**
Arizona-Tucson mine (Ayres, 1924).

faces (Ayres, 1924). Ajo district, where it is particularly abundant in dioritic border facies at the New Cornelia mine; also in the Concentrator Volcanics (Gilluly, 1937), a few fine crystal groups have been produced. Pima district, present in the primary sulfide mineralization of the Twin Buttes mine (Stanley B. Keith, pers. comm., 1973). Present in the majority of the mines and prospects in the Santa Rita Mountains (Schrader and Hill, 1915). Also abundant in the mining districts of the Sierrita Mountains and vicinity. Silver Bell district (Kerr, 1951).

*Pinal County:* Pioneer district, as large crystals from the Belmont mine; Magma mine, as large bodies (Short et al., 1943; Mills and Eyrich, 1966). Ray district, Iron Cap mine, as perfect pyritohedral crystals up to 1 inch in diameter in soft clay gangue (Ransome, 1919; Schwartz, 1947; Clarke, 1953; Rose, 1970). Galiuro Mountains, as excellent crystals in the Childs-Aldwinkle mine (Kuhn, 1941). Mammoth district, San Manuel mine, where it is the most abundant sulfide mineral (Schwartz, 1947, 1949, 1958; Lovering, 1948; Creasey, 1959); Kalamazoo ore body (Lowell, 1968).

*Santa Cruz County:* Patagonia Mountains, as striated and twinned crystals up to 8 inches across and as large crystal aggregates at the

Three R mine, and as large crystal groups near the molybdenite bodies of the Santo Niño mine (Schrader and Hill, 1915; Schrader, 1917; Frondel and Wickman, 1970). A very large crystal from Duquesne is UA 5693. As striated crystals 2 cm on an edge from the Four Metals mine, Patagonia district (James Bleess, pers. comm., 1972).

*Yavapai County:* United Verde mine contains one of the largest pyritic ore bodies of the world (Fearing, 1926; Lausen, 1928; Schwartz, 1938; Anderson and Creasey, 1958; Moxham et al., 1965); also at other properties in the Jerome-Bradshaw Mountains area (Lindgren, 1926).

*Yuma County:* Dome Rock Mountains, Don Welsh prospect, as crystals more than 1 inch in diameter.

## PYROCHLORE-MICROLITE

Sodium calcium niobium tantalum oxide hydroxide fluoride, $(Na,Ca)_2(Nb,Ta)_2O_6(OH,F)$ to sodium calcium tantalum niobium oxide hydroxide fluoride, $(Na,Ca)_2(Ta,Nb)_2O_6(OH,F)$. A mineral series in which there is extensive ionic substitution of the major elements indicated as well as of many others. Typically occurs in granite pegmatites. The nature of a specific sample can be determined only by careful mineralogical and chemical study.

*Yavapai County:* White Picacho district, in the Outpost, Midnight Owl, and Picacho View pegmatites, as tiny olive-green to dark brown and black crystals of microlite with sharply defined octahedral and dodecahedral faces; associated with pyrochlore (Jahns, 1952).

## PYROLUSITE

Manganese dioxide, $MnO_2$. A member of the rutile structural group. A common manganese mineral which apparently is always formed under oxidizing conditions; usually associated with other manganese minerals such as manganite (from which it forms by alteration), hausmannite, braunite, psilomelane, as well as limonite, hematite, and goethite. Widely distributed in small amounts throughout Arizona.

*Cochise County:* Tombstone district, in commercial quantities at the Oregon-Prompter, Lucky Cuss, Telephone, and Bunker Hill mines (Butler et al., 1938; Rasor, 1939; Romslo and Ravitz, 1947; Havens et al., 1954; Hewett and Fleischer, 1960). Warren district, Higgins mine, with psilomelane and braunite (Palache and Shannon, 1920). Texas Canyon area, as acicular crystals in vugs in the quartz-huebnerite veins cutting the Texas Canyon Quartz Monzonite (Cooper and Silver, 1964).

*Gila County:* Globe-Miami district, with manganite and psilomelane, forms the bulk of the gangue in the manganese-zinc-lead-silver deposits of the district (Peterson, 1962). Globe district, as the variety polianite (UA 7756). East of Ripsey Wash, near Kelvin at Sec. 14, T.3N, R.13E; fibrous, pseudomorphous after manganite (Robert O'Haire, pers. comm., 1972). Banner district, as numerous dendritic forms coating fractures in near-surface limestones at the 79 mine (Keith, 1972).

*Greenlee County:* Clifton-Morenci district, in black, sooty masses with iron oxides in metamorphosed limestones (Lindgren, 1905; Guild, 1910).

*Maricopa County:* Bighorn Mountains, Aguila district, with manganite or wad. White Picacho district, as crusts on lithium phosphate minerals in pegmatites (Jahns, 1952). Black Vulture mine, as replacements in limestone, 32 miles south of Wickenburg (Long et al., 1948).

*Mohave County:* Rawhide Mountains, Artillery Peak, in large deposits with wad (Head, 1941; Lasky and Webber, 1944, 1949). Reported in veins 4 miles south of Hoover Dam. Little Chemehuevi Valley, in veins and shear zones with wad on the Arizona Manganese claims (Havens et al., 1947). Near the Colorado River, 18 miles north of Parker Dam. Artillery Mountains, Black Jack, Price, and Priceless veins and Plancha bedded deposit, associated with a variety of oxidized manganese minerals (Mouat, 1962). Black Warrior mine, 6 miles northeast of Alamo Crossing, with psilomelane (Long et al., 1948).

*Navajo County:* Sonsela Buttes area, filling voids between pebbles or rock fragments in sandstone and conglomerate, with pyrolusite, and as nodules (Mayo, 1955).

*Pima County:* Tucson Mountains, in quartz (UA 4486). Stovall mine, near Alamo Crossing (H 108492); good quality crystals have come from this property.

*Pinal County:* Pioneer district, in the outcrops of the Magma vein. At a number of claims in the vicinity of Winkelman (Dean et al., 1952).

*Santa Cruz County:* Patagonia Mountains, Mowry mine, with wad. Also found at the Hardshell, North Mowry, Hermosa, and Black Eagle mines (Schrader and Hill, 1915; Schrader, 1917). Tyndall district, Cottonwood Canyon, Glove mine (Olson, 1966). Harshaw district, Salvador mine (Romslo and Ravitz, 1947).

*Yavapai County:* In the northern part of the Aguila district. In ores of the Bradshaw Mountains. At Castle Creek, 23 miles northeast of Morristown (Hewett and Fleischer, 1960; Hewett et al., 1963). At the Burmeister property, 12 miles southeast of Mayer on the Agua Fria River, with manganite and minor psilomelane (Long et al., 1948). White Picacho district, as thick crusts coating lithium phosphate minerals, in pegmatites (Jahns, 1952). Castle Creek district, north of Castle Hot Springs, in Precambrian granite gneiss with psilomelane (Hewett and Fleischer, 1960; Hewett et al., 1963). Big Bug district, 13 miles from Mayer, in minor amounts with psilomelane in nodules and stratified veins (*Engr. Min. Jour.*, 1918).

*Yuma County:* Dobbins claims, 6 miles east of Bouse, and at a locality 2½ miles west of Bouse. Deposits are reported from Sec. 36, T.10N, R.14E, Sheep Tanks and Silver districts (Wilson, 1933).

## PYROMORPHITE

Lead phosphate chloride, $Pb_5(PO_4)_3Cl$. A member of the apatite group. An uncommon mineral which occurs in the oxidized portion of lead deposits, typically associated with cerussite, limonite, smithsonite, anglesite, malachite, wulfenite, vanadinite, and other oxidized minerals.

*Cochise County:* Tombstone district, as small crystals associated with wulfenite (Butler et al., 1938). Also known from the Warren district. At a locality about 8 miles northeast of Benson, in quartz veins cutting gneiss, in minor amounts with chrysocolla, vauquelinite, and perite (Phelps Dodge Corp., pers. comm., 1972). Huachuca Mountains (UA 6254).

*Gila County:* Banner district, sparingly as clear yellow needle-like crystals on chrysocolla; crystal forms noted include $\{01\bar{1}1\}$, $\{20\bar{2}1\}$, $\{40\bar{4}1\}$, $\{10\bar{1}1\}$, and $\{0001\}$.

*Graham County:* Golondrina property, in pyroclastic rocks, associated with chalcopyrite, chalcocite, and malachite; contains as much as 0.6% uranium (Granger and Raup, 1962).

*Maricopa County:* In the pegmatites of the White Picacho district (Jahns, 1952).

*Pima County:* Cerro Colorado Mountains, at the Cerro Colorado mine. Cababi district, South Comobabi Mountains, at the Mildren and Steppe claims (Williams, 1963).

*Pinal County:* Mammoth district, Mammoth–St. Anthony mine, as olive-green crystals on mottramite, with vanadinite (Peterson, 1938). Slate Mountains, Jackrabbit mine, coating fault breccia and in vugs in silicified limestone (Hammer, 1961).

*Santa Cruz County:* Harshaw district, Trench mine, as incrustations. Patagonia district, Javalina prospect (Schrader and Hill, 1915; Schrader, 1917).

*Yavapai County:* Bradshaw Mountains, at a prospect on the Slate Creek property of Kalium Chemicals Ltd., as tiny transparent, light green, barrel-shaped prismatic crystals, associated with plentiful mottramite and with botryoidal crusts of mimetite (W. C. Berridge, pers. comm., 1973).

*Yuma County:* At the Iber-Plomosa mine near Bouse. Castle Dome Mountains, Castle Dome district, in old workings at Castle Dome where it is associated with wulfenite and grades into vanadinite (Blake, 1881; Wilson, 1933).

## PYROPE

Magnesium aluminum silicate, $Mg_3Al_2(SiO_4)_3$. A member of the garnet group. Occurs typically in ultrabasic rocks; found in blocks in mafic volcanic agglomerates.

*Apache County:* Navajo Indian Reservation, Garnet Ridge, as pebbles of gem quality (Williams, 1936). Also at Buell Park, near Fort Defiance where it is found in alluvium, agglomerate, and as inclusions in igneous rock (Gregory, 1917; Gavasci and Kerr, 1968). At Green Knobs, near Red Lake, as pale violet, wine-colored, and orange xenocrysts in diatremes filled with tuffs, agglomerates, and minor amounts of intrusive and extrusive mafic igneous rocks. Also found in abundance with olivine on the surfaces of ant hills (O'Hara and Mercy, 1966).

## PYROPHYLLITE

Aluminum silicate hydroxide, $Al_2Si_4O_{10}$-$(OH)_2$. An uncommon metamorphic mineral largely formed by hydrothermal alteration of feldspar; commonly associated with quartz and kyanite. Closely resembles talc.

*Cochise County:* Warren district, Warren, with barite (UA 5978); common in the Sacramento stock, Lavender pit.

*Gila County:* South of Christopher Mountain, in Gordon Canyon, where it occurs in an old weathered zone in rhyolite beneath Mazatzal Quartzite (D. L. Livingston, pers. comm., 1972).

*Mohave County:* Williams River, near Alamo at the Cactus Queen mine. Also reported in large quantities from southeast of Yucca.

*Pima County:* From a locality 50 miles southwest of Ajo (UA 8865).

*Yuma County:* Three miles southwest of Quartzsite, with dumortierite, andalusite, and kyanite (Wilson, 1929). Near Alamo Springs, 27 miles southeast of Quartzsite. Also reported from the vicinity of Bouse.

## PYROXENE

A group name for minerals having the general composition calcium iron magnesium aluminum silicate $(Ca,Fe,Mg,Al)SiO_3$, and including a variety of species differing in chemistry and, to some extent, in structural features. Individual species, where recognized, are listed individually. Of common occurrence in igneous rocks, some of which consist almost entirely of pyroxene. Abundant in dark-colored volcanic rocks; certain species are typically formed in the contact metamorphic environment.

## PYRRHOTITE

Iron sulfide, $Fe_{1-x}S$. The hexagonal, iron-deficient modification of iron sulfide. Occurs as a high temperature, early-formed mineral in veins, as a primary mineral in some igneous rocks, and in pegmatites and in contact metamorphic deposits.

*Cochise County:* Cochise district, Johnson Camp, sparse in a drill core taken near the Mammoth mine (Cooper and Silver, 1964).

*Gila County:* Christmas mine, where it is common in the pyrrhotite-chalcopyrite zone of the lower Martin ore body (Knoerr and Eigo, 1963; Perry, 1969; McCurry, 1971). A common constituent disseminated in deposits in the Workman Creek area and at the Brush, Sorrel Horse, and Citation deposits in the Cherry Creek area, northwest part of Gila County, in hornfels and related metamorphic rocks (Granger and Raup, 1969).

*Greenlee County:* Clifton-Morenci district, where it is found locally in an extensive contact metamorphic assemblage in the northern part of the Morenci open pit mine area (Moolick and Durek, 1966).

*Maricopa County:* White Picacho district, scattered in pegmatites, especially the coarse-grained interior portions (Jahns, 1952).

*Mohave County:* Near Littlefield, in mafic dikes with chalcopyrite and pentlandite. At the Copper World mine near Yucca, with sphalerite, chalcopyrite, and loellingite (Rasor, 1946). Hualpai Mountains, Antler mine, where it is coated by covellite (Romslo, 1948).

*Navajo County:* As troilite in stones of the Holbrook meteorite.

*Pima County:* Santa Rita Mountains, with pyrite, Helvetia district (Schrader and Hill, 1915); as blebs in sphalerite, Busterville mine, Helvetia district. Sierrita Mountains, in chalcopyrite ores, Pima district; a primary mineral at the Twin Buttes mine (Stanley B. Keith, pers. comm., 1973).

*Santa Cruz County:* Patagonia district, Duquesne and Washington Camp, in contact zones in Paleozoic limestones (Schrader, 1917).

*Yavapai County:* With gold ores in the Bradshaw Mountains, in massive form in the Rainbow deposit near Turkey Creek station, Black Canyon district (Lindgren, 1926). In pegmatites of the White Picacho district (Jahns, 1952).

# Q

## QUARTZ

Silicon dioxide, $SiO_2$. Low or alpha quartz. By far the most abundant of the polymorphic forms of silica and the most widespread and abundant mineral of the earth's crust. Occurs in igneous, metamorphic, and sedimentary rocks

**Scepter quartz crystal.**
Quartzsite, Yuma County. Julius Weber.

as well as in hydrothermal veins, metasomatic, and hot spring deposits.

*Cochise County:* Little Dragoon Mountains, Cochise district, Johnson Camp area, as well-formed crystals, some of which attain large size. At Hamburg in the Huachuca Mountains, as clear crystals twinned after the Japanese law (H 83142; UA 9692) (Frondel, 1962), and at the Jack Wakefield mine as beautiful clear groups of crystals reminiscent of Hot Springs, Arkansas, material (S 15034). At Russellville, with inclusions of tiny yellow octahedral crystals of scheelite grown on phantoms within the quartz (S 105926). At Courtland (UA 8989).

*Gila County:* Mazatzal Mountains, as amethyst crystals lining vugs in Mazatzal Quartzite at Four Peaks; some gem-quality material was produced from the area. On a mine dump at the Yankee Boy mine, one group of multiple-twinned crystals consisting of a central individual about 5 mm long and doubly terminated, to which are twinned two shorter individual crystals after the Japanese law; other, smaller simple twins have been collected from this locality which is 6 miles east of Mayer (collection of Mrs. Donald C. Sonnenberg) (Bideaux, 1970). Globe-Miami district, Old Dominion mine, where crystals colored blue by associated chrysocolla line cavities in oxidized ores; some quartz is brilliant red from included finely-divided hematite. Payson district, as clear crystals up to 1 inch in length at the Ox Bow mine. As tiger-eye (silicified chrysotile) from the San Carlos Indian Reservation.

*Graham County:* Galiuro Mountains, Crystal Peak (UA 4424), as twinned crystals up to 1 inch in length on Table Mountain (H 106631; UA 9521), and abundant at Stanley Butte as slender tapering crystals several inches long sometimes associated with and including andradite (S 12847). As crystals of good quality in the veins of the Tenstrike group, Aravaipa district.

*Maricopa County:* North of Scottsdale, in a quarry about 7 miles northeast of Curry's Corner. Amethystine at Four Peaks, near Roosevelt Dam, from which some gem-quality material was produced (UA 5722). As pseudomorphs, near Lake Pleasant (UA 223).

*Mohave County:* Black Mountains, rose quartz is locally abundant in the Moss mine, Oatman district; also as amethystine bands in colorless quartz of gold veins in the San Francisco (Oatman and Katherine) district (Roed-

der et al., 1963). Good quality specimens of rose quartz are reported from a locality 40 miles northeast of Kingman. Cerbat Range, as amethystine crystals in Precambrian granite, McConnico district, northeast of Boulder Spring; a crystal from this locality is reported to have been sold to Tiffany's (Guild, 1910). Kingman Feldspar mine, in pegmatite; produced as a by-product of feldspar mining (Heinrich, 1960).

*Pima County:* Tucson Mountains, as geodes in basalt flows near Sentinel Peak ("A" Mountain) (UA 6353); also found with manganese oxide dendrites in the Contzen Pass area (UA 6324); as crystals with copious chrysocolla inclusions from an unspecified mine dump. As scepter crystals from Arivaca (UA 9361). Ajo district, New Cornelia mine, associated with cuprite and native copper and including shattuckite (UA 3431).

*Pinal County:* Galiuro Mountains, Bunker Hill district, Copper Creek Canyon, at the western end of the Aldwinkle and Longstreet claims, as twins after the Japanese law and as individuals up to 2 inches in length in a soft limonite matrix; smaller specimens are typically flattened on the prism, and the tips of crystals have smaller crystals perched upon them. Some crystals show markedly flattened terminations resembling the basal pinacoid. Tourmaline of the red and green varieties is not uncommonly enclosed in quartz crystals (William and Mildred Schupp, pers. comm., 1969; Gary Edson, pers. comm., 1972) (UA 691, 3182). Pioneer district, in geodes in perlite (UA 1813). Sacaton Mountains, Gila River Indian Reservation, as extensive outcrops of white vein quartz (Wilson, 1969). At a locality about 3 miles southwest of Pinal, as hollow, thin-walled crystals in sandstone (Kunz, 1887).

*Santa Cruz County:* Patagonia Mountains, Duquesne, Holland mine, as slender, tapering crystals up to 12 inches long, occasionally forming Japanese twins, one specimen of which is about 7 inches from tip to tip of the individual crystals; these twins and slender individual crystals occurred in a pocket with drusy siderite and chlorite (R. A. Bideaux collection) (UA 1096, 4459). This is probably the finest United States locality for twins after the Japanese law. Also as amethyst in pegmatite near Duquesne. Belmont and Lead King properties, as a body 100 feet wide containing crystals up to 2 feet long (Schrader and Hill, 1915). In geodes at Temporal Gulch (UA 9352). Sierra de Tordillo, Parker Canyon district.

*Yavapai County:* Bradshaw Mountains, Cash mine, Hassayampa district, where clear crystals lining open veins are accompanied by crystals of adularia, calcite, and sulfide ore minerals. Crystals up to 2 inches long twinned after the Japanese law from the Bagdad area along the creek above the open pit mine; commonly stained with iron oxides. As fern-like growths near Hillside (H 106631). Date Creek, near Congress Junction, amethyst scepters on milky quartz crystals up to two inches long, loose in the soil (Arizona-Sonora Desert Museum, Hill collection).

*Yuma County:* As large crystals in pegmatite in Precambrian granites over considerable areas. A chatoyant variety from which gem-quality cats-eye has been produced occurs near Crystal Park, near Quartzsite (Frondel, 1962) (H 104852).

## Chalcedony

Cryptocrystalline quartz having fibrous microtexture, which occurs as transparent to translucent crusts and coatings and as mammillary, botryoidal, nodular, and irregular masses; exhibits greasy or waxy luster. Usually shows banding parallel to a free surface or to the surface upon which it was deposited. Formed under conditions of low temperature by hydrothermal and weathering processes. A number of variety names are applied to chalcedonic silica based mainly on color and textural characteristics.

*Apache and Navajo Counties:* The principal constituent of petrified wood, at a number of localities (see Petrified Wood).

*Gila County:* Globe-Miami district, colored blue to green from included chrysocolla or malachite.

*Graham County:* An unusual and attractive type, locally termed "fire agate," is found in the Duncan area. It exhibits a play of colors reminiscent of fire opal. The optical effect is apparently due to the presence of included turgite (a variety of hematite having absorbed water).

*Greenlee County:* Clifton-Morenci district, in limestone; also loose at Shannon Mountain.

*Maricopa County:* South of Aguila (A 31736). As "fire agate" at Saddle Mountain.

*Pinal County:* Galiuro Mountains, abundant as chalcedonic roses to 12 inches, and as other forms, in the pass between Little and Big Table Top Mountains.

*Santa Cruz County:* Grosvenor Hills, near the old village of Santa Cruz, Canelo Hills. Santa Rita Mountains, Cottonwood Canyon, at the Glove mine, as pesudomorphs after calcite and as casts of hemimorphite crystals (Olson, 1966) (UA 700, 9367).

*Yavapai County:* In the vicinity of Morgan City and Slow Spring Washes, as spherulitic nodules in lavas. Much of the chalcedony is fluorescent. In opalite, on the Agua Fria River 14 miles from Mayer (UA 1660). As abundant roses weathered from volcanic rocks, at Saddle Mountain, about 30 miles northwest of Hassayampa (Rogers, 1958).

*Yuma County:* As excellent specimen material from the Chocolate Mountains. From a locality east of Parker near milepost 87, Santa Fe Railway. Kofa Mountains, found lining geodes in rhyolite, 20 miles southeast of Quartzsite (Walker, 1957), and in geodes just west of the Kofa Mountains and north of the Castle Dome Mountains, about 26 miles south of Quartzsite (Weight, 1949).

## Agate

A term applied to chalcedony in which thin layers are typically accentuated by color differences. All gradations exist between chalcedony and agate, however, and many varieties of variegated and clouded agate are known. Moss agate is a variety in which manganese oxides form patterns in chalcedony.

*Coconino and Mohave Counties:* As nodules and geodes in the Kaibab Limestone (Guild, 1910).

*Greenlee County:* Near the Willow Spring ranch in the Peloncillo Mountains (Dimick, 1957).

*Maricopa County:* Agua Fria River area, as plume agate and green and white fortification agate.

*Pima County:* Tucson Mountains, where geodes of blue and white agate abound in the basaltic rocks near Sentinel Peak ("A" Mountain) (Guild, 1905). Rarely as carnelian and blue and white banded fortification agate, north of Pantano.

*Yavapai County:* In the vicinity of Morgan City and Slow Springs Washes, as spherulitic nodules in lava. On the Agua Fria River, west of New River Station, as "plume" agate, and mottled material (Richards, 1956). As excellent gray, blue, pink and violet material, a few miles north of Castle Hot Springs (Richards, 1956).

*Yuma County:* In an area several miles in extent, from Gila Bend to the north (Richards, 1956).

## Chert and Flint

The only difference between these materials is color. Cherts are usually, but not always, fairly light in color and may be grayish white, gray, yellowish, reddish, brownish; flint is usually darker in color, grays and black being typical. Both are composed principally of fine-grained fibrous chalcedony and occur typically in sedimentary rocks, especially limestones, in which they may be of primary or secondary origin. Beds of chert in limestone of considerable size are known. The shapes of the smaller bodies of chert in limestone assume great variety.

*Cochise and Pima Counties:* Some beds of the Permian-age limestones which are widely distributed in southeastern Arizona contain abundant chert. Small, reddish, jelly bean-like bodies abound in one bed of the Earp Formation.

*Coconino and Mohave Counties:* Chert is abundant in the Kaibab Limestone of Permian age.

## Chrysoprase

Chalcedony colored green by inclusions of nickel silicate (Frondel, 1962). Other green or greenish-blue chalcedonies such as, for example, those from the Live Oak and Keystone mines in the Globe-Miami district (Guild, 1910), are not properly chrysoprase because they are colored by chrysocolla.

*Mohave County:* Reported from the western slopes of the River Range, Weaver district.

## Petrified (Fossil) Wood

Fossil wood, replaced pseudomorphously by cryptocrystalline quartz, largely chalcedony and the red, brown, or yellow variety, jasper; during the replacement process by silica carried in underground waters, even the most minute details of the original woody structures are often preserved.

*Apache and Navajo Counties:* The Petrified Forest National Monument is of world renown for the remarkable amount and quality of the fossil wood which abounds in the Petrified Forest Member of the Chinle Formation over an area of many square miles extent. Tree logs, trunks,

limbs, and fragments are preserved, often in beautiful colors. In the vicinity of Nazlini Canyon, north of Ganado, and at numerous other localities (Gregory, 1917).

*Coconino, Mohave, and Yuma Counties:* Abundant along the banks of the Colorado River (Guild, 1910; Wilson and Butler, 1930).

*Pima County:* Tucson Mountains area, where it is common in the Amole Formation of Cretaceous age (Donald L. Bryant, pers. comm., 1972). In red beds of Cretaceous age south of the Empire Mountains.

*Santa Cruz County:* In red beds of Cretaceous age, Adobe Canyon, Santa Rita Mountains.

# R

## *RAMSDELLITE

Manganese oxide, $\gamma$-$MnO_2$. The orthorhombic dimorph of pyrolusite. Occurs in vein and bedded manganese deposits with a variety of other oxidized manganese minerals.

*Maricopa County:* Black Rock mine, present in hypogene veins in Precambrian crystalline gneisses (Hewett, 1964).

*Mohave County:* Artillery Mountains, Black Jack, Price and Priceless veins, in veinlets along fractures and joints, as small tabular to blocky crystals with hollandite, psilomelane, cryptomelane, coronadite, pyrolusite, and lithiophorite; also at the Plancha bedded manganese deposit (Mouat, 1962) (UA 9404).

*Pinal County:* Mammoth district, Mammoth–St. Anthony mine, as pseudomorphs after groutite microcrystals.

*Yavapai County:* As masses of equant to tabular crystals up to about 5 mm, from a locality east of Octave and northeast of Wickenburg (UA A86).

## RANSOMITE

Copper iron sulfate hydrate, $CuFe_2(SO_4)_4 \cdot 6H_2O$. A rare mineral found originally at Jerome where it formed as a result of a mine fire.

*Cochise County:* Warren district, where it is found as a post-mine mineral associated with voltaite and roemerite in warm (oxidizing) parts of pyritic ores in the Cole and Campbell mines.

*Yavapai County:* Black Hills, United Verde mine, as crusts and small tufts of crystals formed as a result of the burning of pyritic ore. The type locality (Lausen, 1928; Wood, 1970). The chemical analysis made by T. F. Buehrer gave:

$H_2O$   18.82   $Fe_2O_3$   22.57   $Al_2O_3$   1.52
$SO_3$   46.30                      $CuO$   11.29

TOTAL   100.50%

Specific gravity: 2.63.

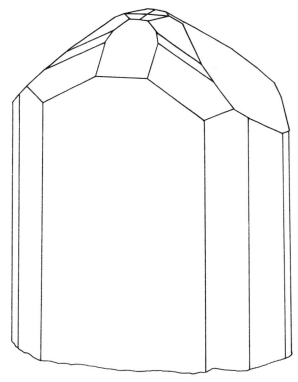

**Ransomite.**
United Verde mine (Lausen, 1928).

## RAUVITE

Calcium uranyl vanadium oxide hydrate, $Ca(UO_2)_2V_{10}O_{28} \cdot 16H_2O$. A secondary mineral found in masses, as crusts and coatings, and as interstitial matter in sandstone with other oxidized uranium-vanadium minerals in Colorado Plateau-type deposits.

*Apache County:* Monument Valley, Monument No. 2 mine (Weeks et al., 1955; Finnell, 1957; Witkind and Thaden, 1963).

*Navajo County:* Monument Valley, Monument No. 1 mine, and the adjoining Mitten No. 2 mine (Witkind, 1961; Witkind and Thaden, 1963).

## REALGAR

Arsenic sulfide, AsS. Occurs as a minor constituent in some gold, silver, and lead veins with orpiment, stibnite, and other arsenic minerals;

also associated with volcanic fumerolic action and with some hot springs; typically associated with orpiment to which it readily alters.

*Pinal County:* In 1915 several pounds of realgar and orpiment were discovered at an unspecified locality near the junction of the Gila River and Hackberry Wash.

*Yavapai County:* Bradshaw Mountains, from the vicinity of Castle Hot Springs.

## RHODOCHROSITE

Manganese carbonate, $MnCO_3$. A member of the calcite group. A common gangue mineral in sulfide mineral deposits formed under a wide range of temperatures; also as a secondary mineral in iron and manganese oxide deposits.

*Cochise County:* Tombstone district, as small grains in oxidized alabandite ore from the Lucky Cuss mine (Hewett and Rove, 1930; Butler et al, 1938; Rasor, 1939). Warren district, Higgins mine, replacing dolomitic limestone, in drusy cavities in alabandite (Hewett and Rove, 1930; Hewett and Fleischer, 1960). Chiricahua Mountains, at the Humboldt mine, with rhodonite, calcite, and quartz, associated with alabandite in lenses in a fissure vein cutting limestone (Hewett and Rove, 1930).

*Gila County:* Banner district, in the London Range shaft. Globe-Miami district, with manganian ankerite, the principal hypogene gangue mineral of the Ramboz (Silver Glance) deposit (Hewett and Fleischer, 1960; Peterson, 1962; Hewett et al., 1963).

*Pima County:* Rincon Mountains, with pyrolusite (UA 7554).

*Santa Cruz County:* Patagonia Mountains, at the Trench mine, where it occurs with alabandite, sphalerite, and galena (Schrader and Hill, 1915; Schrader, 1917; Hewett and Rove, 1930). Santa Rita Mountains, Cottonwood Canyon, Glove mine, with manganese oxide minerals (Olson, 1966).

## *RHODONITE

Manganese iron magnesium silicate, $(Mn,Fe,Mg)SiO_3$. Rhodonite is a member of the pyroxenoid group of minerals. It forms as an igneous mineral as well as by hydrothermal processes. It also occurs in pegmatites and in contact metamorphic deposits by the alteration of rhodochrosite.

*Cochise County:* Chiricahua Mountains, Humboldt mine, with rhodochrosite and alabandite in lenses in a fissure vein in limestone (Hewett and Rove, 1930).

*Pima County:* From an unspecified locality in the Twin Buttes area (UA 4496). Also reported to occur near Sasabe (Lee Hammons, pers. comm., 1974).

*Santa Cruz County:* South central Patagonia Mountains, in quartz gangue in veins cutting Tertiary volcanics (Schrader, 1917).

## RHOMBOCLASE

Acid iron sulfate hydrate, $HFe(SO_4)_2 \cdot 4H_2O$. A secondary mineral formed from the oxidation of pyrite; as an incrustation on mine walls with chalcanthite, roemerite, and epsomite.

*Cochise County:* Warren district, in porous crusts with roemerite, Copper Queen mine (Merwin and Posnjak, 1937).

## *RICHTERITE

Sodium potassium calcium magnesium manganese silicate hydroxide, $(Na,K,Ca)_3(Mg,Mn)_5$-$Si_8O_{22}(OH)_2$. A member of the amphibole group. Found in igneous and in both thermally and contact metasomatically metamorphosed rocks.

*Coconino County:* Canyon Diablo area, associated with krinovite, roedderite, high albite, ureyite, and chromite in the octahedrite meteorite (Olsen and Fuchs, 1968).

## RICKARDITE

Copper telluride, $Cu_4Te_3$. A rare, late-formed mineral found in veins with pyrite, tellurium, and other tellurides.

*Cochise County:* Warren district, as small purple fragments in a sample of sulfide pulp from the 1400-foot level of the Junction mine. The identification was based on a positive qualitative test for tellurium and the visual detection of the characteristic purple-red color of rickardite (Crawford, 1930).

## *RIEBECKITE

Sodium iron silicate hydroxide, $Na_2Fe_3^{2+}Fe_2^{3+}$-$(Si_8O_{22})(OH)_2$. Associated with granitic igneous rocks and with low-grade regionally metamorphosed schists.

*Pima County:* North end of the Sierrita Mountains and the south end of the Roskruge Mountains, as the asbestiform variety, crocidolite (Robert O'Haire, pers. comm.) (UA 3065, 9212).

## *RIPIDOLITE

Magnesium iron aluminum silicate hydroxide, $(Mg,Fe,Al)_6(Si,Al)_4O_{10}(OH)_8$. A member of the chlorite group, formed during thermal metamorphism and by hydrothermal alteration processes.

*Yavapai County:* Verde district, United Verde mine area, where it is a constituent of the "black schist"; associated with other members of the chlorite group (Anderson and Creasey, 1958).

## *ROBERTSITE

Calcium manganese phosphate hydroxyl hydrate, $CaMn_4(OH)_6(H_2O)_3(PO_4)_4$. A rare mineral previously reported only in certain pegmatites near Custer, South Dakota, where it is associated with other dark-colored, late-stage iron and manganese phosphate minerals.

*Yavapai County:* In the pegmatites of the White Picacho district. (Material collected by Joseph E. Urban.)

## Rock Salt (see HALITE)

## *ROEDDERITE

Sodium potassium magnesium iron silicate, $(Na,K)_2(Mg,Fe)_5Si_{12}O_{30}$. Occurs in iron meteorites. Found in octahedrite meteorites from three areas including Canyon Diablo.

*Coconino County:* Canyon Diablo area, with krinovite in graphite nodules in the octahedrite meteorite; also associated with richterite, high albite, ureyite, and chromite (Olsen and Fuchs, 1968).

## ROEMERITE

Iron sulfate hydrate, $Fe^{2+}Fe_2^{3+}(SO_4)_4 \cdot 14H_2O$. A secondary mineral commonly formed by the alteration of pyrite; frequently associated with copiapite and other secondary sulfates. The louderbackite of Lausen (1928) was subsequently shown to be this species (Pearl, 1950).

*Cochise County:* Warren district, in porous crusts, Copper Queen mine (Merwin and Posnjak, 1937).

*Yavapai County:* Black Hills, United Verde mine, formed under fumerolic conditions as a result of burning pyritic ores, as thin crusts on pyrite (Lausen, 1928; Wood, 1970). H 90534 is part of the type louderbackite (Pearl, 1950). An analysis by T. F. Buehrer on the original "louderbackite" is as follows:

$H_2O$ 31.33  $Fe_2O_3$ 20.84  FeO 7.01
$SO_3$ 39.34  $Al_2O_3$ 2.55  $Na_2O$ 0.88

TOTAL 101.95%

Specific gravity: 2.19.

## ROMANECHITE
(see PSILOMELANE)

## ROSASITE

Copper zinc carbonate hydroxide, $(Cu,Zn)_2$-$(CO_3)(OH)_2$. A secondary mineral found with aurichalcite, brochantite, malachite, and other secondary minerals in the oxidized portions of copper-zinc-lead deposits.

*Cochise County:* Tombstone district, as bright green mammillary spherules in siliceous linings of vugs and between hemimorphite crystals at the Toughnut and Empire mines (Butler et al., 1938). With smithsonite and manganese oxides at Gleeson Ridge, Turquoise district. Warren district, where it is not uncommonly found in the gossany areas in underground workings; as small, dense spherules of radiating blue-green fibers in vugs with calcite, aurichalcite, and hemimorphite.

*Gila County:* Banner district, 79 mine, in the oxidized portion of the mine as "deep blue-green velvety mats or warty crusts encrusting manganese oxides and smithsonite. Some of the mats are associated with aurichalcite, smithsonite, wulfenite, calcite, malachite, and mimetite"; formed contemporaneously with aurichalcite (Keith, 1972).

*Pima County:* Waterman Mountains, Silver Hill mine (UA 6937, 5537). Cababi district, South Comobabi Mountains, Mildren and Steppe claims (Williams, 1963). Pima district, San Xavier West mine, intergrown with flat bladed calcite (Arnold, 1964).

*Santa Cruz County:* Santa Rita Mountains, Cottonwood Canyon, Glove mine (Olson, 1966).

## ROSCOELITE

Potassium vanadium aluminum magnesium silicate hydroxide, $K(V,Al,Mg)_3(Al,Si)_3O_{10}$-$(OH)_2$. A rare vanadian member of the mica group.

*Apache County:* Monument Valley, Monument No. 2 mine, with many other uranium-vanadium minerals (Gruner and Gardiner, 1952).

*Navajo County:* Monument Valley, Mitten No. 2 mine (Witkind and Thaden, 1963); Monument No. 1 mine (Holland et al., 1958).

## ROSSITE

Calcium vanadium oxide hydrate, $CaV_2O_6 \cdot 4H_2O$. A rare mineral associated with the oxidized uranium-vanadium deposits of the Colorado Plateau; intimately associated with metarossite to which it alters.

*Apache County:* Lukachukai Mountains, reported from the Mesa No. 1 mine.

## RUTILE

Titanium dioxide, $TiO_2$. Widespread as an accessory mineral in granitic igneous rocks; also common in metamorphic rocks and as a product of alteration of other titanium-bearing silicates. Common as a detrital mineral in sands and sedimentary rocks.

*Apache County:* Monument Valley area, Garnet Ridge, found as mineral fragments in a breccia dike piercing the sedimentary rocks; commonly included as needle-like crystals in garnet (Gavasci and Kerr, 1968).

*Cochise County:* Warren district, as an accessory mineral in granite, northwest of Bisbee; Sacramento Hill, in a hydrothermally altered granite porphyry, as clusters of tiny crystals in pseudomorphs after biotite (Schwartz, 1947).

*Gila and Pinal Counties:* As an accessory mineral in Madera Diorite, Pinal Schist, and the Solitude and Ruin Granites (Peterson, 1962). Globe-Miami district, Miami ore body, as a product of hydrothermal alteration of granite porphyry and, in the vicinity of the ore body, in schist (Schwartz, 1947, 1958); Castle Dome mine, as a product of the clay mineral phase of alteration, recrystallized from sphene and ilmenite (Peterson, et al., 1946; Schwartz, 1947).

*Greenlee County:* Morenci district, as clusters replacing ferromagnesian minerals in an intensely altered porphyry copper ore body; associated with a suite of hydrothermal alteration minerals (Reber, 1916; Schwartz, 1947, 1958).

*Mohave County:* Cerbat Mountains, Wallapai district, present in a hydrothermally altered disseminated sulfide deposit in a porphyritic intrusive, as sagenite webs and in clusters of crystals; also replacing sphene (Thomas, 1949).

*Pima County:* Ajo district, as stumpy crystals which are an alteration product of biotite (Schwartz, 1958; Hutton and Vlisidis, 1960). Silver Bell area, as small, acicular crystals associated with quartz, and as small grains and prismatic crystals (Kerr, 1951).

*Pinal County:* Mineral Creek district, Ray area, found in hydrothermally altered quartz monzonite porphyry, in pseudomorphs after biotite, with chlorite, sericite, and hydromica (Schwartz, 1952). Mammoth district, San Manuel mine, in hydrothermally altered monzonite and quartz monzonite porphyries associated with chlorite pseudomorphs after biotite (Schwartz, 1947, 1949, 1958); found in the propylitic, potassium silicate, and argillic phases of alteration (Creasey, 1959). Sacaton Mountains, Sec. 12, T.5S, R.5E, with corundum and quartz, as small, irregular masses in felsite (Wilson, 1969).

*Santa Cruz County:* Patagonia Mountains, Duquesne, as slender crystals and reticulated masses to several inches in altering granite at the Santo Niño mine (J. H. Courtright, pers. comm.).

*Yavapai County:* Black Hills, as well-developed crystals at the United Verde mine. Bradshaw Mountains, with tourmaline in the gangue of the Howard Copper property, Black Canyon district. Eureka district, Bagdad, as stubby crystals or granular aggregates derived from sphene and biotite in an altered quartz monzonite stock, associated with quartz and orthoclase, more rarely with sericite (Anderson, 1950).

*Yuma County:* East margin of the Dome Rock Mountains, 3 miles southwest of Quartzsite, as small grains scattered throughout Precambrian schist; some well-formed crystals occur in fractures in quartz veins (Wilson, 1929).

# S

## SABUGALITE

Acid aluminum uranyl phosphate hydrate, $HAl(UO_2)_4(PO_4)_4 \cdot 16H_2O$. A rare secondary mineral found in vein-type uranium deposits as an alteration product of uraninite and in deposits of the Colorado Plateau type.

*Apache County:* Black Water mine, Black Mesa Basin.

*Coconino County:* Cameron district, O'Jaco, Huskon No. 5, and Arrow Head Nos. 1, 3, and 7 claims (Holland et al., 1958).

## *SALEEITE

Magnesium uranyl phosphate hydrate, $Mg(UO_2)_2(PO_4)_2 \cdot 8H_2O$. A secondary mineral of the autunite group associated with oxidized uranium deposits.

*Gila County:* Cherry Creek area, Sue mine, where it is intimately associated with bassetite (Granger and Raup, 1969). In weathered deposits in sandstone in the Dripping Spring Quartzite in the north-central portion of the county (Granger and Raup, 1962).

## *SALMONSITE

Manganese iron phosphate hydrate, $Mn_9$-$Fe_2^{3+}(PO_4) \cdot 14H_2O$ (?). A rare secondary mineral formed as an alteration product of hureaulite in pegmatite; inadequately characterized species.

*Maricopa County:* In lithiophilite, White Picacho district, Midnight Owl mine, near Wickenburg (UA 6916).

## SAMARSKITE

Yttrium cerium uranium calcium lead columbium tantalum titanium tin oxide, (Y,Ce,U,Ca,-Pb)(Nb,Ta,Ti,Sn)$_2O_6$. Generally occurs in granite pegmatites in small amounts with other rare earth minerals.

*Mohave County:* Aquarius Range, in granite pegmatites with allanite. Also reported from northeast of Kingman.

*Pima County:* Sierrita Mountains, as shiny black crystals of what is probably samarskite, in the New Years Eve mine.

*Yavapai County:* Black Hills, reported from near Jerome.

## SANIDINE

Potassium sodium aluminum silicate, (K,Na)-$AlSi_3O_8$. A member of the alkali feldspar group which may be described as a structurally disordered orthoclase. A constituent of alkali and felsic volcanic rocks such as rhyolites and trachytes, as transparent, glassy crystals. The mineral is no doubt much more common in the state than suggested by the few localities noted.

*Cochise County:* Both in the groundmass and as phenocrysts in welded rhyolite tuffs in the Chiricahua National Monument; associated with quartz and magnetite (Enlows, 1955). Also reported from the Warren district. Common as phenocrysts, some showing a blue play of colors, in rhyolites in the Peloncillo Mountains, Cottonwood Canyon.

*Pima County:* Gunsight Mountain, northern Sierrita Mountains, as the barian variety, in a pulaskite dike (Bideaux et al., 1960). As clear round phenocrysts in rhyolite welded tuffs from the Batamote Mountains, near Ajo.

## *SAPONITE

Calcium sodium magnesium iron aluminum silicate hydroxide hydrate, $(Ca/2,Na)_{0.33}(Mg,Fe)_3(Si,Al)_4O_{10}(OH)_2 \cdot 4H_2O$. A member of the smectite (montmorillonite) group.

*Maricopa County:* As smears along fractures in vein quartz at the Tonopah-Belmont mine, south of Wickenburg.

*Pima County:* From an unspecified locality near Pantano (UA 6486).

## *SAPPHIRINE

Magnesium aluminum silicate, $(Mg,Al)_8(Al,Si)_6O_{20}$ (Fleischer, 1975); $(Mg,Fe)_2Al_4O_6(SiO_4)$ (Deere et al., 1962). A relatively rare mineral which occurs in metamorphic rocks high in alumina and low in silica.

*Yuma County:* Three miles southwest of Quartzsite, as irregular grains which occur sparingly in schist. Tentative identification by E. D. Wilson by optical means (Wilson, 1929).

## SAUCONITE

Sodium zinc aluminum silicate hydroxide hydrate, $Na_{0.33}Zn_3(Si,Al)_4O_{10}(OH)_2 \cdot 4H_2O$. A member of the smectite (montmorillonite) group. Previously termed "tallow clay."

*Gila County:* Globe-Miami district, Castle Dome area, as purplish, waxy lumps in manganiferous material. Banner district, 79 mine, associated with chrysocolla and hemimorphite in a stope system above the 470 level (Keith, 1972).

*Pinal County:* Pioneer district, Magma mine, as soft, waxy gouge-like material from near the lower limit of oxidation, associated with coronadite. A partial analysis gave:

| | | |
|---|---|---|
| Zn 14.6 | Al$_2$O$_3$ 22.0 | H$_2$O 14.9 |
| SiO$_2$ 33.6 | | |

## SCAPOLITE

A complex silicate chloride fluoride hydroxide carbonate and sulfate of sodium, calcium, potassium, and aluminum. The scapolite group which can broadly be expressed by the formula $(Na,Ca,K)_4Al_3(Al,Si)_3Si_6O_{24}(Cl,F,OH,CO_3,SO_4)$, ideally includes two end-members, marialite and meionite. The exact composition of the mineral at the occurrences noted below is not known. Mainly confined to regional or contact meta-morphosed rocks such as schists, gneisses, and crystalline limestones.

*Gila County:* Reported from metamorphosed Precambrian rocks in the Diamond Butte quadrangle (Gastil, 1958).

*Yavapai County:* Bradshaw Mountains, Black Canyon district, 6 miles south of Cleator, in Yavapai schist.

## SCHEELITE

Calcium tungstate, $CaWO_4$. A widespread mineral, often an important ore of tungsten; occurs typically in high temperature environments: in granite pegmatites, in contact metamorphic deposits formed in limestone, and in quartz-rich high-temperature veins. Commonly associated with wolframite. Widely distributed in small amounts in Arizona.

*Cochise County:* In small quantities in quartz veins cutting granite, about 4 miles north of Dragoon (Guild, 1910). Cochise district, Johnson Camp area, near the Republic mine, crystals of scheelite embedded in quartz crystals (Romslo, 1949; Cooper and Huff, 1951; Cooper, 1957; Baker, 1960; Cooper and Silver, 1964). At the Cohen Tungsten mine, 10 miles east of Willcox, as light brown crystals up to fifteen pounds in weight, and as smaller crystals embedded in gray, doubly-terminated, quartz crystals. Whetstone Mountains, in quartz veins and as replacements in granite on the eastern slope of the range (Guild, 1910; Dale et al., 1960). Huachuca Mountains, Reef mine, in a quartz vein associated with chalcopyrite, sphalerite, huebnerite, fluorite, galena, pyrite, and stolzite (Palache, 1941a); also at the James, Harper, and other properties (Wilson, 1941; Dale et al., 1960). Swisshelm Mountains (Dale, 1959). In quartz veins in silicified limestones in the Paradise area, Chiricahua Mountains (Dale et al., 1960).

*Gila County:* Globe-Miami district, low on the east slope of Day Peaks, Lost Gulch area, in a mineralized zone in a fault cutting diabase, with stolzite, as sparse, isolated crystals (Faick and Hildebrand, 1958).

*Graham County:* Southwest of the Graham Mountains, SE ¼ Sec. 35, and SW ¼ Sec. 36, T.8S, R.22E, with tourmaline from the Black Beauty group (Dale, 1959).

*Maricopa County:* Mazatzal Mountains, northwest of Four Peaks. Near Morristown, on upper Santo Domingo Wash, with powellite. Cave Creek district, at an unspecified locality, with cuprotungstite and ferberite (Schaller, 1932).

In several properties northwest of Morristown (Dale, 1959). Vulture Mountains, with powellite as disseminations in granitic rocks in the Flying Saucer group of claims, Sec. 12, T.6N, R.6W (Dale, 1959).

*Mohave County:* Hualpai Mountains, with wolframite, in quartz veins, of the Boriana (Hobbs, 1944; Dale, 1961), Telluride Chief, and other properties. Aquarius Range, sparingly at the Williams mine, Boner Canyon (Hobbs, 1944; Dale, 1961). In small amounts with wolframite, in the Cottonwood and Greenwood areas (Dale, 1961). Mohave Mountains, Dutch Flat (Wilson, 1941).

*Pima County:* Las Guijas Mountains (Dale et al., 1960). Santa Rita Mountains, Helvetia district (Schrader and Hill, 1915). Sierrita Mountains, in contact zones in the Twin Butte mine, Pima district (S. B. Keith, pers. comm., 1973). Gunsight Hills (Wilson, 1941; Dale et al., 1960). On a number of claims in the eastern part of the Papago Indian Reservation, about 50 miles west of Tucson (Dale et al., 1960). Santa Catalina Mountains, near Marble and Piety Peaks (Dale et al., 1960). San Luis Mountains, southwest of Arivaca, in quartz veins in the Easter prospect (Dale et al., 1960).

*Pinal County:* Campo Bonito area, Maudina and other properties (Guild, 1910). Northwest of Mammoth, Tarr and Antelope Peak areas (Wilson, 1941).

*Santa Cruz County:* San Cayetano district, near Calabasas, with wolframite (Dale et al., 1960). Patagonia Mountains, with molybdenite, 4 miles south of Duquesne (Schrader and Hill, 1915; Wilson, 1941). Nogales district, Reagan Camp, with wolframite (Schrader, 1917).

*Yavapai County:* Bradshaw Mountains, Tip Top district (Dale, 1961). Hassayampa district. Wickenburg Mountains, White Picacho district (Jahns, 1952), disseminated in garnet-epidote schist on the upper Santo Domingo and Little Santo Domingo Washes. Silver Mountain area (Wilson, 1941).

*Yuma County:* Trigo Mountains, in a sheared quartz vein at the Gold Reef claims, Silver district (Wilson, 1941). Granite Wash Mountains, Three Musketeers mine in Sec. 24, T.6N, R.15W, as crystals up to 4 inches across, in white quartz; the crystals show brilliant blue-white fluorescence (David Shannon, pers. comm., 1971). From a placer near Quartzsite (H 97745). Sporadic occurrences have been noted in the Dome Rock, Plomosa, Kofa, Gila, and Little Harquahala Mountains (Dale, 1959).

## *SCHOEPITE

Uranium oxide hydrate, $UO_3 \cdot 2H_2O$. A secondary mineral commonly associated with uraninite from which it may form by alteration.

*Apache County:* Monument Valley, Monument No. 2 mine, as a surficial alteration product of uraninite, with becquerelite, and small amounts of fourmarierite (Frondel, 1956).

*Coconino County:* Cameron area, Black Point-Murphy mine, sparsely distributed in Pleistocene gravels (Austin, 1964).

*Yavapai County:* Abe Lincoln mine, northeast of Wickenburg, the principal secondary uranium mineral, as a coating on pyrite grains (Raup, 1954; Granger and Raup, 1962).

## SCHORL (see TOURMALINE)

## SCHREIBERSITE

Iron nickel phosphide, $(Fe,Ni)_3P$. Present in the metallic meteorites of the state.

## SCHROECKINGERITE

Sodium calcium uranyl carbonate sulfate fluoride hydrate, $NaCa_3(UO_2)(CO_3)_3(SO_4)F \cdot 10H_2O$. Occurs as a late secondary mineral with other oxidized uranium minerals; known to form as a post-mine opening mineral.

*Coconino County:* Cameron area, coating fractures in sandstone at the Jack Daniels No. 1 mine, possibly of post-mine age; Foley Bros. No. 5 mine (Austin, 1964).

*Yavapai County:* Hillside mine, as $1/8$-inch-thick coatings on gypsum, with andersonite, bayleyite, and swartzite. Johannite and pitchblende are reported from the same mine (Axelrod et al., 1951).

## SCOLECITE

Calcium aluminum silicate hydrate, $CaAl_2Si_3O_{10} \cdot H_2O$. A member of the zeolite group. Occurs as seams and amygdule fillings in mafic volcanic rocks; also less commonly of hydrothermal origin in metamorphic rocks.

*Graham County:* At Black Point, about 4 miles below Geronimo on the Gila River, as amygdule fillings up to $1/2$ inch in diameter in basalt; highly fluorescent.

*Pinal County:* Copper Creek district, in brecciated zones in the Copper Creek Granodiorite with calcite, as well-formed prisms up to 1 inch in length.

## SCORODITE

Iron arsenate hydrate, $FeAsO_4 \cdot 2H_2O$. A relatively common mineral which forms under oxidizing conditions in gossans associated with arsenic-bearing mineralization, often the first mineral to replace arsenopyrite; rarely a primary hydrothermal mineral.

*Coconino County:* Horseshoe Mesa, Grand Canyon National Park, Grandview (Last Chance) mine, as light brown to gray mammillary masses having radiating internal structure, associated with metazeunerite and olivenite (Leicht, 1971; John S. White, Jr., pers. comm.).

*Mohave County:* Cerbat Mountains, a common secondary mineral in the Wallapai district; Mollie Gibson mine (Thomas, 1949).

## SELENIUM

Selenium, Se. A rare secondary mineral. The Jerome occurrence marks its first recognition as the mineral.

*Yavapai County:* Verde district, United Verde mine, formed as a coating of needle-like crystals on rock above the burning pyrite ore body. Crystals are up to 2 cm in length and are bounded by first and second order rhombohedrons (Palache, 1934) (H 92678, 92679). An analysis by F. A. Gonyer showed the mineral to contain no tellurium and only a trace of sulfur.

## SENGIERITE

Copper uranyl vanadate hydrate, $Cu(UO_2)_2(VO_4)_2 \cdot 8\text{-}10H_2O$. A rare secondary mineral. The occurrence noted here was the first of the mineral in the United States.

*Cochise County:* Originally found in one isolated pocket in the Cole mine at Bisbee in 1935 where it was associated with malachite and chalcocite, covellite, and chlorargyrite. Later found to be more widely distributed in the Cole mine; it "occurred in one of a series of five sulfide-oxide veins which extended from the 800 level to the 1300 level of the Cole Shaft" (Hutton, 1957).

## SEPIOLITE (Meerschaum)

Magnesium silicate hydroxide hydrate, $Mg_4Si_6O_{15}(OH)_2 \cdot 6H_2O$. Formed by alteration of magnesian rocks and generally associated with serpentine or magnesite. A component of certain clays; sometimes found in hydrothermal veins. A compact variety constitutes the meerschaum of industry.

*Greenlee County:* Clifton area (UA 8847).

*Maricopa County:* From a locality 42 miles north of Phoenix and 2 miles east of Highway 69.

*Pima County:* San Xavier West mine, Pima district (UA 4164).

*Yavapai County:* In the basin of the Santa Maria River, west of the McCord Mountains, as crystalline, fibrous material (Kauffman, 1943). An analysis by Kauffman gave:

| | | | | | |
|---|---|---|---|---|---|
| SiO₂ | 54.83 | CaO | 0.55 | K₂O | 0.03 |
| Al₂O₃ | 0.28 | MgO | 24.51 | H₂O(−) | 8.18 |
| Fe₂O₃ | 0.45 | Na₂O | 0.35 | H₂O(+) | 10.74 |

TOTAL 99.92%

## Serpentine

Magnesium silicate hydroxide, $Mg_3Si_2O_5(OH)_4$, with variable small amounts of ferric iron and aluminum in substitution for magnesium. Serpentines usually consist of mixtures of the species chrysotile and antigorite which are members of the group (Nagy and Faust, 1956). The fibrous habit of chrysotile has been shown to be the result of the "rolled-up tube" nature of its kaolinite-like sheet structure. The nomen-

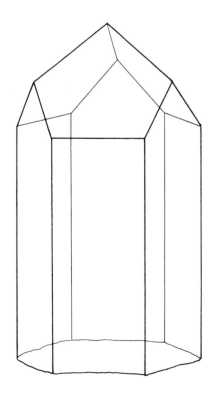

**Selenium.**
United Verde mine (Palache, 1934).

clature of the serpentine minerals is further complicated by stacking polymorphism. The most notable serpentine mineral in Arizona is chrysotile which has been extensively mined as asbestos. Antigorite is abundant, however, as a major constituent of the non-asbestiform serpentine rock associated with the asbestos ore. Of secondary origin derived from alteration of non-aluminous magnesian silicates, particularly olivine, amphibole, or pyroxene, in places as large masses derived from peridotite or other mafic igneous rocks. A common product of contact metamorphism of magnesian limestones; also found in hydrothermal veins.

Extensive lists of asbestos occurrences will be found in Arizona Bureau of Mines Bulletins 126 and 180, and in Funnell and Wolfe (1964).

*Apache County:* Garnet Ridge, abundant in a boulder, pseudomorphous after an orthopyroxene; related to a breccia dike which pierces sedimentary rocks (Gavasci and Kerr, 1968).

*Cochise County:* Tombstone district, in altered limestone, Lucky Cuss mine (Butler et al., 1938). Chiricahua Mountains, as chrysotile, with contact metamorphic ores. Dos Cabezas Mountains, in metamorphosed limestones (Guild, 1910). Little Dragoon Mountains, Johnson Camp area, as a retrograde product of higher temperature metamorphism of forsterite, tremolite, and diopside in the Moore ore body (Cooper and Huff, 1951; Cooper and Silver, 1964).

*Coconino County:* Grand Canyon region, at the Bass and Hance properties where the Precambrian Bass Limestone has been altered adjacent to diabase sills (Selfridge, 1936); crossfiber asbestos is locally up to 4 inches long, but is usually shorter (Funnell and Wolfe, 1964).

*Gila County:* The most extensive and commercially important deposits of chrysotile in Arizona are north and northeast of Globe along the Salt River in the vicinity of Chrysotile and along Cherry and Ash Creeks (*Engr. Min. Jour.,* 1915; Melhase, 1925; Stewart and Haury, 1947; Stewart, 1956). The deposits originated through metamorphic action of diabase intrusives upon Precambrian Mescal Limestone. Typical analyses follow:

R. E. Zimmerman

(1)

| | | | |
|---|---|---|---|
| MgO 42.05 | Al$_2$O$_3$ 1.27 | FeO | 0.64 |
| SiO$_2$ 41.56 | | H$_2$O+ | 14.31 |
| | | TOTAL | 99.83% |

(2)

| | | | |
|---|---|---|---|
| MgO 41.85 | Al$_2$O$_3$ 0.91 | CaO | 0.07 |
| SiO$_2$ 41.35 | FeO 0.69 | H$_2$O+ | 13.34 |
| | | TOTAL | 98.21% |

Geo. T. Faust (Nagy and Faust, 1956)

| | | | |
|---|---|---|---|
| MgO 41.44 | FeO 0.11 | MnO | 0.03 |
| SiO$_2$ 42.02 | Fe$_2$O$_3$ 0.19 | H$_2$O− | 1.64 |
| Al$_2$O$_3$ 0.52 | | H$_2$O+ | 14.04 |
| | | TOTAL | 99.99% |

Other deposits occur in the Pinal and Mescal Mountains and in the Pinto Creek region. Sierra Ancha Mountains, at the head of Pocket Creek (Bateman, 1923; Sampson, 1924). Banner district, Christmas mine, replacing dolomite and forsterite in the ore bodies (Perry, 1969). San Carlos Indian Reservation, Emsco and Bear mines (Bromfield and Schride, 1956). Globe district, Old Dominion mine, in Mescal Limestone near diabase sills.

*Greenlee County:* Clifton-Morenci district, of local occurrence with other calc-silicate minerals in an extensive contact metamorphic assemblage in the northern part of the Morenci open pit area (Moolick and Durek, 1966); on the ridge just west of Morenci at the Thompson mine, as green banded material associated with magnetite (Lindgren, 1905; Reber, 1916; Creasey, 1959).

*Pima County:* Pima district, developed in the metamorphic rocks at the Twin Buttes open pit mine (Stanley B. Keith, pers. comm., 1973).

*Pinal County:* Mammoth district, Mammoth–St. Anthony mine, antigorite, as shining white balls massed in aggregate as a matrix for wulfenite and descloizite. Hewett Wash area (Stewart, 1956).

*Yavapai County:* Copper Basin district, where it is an alteration product in ores in breccia pipes (Johnston and Lowell, 1961).

## SHATTUCKITE

Copper silicate hydroxide, Cu$_5$(SiO$_3$)$_4$(OH)$_2$. A secondary mineral which occurs as an alteration of other secondary copper minerals. First described from Bisbee by Schaller (1915).

*Cochise County:* Warren district, Shattuck mine, as pseudomorphs after malachite and as small blue spherules; associated with bisbeeite which replaces it (Schaller, 1915). Analyses by W. T. Schaller (reported by Wells, 1937) are as follows:

(1)

| | | |
|---|---|---|
| SiO$_2$ 39.68 | H$_2$O  5.94 | FeO 0.16 |
| CaO   0.05 | CuO 54.80 | ZnO   tr. |

TOTAL  100.63%

(on mineral from heavy liquid separation)

(2)

| | | |
|---|---|---|
| SiO$_2$ 37.91 | H$_2$O  5.83 | FeO 0.43 |
| CaO   — | CuO 55.51 | ZnO   — |

TOTAL  99.68%

(on deep blue spherules)

*Greenlee County:* Clifton-Morenci district, in Gila Conglomerate (UA 1036).

*Maricopa County:* South of Wickenburg, at several localities including the Moon Anchor mine and Potter-Cramer property, associated with a number of oxidized minerals (Williams, 1968).

*Pima County:* Ajo district, in veins at the New Cornelia open pit mine (Schaller and Vlisidis, 1958; Newberg, 1964; Mrose and Vlisidis, 1966; Vlisidis and Schaller, 1967). An analysis by Vlisidis and Schaller gave:

| | | |
|---|---|---|
| SiO$_2$ 36.06 | CaO  0.01 | MnO   0.03 |
| CuO 59.39 | MgO  0.02 | H$_2$O(+) 2.62 |
| FeO   0.19 | | TOTAL 98.32% |

Specific gravity (corrected): 4.09. An analysis by H. B. Wiik (Sun, 1961) on Ajo material yielded the following:

| | | |
|---|---|---|
| SiO$_2$ 35.90 | CaO  0.00 | CO$_2$    0.00 |
| CuO 55.27 | TiO$_2$  0.00 | H$_2$O(+) 6.31 |
| Fe$_2$O$_3$ 0.31 | Al$_2$O$_3$ 0.79 | H$_2$O(−) 0.08 |

TOTAL 98.66%

**Shattuckite.**
Ajo, Pima County. Julius Weber.

At an unspecified locality in the Tortolita Mountains. Santa Catalina Mountains, Molino Basin, with flecks of gold on chrysocolla-stained quartz.

*Pinal County:* Mammoth district, San Manuel mine, as veinlets in chrysocolla. At the Poston Butte porphyry copper deposit. Associated with fornacite and cerussite in dumps near the Roadside mine, Slate Mountains.

## *SHERWOODITE

Calcium vanadium oxide hydrate, Ca$_3$V$_8$O$_{22}$·15H$_2$O. A secondary oxidation product of lower valence vanadium minerals, commonly associated with native selenium, metatyuyamunite, and melanovanadite.

*Apache County:* Joleo mine (AEC).

## SICKLERITE

Lithium manganese iron phosphate, (Li,Mn,-Fe)PO$_4$. A rare secondary mineral formed by the alteration of lithiophilite and triphylite in pegmatites.

*Yavapai County:* White Picacho district, as thin discontinuous rims on lithiophilite and triphylite (Jahns, 1952).

## SIDERITE

Iron carbonate, FeCO$_3$. A member of the calcite group. Commonly occurs in bedded sedimentary rocks where it may be an important ore mineral; it also commonly occurs as a hydrothermal vein mineral.

*Cochise County:* Warren district, where boxwork siderites have proved to be a useful guide to ore (Trischka et al., 1929); Campbell mine, fine granular, dark brown, stalactitic, coated by small iridescent crystals.

*Gila County:* Dripping Spring Mountains, with wulfenite and vanadinite, McHur prospect, Banner district.

*Greenlee County:* Clifton-Morenci district, as a weathering product found in massive form in highly mineralized limestone (Reber, 1916).

*Mohave County:* Cerbat Range, a common gangue mineral in the Wallapai and Gold Basin districts (Thomas, 1949).

*Pima County:* Santa Rita Mountains, Iron Mask mine, Old Baldy district, with magnetite and schorl. Empire Mountains, Hilton mines (Schrader and Hill, 1915). Cababi district, Mildren and Steppe claims (Williams, 1963). Silver Bell district, as small crystals (Stewart, 1912).

*Pinal County:* Mammoth district, San Manuel mine (UA 8836).

*Santa Cruz County:* Tyndall district, Cottonwood Canyon, Glove mine, replacing limestone (Olson, 1966). Patagonia Mountains, Duquesne district, Mowry mine; Harshaw district, Flux mine (Schrader, 1917).

*Yavapai County:* Bradshaw Mountains, Lynx Creek, in veins with chlorite and tourmaline; Gold Note group, Turkey Creek district; Peck and Swastika mines, Peck district, where it is associated with native silver and bromargyrite. As large crystalline nodules from the vicinity of Yarnell. Stonewall Jackson mine, near Prescott, where it constitutes the gangue with which native silver, chlorargyrite, and chalcanthite are associated (Guild, 1917). Black Hills, at the Shea, Brindle Pup, and Mingus Mountain mines, in quartz veins with sulfides (Lindgren, 1926).

*Yuma County:* Dome Rock Mountains, in cinnabar veins with tourmaline. Harcuvar Mountains, as nearly jet-black cleavable material with chalcopyrite, in the Cunningham Pass area. Harquahala Mountains (UA 7586).

## SILLIMANITE

Aluminum silicate, $Al_2SiO_5$. Characteristically occurs in high-grade thermally metamorphosed argillaceous rocks; commonly associated with andalusite or cordierite.

*Cochise County:* Cochise district, as acicular porphyroblasts up to 1 cm long in hornfelsed shale near the Texas Canyon stock, about 2 miles southwest of Adams Peak (Cooper and Silver, 1964).

*Coconino County:* Grand Canyon, abundant in the Inner Gorge one-half mile downstream from Monument Creek (Campbell, 1936).

*Gila and Pinal Counties:* In Pinal Schist near post-Cambrian granite contacts, in the Pinal Ranch Quadrangle area (Peterson, 1962).

*Mohave County:* Hualpai Mountains, Maynard district, in quartz veins cutting schist. Cerbat Mountains, principally as the variety fibrolite in elongate masses of interlacing needles intergrown with biotite, garnet, quartz, and feldspar in migmatite (Thomas, 1953).

*Pinal County:* Gila Indian Reservation, Sacaton Mountains, in metamorphosed sediments encased in granodiorite, associated with titanandalusite, corundum, and cordierite (Bideaux et al., 1960).

*Yavapai County:* Santa Maria Mountains, Eureka district, near Camp Wood, as veins and nodules in schist; in the Hillside mica schist along Copper Creek in the vicinity of Bagdad (Anderson et al., 1955).

*Yuma County:* Near the eastern margins of the Dome Rock Mountains, 3 miles southwest of Quartzsite, in Precambrian schists, associated with dumortierite, kyanite, and andalusite (Wilson, 1929).

## SILVER

Silver, Ag. Most commonly occurs in the upper portions of silver-bearing deposits and in the zone of sulfide enrichment of copper deposits with chalcocite. Also, but less commonly, of primary origin, associated with galena or tetrahedrite.

*Cochise County:* Tombstone district, as disseminated flakes at the Empire mine, and as small masses of wire silver at the Flora Morrison mine (Guild, 1917; Butler et al., 1938). Warren district, notably in the Campbell and Cole mines, in small amounts commonly associated with secondary chalcocite which it has been observed to coat (H 94731, 94733; S 97831; UA 1249); more rarely with halloysite, Holbrook mine (Schwartz and Park, 1932). Pearce district, Commonwealth mine, with chlorargyrite, embolite, jarosite, and acanthite (Endlich, 1897).

*Gila County:* Globe district, as minute flakes in calcite at the Continental mine; as stout wires in the oxidized ores of the Old Dominion mine. Fine specimens were also recovered from placers about 4 miles north of Globe (Ransome, 1903). At Richmond Basin, as one of the chief ore minerals, in fairly large masses (Peterson, 1962). At Payson, as wire silver in the oxidized ore of the Silver Butte mine (Lausen and Wilson, 1927). Saddle Mountain district, Little Treasure mine, as wire silver (Ross, 1925b).

*Graham County:* Aravaipa district, at the La Clede mine.

*Greenlee County:* At the Detroit, Arizona-Copper, Shannon, Silver King, Gold Bar, Capote, Silver Bonanza, and other mines.

*Maricopa County:* From pegmatites in the White Picacho district (Jahns, 1952). Wickenburg Mountains, Monte Cristo mine, where it formed contemporaneously with nickel-skutterudite, chalcopyrite, tennantite, enargite, and acanthite (Bastin, 1922).

*Mohave County:* Cerbat Mountains, Distaff mine, as chunks weighing several pounds, with acanthite in the deeper workings (Bastin, 1925); Chloride district, at the Lucky Boy and Samoa mines (Thomas, 1949); Golden Bee, Queen Bee, and Rural mines, Mineral Park district (Bastin, 1925); Stockton Hill district, in solid chunks and as masses of wire silver at the Banner group of claims; Wallapai district, Tennessee-Schuylkill mine (Schrader, 1909). Buckeye mine, noted for large masses of the solid mineral and for beautiful specimens of wire silver (Newhouse, 1933).

*Pima County:* Tortolita Mountains, Apache property, with chalcocite and chlorargyrite. Cerro Colorado Mountains, Cerro Colorado mine, in part as wire silver, with stromeyerite and tetrahedrite (UA 2873) (Guild, 1917). Le Conte (1852) noted silver from a small mountain chain about 40 miles southeast of Tucson; it occurred with galena and "blende." Rincon Mountain foothills, Tanque Verde district, as flakes "in considerable quantity" in copper ore (*Engr. Min. Jour.*, 1897a). Silver Bell district (Stewart, 1912). Prince Rupert mine near Crittenden, reported to occur as fine specimens of wire silver (*Engr. Min. Jour.*, 1892b). Slate Mountains, Jackrabbit mine, as flakes and thin sheets in fractures (Hammer, 1961).

*Pinal County:* Pioneer district, as fine specimens in the Silver King mine where it fills cracks in stromeyerite, bornite, and chalcopyrite in masses of considerable size, also in "beautiful filiform, the branches of which envelop individual chalcocite grains, some of the finer filaments even extending into fractures and cleavage cracks of the chalcocite" (Guild, 1917) (S 65154, 65155; UA 616; H 99620); in the upper portions of the Magma mine (UA 884, 9524). Galiuro Mountains, Little Treasure and Adjust mines, Saddle Mountain district, as wire silver. Mineral Creek district, Ray mines, with secondary copper minerals in the west side of the Pearl Handle pit (Metz and Rose, 1966).

*Santa Cruz County:* Patagonia Mountains, Domino mines, Palmetto district, with crystallized cerussite and wulfenite; Harshaw district, World's Fair mine, with tetrahedrite (UA 211). Santa Rita Mountains, as small crystals surrounded by magnetite in diorite, on the southern slopes of the range (Schrader and Hill, 1915). Oro Blanco district, Eureka-Mabel mine (Blake, 1904).

*Yavapai County:* Verde district, United Verde mine, as a thin layer of high-grade ore in gossan immediately above the sulfide ore body (Anderson and Creasey, 1958). Bradshaw Mountains, at several properties, for example the Dos Oris mine, Hassayampa district, with acanthite and chlorargyrite. Big Bug district, Arizona-National mine, as wire silver in cavities with acanthite (Anderson and Creasey, 1958). Goodwin properties, Turkey Creek district, with chlorargyrite. Thunderbolt mine, Black Canyon district, with

proustite. Tip Top mine, Tip Top district, with ruby silver and chlorargyrite (Lindgren, 1926). Prescott area, Stonewall Jackson mine, in siderite gangue with acanthite and chlorargyrite (Guild, 1917).

## SKUTTERUDITE and NICKEL–SKUTTERUDITE (CHLOANTHITE)

Cobalt nickel arsenide, $(Co,Ni)As_{2-3}$. A mineral series which varies in composition between cobalt to nickel end-members; most representatives are arsenic deficient. The name skutterudite is applied to the cobalt end-member and the nickel end-member is termed either nickel-skutterudite or chloanthite. Iron is a common substitutional element in the mineral series. Found in veins of moderate temperature of formation with other cobalt and nickel minerals.

*Graham County:* Santa Teresa Mountains, Blue Bird mine, 15 miles west of Fort Thomas, as skutterudite.

*Maricopa County:* Wickenburg Mountains, as nickel-skutterudite, with native silver, at the Monte Cristo mine (Bastin, 1922).

## SMITHSONITE

Zinc carbonate, $ZnCO_3$. A secondary mineral commonly found in the oxidized zone of zinc deposits in limestone gangue; formed from the alteration of sphalerite and commonly associated with hemimorphite, cerussite, malachite, azurite, anglesite, and other oxidized lead and zinc minerals.

*Cochise County:* Tombstone district, as tiny rhombohedral crystals at the Toughnut mine (Butler et al., 1938) (UA 35). Turquoise district, as blue to green incrustations and crystalline masses at the Mystery and Silver Bill mines (UA 6517); San Juan mine (Trischka et al., 1929). Huachuca Mountains (UA 1430). Warren district, where it is frequently associated with siderite in the typical boxwork gossan areas (Trischka et al., 1929).

*Coconino County:* Horseshoe Mesa, Grand Canyon National Park, Grandview (Last Chance) mine, in small amounts, associated with malachite and azurite lining vugs in clay in limestone (Leicht, 1971).

*Gila County:* At the Curtin and Humphrey mine, with cerussite and anglesite. Banner district, at the 79 mine, a cuprian variety, as crystals lining vugs, associated with wulfenite, hemimorphite, and aurichalcite (Keith, 1972).

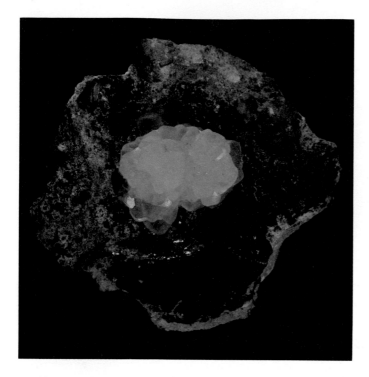

**Smithsonite.**
Silver Bill mine, near Gleeson,
Cochise County. Les Presmyck
collection. Jeff Kurtzeman.

*Graham County:* Aravaipa district, at the Head Center mine (Simons, 1964).

*Greenlee County:* Shannon Mountain, Clifton-Morenci district.

*Pima County:* Empire Mountains, at the Hilton mines. Santa Rita Mountains, Old Baldy district (Schrader, 1917). Sierrita Mountains, as "dry bone ore," Pima district; Papago district, at the Yellow Bird mine (UA 1338, 1363). Silver Bell district, as earthy mixtures of smithsonite and cerussite. Waterman Mountains, Silver Hill mine, as small rhombohedra and as pale blue botryoidal masses (similar to material from the Kelly mine, New Mexico), and also as "dry bone ore," associated with rosasite (UA 1186). Tucson Mountains, at the Thunderbird mine (UA 1710) and the Gila Monster mine (UA 2667). Cababi district, South Comobabi Mountains, at the Mildren and Steppe claims (Williams, 1963).

*Pinal County:* Mammoth district, Mammoth–St. Anthony mine, as crusts and porous masses (UA 6569), and as crude crystals, sometimes with wulfenite.

*Santa Cruz County:* Patagonia Mountains, at the Westinghouse property, Duquesne, associated with cerussite, anglesite, chrysocolla, and cuprite (Schrader and Hill, 1915; Schrader, 1917). Santa Rita Mountains, Tyndall district, at the Glove mine in Cottonwood Canyon, where it is abundant in the form of "dry bone ore" incrustations in the intermediate zone of the replacement ore bodies in limestone; rarely as gray balls (Olson, 1966).

*Yavapai County:* Eureka district, Hillside mine, in the oxidation zone of sulfide-bearing veins in mica schist, associated with cerussite, anglesite, and hemimorphite (Axelrod et al., 1951).

*Yuma County:* Trigo Mountains, associated with cerussite and yellow lead oxide in cellular to crystalline masses. Castle Dome district, where it is found in channels and vugs, associated with hydrozincite, wulfenite, vanadinite, and mimetite (Batty et al., 1947). Silver district (Wilson, 1933).

## SODA-NITER

Sodium nitrate, $NaNO_3$. A water-soluble mineral which occurs in arid environments as coatings and efflorescences on rocks and soil.

*Mohave County:* Rawhide Mountains, 3 miles south of Artillery Peak.

*Pinal County:* Superstition Mountains (UA 7838). Gila River Indian Reservation, presumed to be a minor constituent of sodium nitrate- and sodium chloride-bearing crusts in salt flats of Santa Cruz Wash (Wilson, 1969).

# SPANGOLITE

Copper aluminum sulfate hydroxide chloride hydrate, $Cu_6Al(SO_4)(OH)_{12}Cl \cdot 3H_2O$. A rare secondary mineral associated with other oxidized copper minerals in the oxidized portions of copper deposits. First described by S. L. Penfield in 1890 on a specimen from an uncertain locality thought to be in the vicinity of Tombstone. Palache and Merwin (1909), however, subsequently argued that the type specimen was from Bisbee. A specimen at Yale University (No. 3482) containing spangolite on cuprite with connellite and stated to be the type specimen from Tombstone strongly resembles known Bisbee material. A chemical analysis of the type material by Penfield gave:

| | | |
|---|---|---|
| SO$_3$  10.11 | Al$_2$O$_3$  6.60 | H$_2$O  20.41 |
| Cl    4.11 | CuO  59.51 | (O=Cl  0.92) |

TOTAL 99.82%

Specific gravity: 3.14.

**Spangolite.**
Bisbee, Cochise County. R. A. Bideaux.

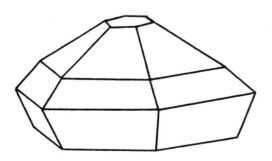

**Spangolite.**
Bisbee (Frondel, 1949).

*Cochise County:* Warren district, Czar mine, in cuprite-azurite matrix with readily identifiable spots of spangolite up to ⅛ inch in diameter (Frondel, 1949); Copper Queen mine (Ford, 1914); crystal cleavages up to an inch across are known.

*Greenlee County:* Clifton-Morenci district, as a scaly coating on chrysocolla from Metcalf (Lindgren and Hillebrand, 1904; Lindgren, 1905).

# SPESSARTINE

Manganese aluminum silicate, $Mn_3Al_2(SiO_4)_3$. A less common member of the garnet group which occurs in skarns and other manganese-rich associations of metamorphic origin; also in granite pegmatites.

*Mohave County:* Near Wright Creek ranch, 15 miles south of Peach Springs, as a salmon-pink or wine-red accessory mineral in several pegmatite dikes in schist and granite (Schaller et al., 1962).

*Pinal County:* Near Saddle Mountain, Winkelman area, as highly perfect crystals, with pseudobrookite, topaz, and bixbyite (Richard L. Jones, pers. comm.).

*Yavapai County:* In nearly all of the pegmatites of the White Picacho district, as grains up to ½ inch in diameter (Jahns, 1952).

# SPHALERITE

Zinc sulfide, ZnS. The most abundant and important ore mineral of zinc. A widespread sulfide mineral common in the base metal deposits in which it is formed under a wide range of conditions. Far more abundant and widespread than suggested by the listing here.

*Cochise County:* Warren district, Campbell mine, from which large quantities were mined, particularly during the years 1945–1949 (Trischka et al., 1929; Hewett and Rove, 1930; Schwartz and Park, 1932; Wilson et al., 1950; Bain, 1951); triboluminescent at the Junction mine (Mitchell, 1920) (S 11794). Tombstone district, Silver Thread and Sulfuret mines, less abundant elsewhere in the district (Butler et al., 1938; Rasor, 1939). Little Dragoon Mountains, in copper ores, Cochise district, Johnson Camp area (Kellogg, 1906; Wilson et al., 1950; Cooper, 1957; Cooper and Silver, 1964); Keystone and St. George deposits (Romslo, 1949). Turquoise district, as scattered bunches in pyritic ore, locally well crystallized (Wilson et al., 1951). Huachuca Mountains, State of Texas mine (Wilson et al., 1951). Dragoon Mountains, the most abundant sulfide ore mineral at the Abril mine (Perry, 1964).

*Gila County:* Banner district, as an abundant ore mineral in the lower levels of the 79 mine; as well-formed tetrahedral crystals (Kiersch, 1949; Wilson et al., 1951; Keith, 1972); Christmas mine, as a moderately common sulfide in skarn (Knoerr and Eigo, 1963; David Perry, pers. comm., 1967). Globe-Miami district, where it is sparingly but widely distributed throughout the area (Peterson, 1962).

*Graham County:* Aravaipa district, at the Iron Cap mine (UA 5829) and other properties (Denton, 1947b; Simons and Munson, 1963); Stanley district (Wilson et al., 1950).

*Greenlee County:* Clifton-Morenci district, in large quantities in the deeper levels of many of the older mines in the region (Guild, 1910; Lindgren, 1905; Reber, 1916).

*Mohave County:* Chloride district, at the Tennessee and Towne mines; Mineral Park district, Keystone mine; Cerbat Range, Cerbat district; Wallapai district (Haury, 1947); in most of the ores of the region, especially at the Vanderbilt and Flores mines. Union Pass district (Schrader, 1909; Thomas, 1949; Wilson et al., 1950). Ithaca Peak, present in veins with chalcopyrite, pyrite, and galena in a quartz monzonite stock (Eidel, 1966). Near Yucca, the most abundant sulfide in a lenticular ore body, with tremolite gangue and chalcopyrite, pyrrhotite, and minor loellingite (Rasor, 1946).

*Pima County:* In many mines and prospects of the Santa Rita and Empire Mountains (Schrader and Hill, 1915). Sierrita Mountains, Pima district, as sulfide ore bodies in limestone (Guild, 1910); Paymaster mine, Olive Camp (Ransome, 1922; Wilson et al., 1951; Nye, 1961); San Xavier mine (UA 7220). Silver Bell district (Stewart, 1912). Twin Buttes mine, near Continental, in a drill hole in the Arkose ore body (Stanley B. Keith, pers. comm., 1973).

*Pinal County:* Pioneer district, at the Belmont and Magma mines (Short et al., 1943). At the Silver King mine it was the most abundant sulfide in the ore, and cleavable masses are described as being "held together by threads of native silver" (Guild, 1917; Short et al., 1943; Wilson et al., 1950). Galiuro Mountains, Adjust, Saddle Mountain, and Little Treasure properties, Saddle Mountain district. Mammoth district, Mammoth–St. Anthony mine, on the lower levels, but extensively altered to smithsonite and hemimorphite in the oxidized zone (Peterson, 1938; Wilson et al., 1950; Fahey et

al., 1950). Vekol district, marmatitic, with pyrrhotite, malachite, chrysocolla, pyrite, and chalcopyrite at the Reward (Vekol) mine (Denton and Haury, 1946).

*Santa Cruz County:* Common in the copper and silver ores of the Santa Rita and Patagonia Mountains (Schrader and Hill, 1915; Schrader, 1917). As magnificent crystal groups at the Westinghouse property at Duquesne, where a single crystal measured nearly 2½ inches in diameter; Callaghan Lead-Zinc mine (UA 1099), and at the Indiana mine (Stanley B. Keith, pers. comm., 1973). Oro Blanco Mountains, Montana mine, with galena (Guild, 1910; Wilson et al., 1951). Idaho and Montana mines, Ruby district (Warren and Loofburrow, 1932; Anderson and Kurtz, 1955). Tyndall district, Cottonwood Canyon, Glove mine (Olson, 1966).

*Yavapai County:* In the pyritic ores of the United Verde mine (Fearing, 1926; Lausen, 1928; Moxham et al., 1965); also found at the Copper Chief mine, Verde district. Present in most of the districts in the Bradshaw Mountains. Davis mine, Hassayampa district, as an unusual golden yellow variety. Big Bug district, Iron King mine (Lindgren, 1926; Wilson et al., 1950; Creasey, 1952; Anderson and Creasey, 1958).

## Sphene (see TITANITE)

## SPINEL

Magnesium aluminum oxide, $MgAl_2O_4$. A relatively common mineral, the structure of which is assumed by several other oxides. Members of the spinel series which occur in Arizona are discussed individually. Spinel *sensu strictu* is formed under high-temperature conditions in thermal and contact metamorphic rocks.

*Apache County:* From an unspecified locality near McNary (UA 3688).

*Gila County:* Crystals of spinel, probably the variety picotite, occur in olivine volcanic bombs near Peridot.

*Maricopa and Yavapai Counties:* Found in the pegmatites of the White Picacho district as greenish-black crystals of the ferroan variety, pleonaste (Jahns, 1952).

## SPODUMENE

Lithium aluminum silicate, $LiAl(SiO_3)_2$. An important ore of lithium which occurs in lithium-bearing granite pegmatites where it is typically associated with quartz, albite, beryl, tourmaline, and lepidolite.

*Maricopa County:* In some of the pegmatites of the White Picacho district, especially the Morning Star and Midnight Owl properties (Jahns, 1952, 1953). From Mitchell Wash, northeast of Morristown (UA 5625).

## *STARKEYITE (Leonhardtite)

Magnesium sulfate hydrate, $MgSO_4 \cdot 4H_2O$. A water soluble secondary mineral forming efflorescences in mine openings.

*Pinal County:* Mammoth district, San Manuel mine, one of several secondary sulfate minerals formed in drifts; on the 2015 and 2700 foot levels with epsomite. The starkeyite may have formed by dehydration of epsomite upon removal from the humid environment of the mine. (Material collected by Joseph Urban) (Anthony and McLean, 1975.)

## STAUROLITE

Magnesium iron aluminum silicate hydroxide, $(Fe,Mg)_2Al_9Si_4O_{23}(OH)$. A relatively common mineral formed in schists during intermediate-grade metamorphism of argillaceous sediments; commonly associated with garnet, kyanite, andalusite or sillimanite, and tourmaline in mica schists.

*Coconino County:* Grand Canyon, as brownish-red, stout, prismatic crystals, with garnet in metamorphic rocks at Lone Tree Canyon (Campbell, 1936).
*Yavapai County:* Bradshaw Mountains, in schist near contacts with intrusive granite bodies; as twinned crystals from Cleator (AM 35853).

## STEIGERITE

Aluminum vanadate hydrate, $AlVO_4 \cdot 3H_2O$. A rare secondary mineral which occurs in sandstone with gypsum, corvusite, and secondary uranium and vanadium minerals.

*Apache County:* Monument Valley, where it is found with navajoite and tyuyamunite in the Monument No. 2 mine (Weeks et al., 1955; Witkind and Thaden, 1963) (AM 29436).

## *STELLERITE

Calcium aluminum silicate hydrate, $Ca(Al_2Si_7)O_{18} \cdot 7H_2O$. A member of the zeolite group closely resembling stilbite and, like it, most commonly found in cavities and vesicles in flow rocks, especially basalt.

*Cochise County:* From an unspecified locality near Douglas (UA 6337).

## STEPHANITE

Silver antimony sulfide, $Ag_5SbS_4$. A late-forming vein mineral in hydrothermal deposits where it is associated with galena, tetrahedrite, and other silver minerals.

*Gila County:* Richmond Basin district, at the Mack Morris mine, with stromeyerite (UA 1358).
*Santa Cruz County:* Patagonia Mountains, Golden Rose mine, Patagonia district (Schrader, 1917).
*Yavapai County:* Bradshaw Mountains, Tuscumbria mine, Bradshaw district.

## *STERNBERGITE

Silver iron sulfide, $AgFe_2S_3$. A rare mineral found in silver ores, with the ruby silvers.

*Cochise County:* At the Leroy mine, Dos Cabezas Mountains, as euhedral crystals up to 2 mm long enclosed in galena and sphalerite, from a silver-rich ore shoot on the 150 level.

## *STETEFELDTITE

Silver antimony oxide hydroxide, $Ag_2Sb_2(O,OH)_7(?)$. A rare secondary mineral regarded by some workers as the silver analogue of bindheimite; other authorities believe it to be a mixture.

*Yuma County:* Silver district, Trigo Mountains, Red Cloud mine as a yellow powder or crust on wulfenite (Gary M. Edson, pers. comm., 1973).

## *STEVENSITE

Magnesium silicate hydroxide, $Mg_3Si_4O_{10}(OH)_2$. A member of the montmorillonite group of minerals; of hydrothermal origin.

*Cochise County:* Warren district, from the oxide ores in the Holbrook pit; intimately intergrown with clinochrysotile.

## STEWARTITE

Manganese phosphate hydrate, $Mn_3(PO_4)_2 \cdot 4H_2O$. A rare secondary mineral formed from the alteration of primary phosphate minerals in granite pegmatites.

*Maricopa and Yavapai Counties:* White Picacho district, as numerous pale yellow films, as finely crystallized aggregates cementing microbreccias of hureaulite and lithiophilite, and as thin, sub-parallel fibers along fractures in other minerals, principally strengite and purpurite (Jahns, 1952).

## STIBICONITE

Antimony oxide hydroxide, $Sb_3O_6(OH)$. A secondary mineral formed by the oxidation of stibnite and other antimony minerals.

*Cochise County:* Warren district, from the 1,000-foot level of the Cole mine, with chalcocite (AM 28997).

*Pima County:* South Comobabi Mountains, Cababi district, Silver-Lead Claim, as a granular white powder in small pockets with lead oxides (Williams, 1962).

*Pinal County:* Vekol Mountains, associated with galena and sphalerite.

*Yavapai County:* Bradshaw Mountains, in a 2- to 3-foot vein in the vicinity of Cañon.

*Yuma County:* Dome Rock Mountains, as radiating blades of stibnite which are partly altered to cervantite and stibiconite, in veins.

## STIBNITE

Antimony sulfide, $Sb_2S_3$. Typically formed in low-temperature hydrothermal veins with other antimony minerals. The most important ore mineral of antimony.

*Gila County:* Near Payson, small amounts occur in some copper ores. Reported on Slate Creek, Mazatzal Mountains.

*Graham County:* Stanley district, in contact metamorphic ores, Cold Springs prospect (Ross, 1925a).

*Mohave County:* Cerbat Range, with galena, sphalerite, and pyrite, at the Golden Gem and Vanderbilt mines, Cerbat district (Schrader, 1909; Thomas, 1949).

*Pima County:* Pima district, from the Whitcomb property, near the Olivette mine (UA 1718). Tucson Mountains (UA 1296).

*Santa Cruz County:* Patagonia Mountains, Dura mine, Nogales district (Schrader, 1919).

*Yavapai County:* Bradshaw Mountains, Seventy-six, Swastika (S 116870), and other properties of the Tip Top district; near Tuscumbria mine, and at Malley Hill mine on Lynx Creek, Bradshaw district. Robinson property, Walker district (Lindgren, 1926). Turkey Creek district.

*Yuma County:* Dome Rock Mountains, 8 miles southwest of Quartzsite, as radiating blades, with cervantite and stibiconite.

## *STILBITE

Sodium calcium aluminum silicate hydrate, $NaCa_2Al_5Si_{13}O_{36} \cdot 14H_2O$. A widespread member of the zeolite group, most commonly found in vesicles and cavities in lavas, especially basalt.

*Gila County:* Banner district, Christmas mine, the most common zeolite in hydrothermally altered diorite, in veinlets with associated sulfides; locally replaces diorite pervasively; also as radiating clusters of prismatic crystals lining vugs in diorite; typically light pink in color (David Perry, pers. comm., 1967; 1969). Globe-Miami district, sparse in altered quartz monzonite at the Castle Dome mine (Peterson et al., 1951; Peterson, 1962).

*Pima County:* Esperanza mine, found in drill core in 1967. Duval-Sierrita open pit mine, in gypsum (UA 9588).

*Pinal County:* Tortolita Mountains, with laumontite and copper silicates in quartz veins at the Azurite mine (Bideaux et al., 1960).

*Santa Cruz County:* Patagonia Mountains, Washington Camp, on adularia (UA 8479).

## *STISHOVITE

Silicon dioxide, tetragonal $SiO_2$. A high density polymorph of $SiO_2$; notable for being the only mineral in which silicon occurs in 6-coordination with oxygen. A product of the high-pressure shock waves produced by meteor impact.

*Coconino County:* Meteor Crater, where it occurs in shock impact-metamorphosed Coconino Sandstone with coesite. This is the type locality (Chao et al., 1962; Fahey, 1964; Bohn and Stöber, 1966; Gigl and Dachille, 1968).

## STOLZITE

Lead tungstate, $PbWO_4$. A rare secondary member of the wulfenite group found in oxidized lead deposits with cerussite, wulfenite, vanadinite, mimetite, and limonite.

*Cochise County:* Dragoon Mountains, Boerich claims, and also at the Primos mine where it occurs as small, highly complex, pale yellow crystals in cavities in a quartz vein containing scheelite, chalcopyrite, sphalerite, huebnerite, fluorite, and galena (Palache, 1941a) (H 101780). Huachuca Mountains, at the Reef mine as pale yellow crystals, 1-2 mm in length, on the walls of cavities in a quartz vein, associated with scheelite, galena, chalcopyrite, pyrite, and limonite (Palache, 1941a) (H 94680). Dos Cabezas district, on the northwest flank of the Dos Cabezas Mountains, as small white tablets up to 3 mm across, in quartz veins bearing scheelite, galena, and pyrite. Unit cell parameters are $a = 5.40$, $c = 12.054$ Å.

*Gila County:* Globe-Miami district, Lost

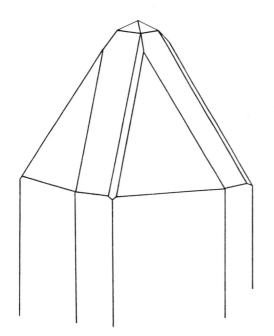

**Stolzite.**
Primos mine (Palache, 1941).

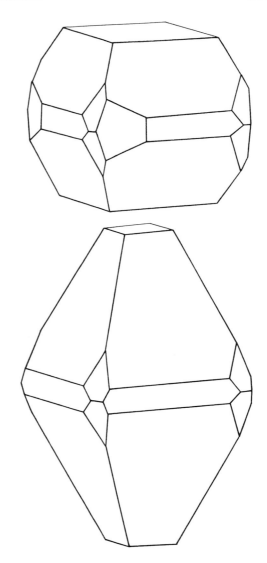

**Stolzite.**
Reef mine (Palache, 1941).

Gulch area, on the east flank of Day Peaks, as a molybdenian variety (Dale, 1961). An analysis by F. G. Hawley (Peterson, 1962) follows:

| | | |
|---|---|---|
| PbO 56.27 | As$_2$O$_5$ 4.56 | CaO 0.80 |
| WO$_3$ 25.40 | P$_2$O$_5$ 2.14 | Fe$_2$O$_3$ 0.28 |
| MoO$_3$ 8.62 | | TOTAL 98.07% |

At a locality about ¾ mile southwest of the Copper Cities mine, in cavities in quartz and disseminated in limonite, as imperfectly formed crystals, associated with scheelite (Faick and Hildebrand, 1958).

*Maricopa County:* Near the mouth of Amazon Wash, east of Wickenburg.

*Yuma County:* On the Livingstone claims, south of Quartzsite.

## STRENGITE

Iron phosphate hydrate, FePO$_4$·2H$_2$O. A member of the variscite group; forms a solid solution series with the aluminum end-member, variscite. Occurs under surface or near-surface conditions as a product of the alteration of iron-bearing phosphate minerals.

*Maricopa and Yavapai Counties:* White Picacho district, as crusts and cavity fillings which usually appear as small felted aggregates with distinctive pinkish-lavender or deep red color, associated with purpurite (Jahns, 1952).

## *STRINGHAMITE

Calcium copper silicate hydrate, CaCuSiO$_4$·2H$_2$O. Probably a late stage hydrothermal mineral and known from only a few localities.

*Pinal County:* Occurs in the open pit mine at Christmas, Banner district, in tactite ores with kinoite and apophyllite. The crystals are a distinctive lavender color ("cornflower blue"), but minute, and invariably occur only on fractures in retrogressively altered tactites formerly rich in wollastonite.

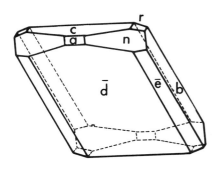

**Stringhamite.**
Christmas mine, Gila County.

## STROMEYERITE

Silver copper sulfide, AgCuS. An uncommon secondary mineral found in zones of sulfide enrichment; associated with argentian tetrahedrite and bornite with which it is frequently intimately intergrown.

*Cochise County:* Tombstone district (Romslo and Ravitz, 1947), where it was probably an important source of silver in the ores of the Empire and Toughnut mines (Guild, 1917; Butler et al., 1938). Warren district, Campbell mine (Schwartz and Park, 1932), and at the Cole mine (Bideaux et al., 1960).

*Gila County:* Globe district, in the ores of the Old Dominion mine; also reported from the Mack Morris mine, Richmond Basin.

*Pima County:* Cerro Colorado Mountains, Cerro Colorado (Heintzelman) mine, associated with tetrahedrite and native silver. South Comobabi Mountains, Cababi district, on the Mildren and Steppe claims (Williams, 1963).

*Pinal County:* Silver King mine, Pioneer district, where it was the most important silver mineral of the ores (Guild, 1910, 1917); found sparingly in the hypogene ores at the Magma mine (Short et al., 1943; Hammer and Peterson, 1968) (H 108641). Galiuro Mountains, associated with tennantite in the lower levels of the Blue Bird mine (Kuhn, 1951).

*Santa Cruz County:* Patagonia Mountains, Tyndall district, at the Ivanhoe mine near Squaw Gulch, associated with proustite, polybasite, native silver, and galena (*Engr. Min. Jour.*, 1912).

## *STRONTIANITE

Strontium carbonate, $SrCO_3$. A low temperature hydrothermal mineral in veins with sulfides as a gangue mineral and with barite, calcite, and celestite in veins in limestone.

*Mohave County:* From a locality ten miles south of Lake Mead City (Robert O'Haire, pers. comm., 1972).

## SULFUR

Sulfur, S. Occurs in a variety of ways: as a result of volcanic activity; in the oxidized portions of sulfide deposits; as a product of mine fire activity; as a result of reduction of gypsum and other sulfates; and from the partial oxidation of pyrite under special conditions as from pyritic waste in old mine dumps.

*Apache County:* Monument Valley, Monument No. 2 mine, as rare crystals in silicified wood (Jensen, 1958; Witkind and Thaden, 1963).

*Cochise County:* Tombstone district, as resinous amber-yellow material somewhat resembling sphalerite, in small crystals replacing anglesite and galena in the Skipjack shaft fissure on the fourth level of the Empire mine (Butler et al., 1938). Warren district, Mary Jo mine; also as small crystals with sphalerite from the 1,500-foot level of the Junction mine; in the Lavender pit, in cavities left by leaching of pyrite.

*Coconino County:* San Francisco Mountains, in small amounts at Sunset Crater and other nearby craters (Guild, 1910).

*Gila County:* Globe-Miami district, Castle Dome mine, as small well-formed crystals in open spaces in veins, formed as a result of the oxidation of galena and sphalerite (Peterson et al., 1951; Peterson, 1962).

*Maricopa County:* As crystals in cavities in quartz, as the result of the decomposition of pyrite, on the Surprise claims, northeast of Morristown.

*Pinal County:* About 2½ miles east of Winkelman, as tiny crystals lining small vugs in a quartz vein. Mammoth district, in small amounts in the oxidized zone at the Mammoth–St. Anthony mine.

*Yavapai County:* United Verde mine, deposited under solfataric conditions caused by the burning of a portion of the pyritic ore body. A variety containing arsenic and selenium (jeromite) occurs as an amorphous black globular coating on rock fragments below iron hoods placed over vents in the burning ore body (Anderson, 1927; Lausen, 1928; Hutton, 1959).

## *SVANBERGITE

Strontium aluminum phosphate sulfate hydroxide, $SrAl_3PO_4SO_4(OH)_6$. A member of the beudantite group, found in aluminous schists; also formed as a product of hydrothermal alteration of igneous rocks.

*Pima County:* Cababi district, South Comobabi Mountains, on the Mildren and Steppe claims (Williams, 1963).
*Yuma County:* In pyrophyllite with hematite at an unspecified locality near Quartzsite (UA 9725; collected by Richard L. Jones).

## SWARTZITE

Calcium magnesium uranyl carbonate hydrate, $CaMg(UO_2)(CO_3)_3 \cdot 12H_2O$. A rare secondary mineral formed as efflorescences on the walls of mine workings. The Hillside mine is the type locality.

*Yavapai County:* With gypsum, schroeckingerite, bayleyite, andersonite, johannite, and uraninite, as an efflorescence on walls in the Hillside mine (Axelrod et al., 1951). A chemical analysis by F. S. Grimaldi on 0.25 gram gave the following result:

| | | | |
|------|-------|--------|-------|
| MgO | 5.24 | UO$_3$ 37.19 | H$_2$O 29.31 |
| CaO | 8.40 | CO$_2$ 17.16 | Acid |
| Na$_2$O | 0.25 | SO$_3$ 1.98 | Insol., } 0.30 |
| K$_2$O | 0.47 | | Ignited |

TOTAL 100.30%

Specific gravity: 2.3.

## *SYLVITE

Potassium chloride, $KCl$. A member of the halite group with which it commonly occurs as a widespread precipitate from oceanic waters.

*Apache and Navajo Counties:* East-central Arizona, encountered in drill holes which delineate a northeast-trending potash zone beneath an area of about 300 square miles, in Permian evaporites (Peirce, 1969). The log of a hole drilled at Sec. 24, T.18N, R.25E, showed, in addition, carnallite, polyhalite, halite, anhydrite, and gypsum (H. Wesley Peirce, pers. comm., 1972).

## *SZOMOLNOKITE

Iron sulfate hydrate, $FeSO_4 \cdot H_2O$. A water-soluble secondary mineral precipitated from highly acidic solutions derived from the breakdown of pyrite; associated with other sulfates.

*Cochise County:* Warren district, Lavender pit, as brown, warty crusts enclosing corroded pyrite grains, especially in intensely pyritized areas.

# T

## TAENITE

Nickel iron, gamma-iron (face-centered cubic structure) containing variable amounts (from about 27 to 65%) of nickel. With kamacite, a major constituent of the iron meteorites of the state.

## TALC

Magnesium silicate hydroxide, $Mg_3Si_4O_{10}(OH)_2$. A layer-structured, mica-like mineral, talc is formed during low-grade metamorphism of siliceous dolomites and as a product of hydrothermal alteration of ultramafic rocks. Commonly associated with serpentine.

*Apache County:* Monument Valley, Garnet Ridge, where it is found in mineral fragments as well as in a breccia dike piercing sedimentary rocks (Gavasci and Kerr, 1968).
*Cochise County:* Gunnison Hills, as an alteration product of silty dolomite pebbles in Glance Conglomerate (Cooper and Silver, 1964). In contact metamorphosed dolomites around the Texas Canyon stock, near Johnson Camp, Little Dragoon Mountains (Cooper, 1957).
*Gila County:* Banner district, Christmas mine, associated with tremolite and phlogopite in diopside hornfels in the footwall of the lower Martin ore body (Perry, 1969).
*Greenlee County:* Clifton-Morenci district, of local occurrence with other calc-silicate minerals in an extensive contact metamorphic assemblage in the northern part of the Morenci open pit area (Moolick and Durek, 1966).
*Maricopa County:* As massive green, waxy material in pegmatite, about 10 miles south of Wickenburg.

*Pinal County:* Campo Bonito area, Santa Catalina Mountains (UA 8842).

*Yavapai County:* Relatively pure material is reported to occur in appreciable amounts in the Eureka district.

## TANTALITE
### (see COLUMBITE-TANTALITE)

## TAPIOLITE

Iron tantalum niobium oxide, $Fe(Ta,Nb)_2O_6$. A rare mineral which occurs in granite pegmatites and as a detrital mineral derived therefrom.

*Yavapai County:* Bradshaw Mountains, in stream gravels on Castle Creek.

## *TEINEITE

Copper tellurite hydrate, $Cu(Te,S)O_3 \cdot 2H_2O$. A very rare secondary mineral previously reported only from the Teine mine, Hokkaido, Japan, where it is associated with native tellurium and sylvanite.

*Cochise County:* Warren district, 1200 level of the Cole shaft, where it is associated with cuprite, malachite, and graemite in a specimen collected from an ore car in 1959. One teineite crystal at least 20 mm long was totally replaced by graemite (Williams and Matter, 1975). Analysis gave:

CuO 27.4      $TeO_2$ 56.2      $H_2O$ 16.7
                                    TOTAL 100.3%

## TELLURIUM

Tellurium, Te. The native element is of rare occurrence in nature; found in hydrothermal quartz veins with native gold and telluride minerals.

*Cochise County:* Tombstone district, as microscopic blebs in galena (Butler et al., 1938); with mackayite and emmonsite (Bideaux et al., 1960).

*Yuma County:* Granite Wash Hills, 4 miles north of Vicksburg.

## TENNANTITE

Copper iron arsenic sulfide, $(Cu,Fe)_{12}As_4S_{13}$. A common sulfosalt mineral of widespread occurrence, in hydrothermal vein deposits with other sulfosalts and sulfides. Tennantite forms a complete solid solution series with the antimony end-member, tetrahedrite (which see).

*Cochise County:* Warren district, Campbell mine, with tetrahedrite, chalcocite, bornite, enargite, and famatinite in Escabrosa Limestone, probably of hypogene origin (Schwartz and Park, 1932).

*Gila County:* Richmond Basin, Helene vein, silver-bearing. East slope of the Mazatzal Mountains, at the Ord mine as the mercurian variety (Faick, 1958). An analysis by R. E. Stiles gave:

| | | | |
|---|---|---|---|
| Hg 16.8 | As 9.6 | Fe | 3.7 |
| Cu 31.8 | Sb 6.8 | MgO | 1.1 |
| S 20.0 | | Insol. | 3.5 |
| | | TOTAL | 93.3% |

(Recalculation assuming Fe and MgO to occur as siderite contaminant gives a total approaching 100 percent.) Globe-Miami district, Miami mine, with aikinite and enargite in veinlets cutting chalcopyrite (Legge, 1939).

*Mohave County:* Cerbat Mountains, a common constituent of the high grade silver ores, commonly associated with proustite (Bastin, 1925; Thomas, 1949).

*Pima County:* Ajo district, rare (Gilluly, 1937). As a minor primary mineral at the Pima open pit mine, Pima district (Himes, 1972).

*Pinal County:* Tortilla Mountains, Florence Lead-Silver mine (Williams and Anthony, 1970). Galiuro Mountains, Copper Creek district, Childs-Aldwinkle mine, with bornite and chalcopyrite (UA 543).

*Yavapai County:* Wickenburg Mountains, argentiferous, with tetrahedrite, enargite, nickeline, and native silver, Monte Cristo mine (Bastin, 1922). Verde district, United Verde mine (Anderson and Creasey, 1958). An analysis gave:

(1)

| | | | |
|---|---|---|---|
| Cu 41.6 | Sb tr. | Ag (oz. | |
| Fe 0.3 | As 17.4 | per ton) | 38.08 |
| Zn 10.9 | S 27.3 | Au (oz. | |
| | | per ton) | 0.075 |

(2)

| | | | |
|---|---|---|---|
| Cu 36.95 | As 16.0 | Ag (oz. | |
| Sb 2.2 | S 24.4 | per ton) | 31.76 |
| | | Au (oz. | |
| | | per ton) | 0.14 |

Big Bug district, Iron King mine (Creasey, 1952).

## TENORITE

Cuprous oxide, CuO. A relatively common mineral of secondary origin found in the oxidized zones of copper deposits, characteristically with chrysocolla, malachite, cuprite, limonite, and hematite. Only a few representative localities can be listed.

*Cochise County:* Warren district, as earthy material mixed with manganese oxides (Ransome, 1904; Schwartz, 1934; Frondel, 1941). Dragoon Mountains, Maid of Sunshine and other mines of the Turquoise district.

*Gila County:* Globe-Miami district, sparingly in oxidized copper ores (Schwartz, 1921, 1934); Van Dyke mine, with chrysocolla (Peterson, 1962). Banner district, 79 mine, where it is associated with malachite and chrysocolla (Keith, 1972).

*Greenlee County:* Clifton-Morenci district.

*Mohave County:* Kingman district, Emerald Isle mine, with chrysocolla.

*Pima County:* Santa Rita Mountains, Rosemont area. Pima district, present in the oxidized portions of the Twin Buttes open pit mine (Stanley B. Keith, pers. comm., 1973).

*Pinal County:* Mammoth district, Mammoth–St. Anthony mine, as coal-black nodules surrounded by thin shells of chrysocolla (Peterson, 1938); associated in lesser quantities with the more abundant chrysocolla, cuprite, malachite, and native copper in the oxidized portions of the San Manuel ore body (Thomas, 1966). Mineral Creek district, with chrysocolla and malachite, cementing part of the White Tail Conglomerate south of Ray (Phelps, 1946; Clarke, 1953); found in Holocene gravels with jarosite, goethite, malachite, and azurite (Phillips et al., 1971).

*Yavapai County:* Walnut Grove district, Zonia Copper mine, with malachite, cuprite, and chrysocolla (Kumke, 1947).

*Yuma County:* At the Success mine near Quartzsite, altering from chalcopyrite (UA 6898).

## *TEPHROITE

Manganese silicate, $Mn_2SiO_4$. A member of the olivine group. Of metamorphic origin, in iron-manganese ore deposits and in their associated skarns.

*Mohave County:* Near Alamo Crossing (AM 34347).

## TETRADYMITE

Bismuth tellurium sulfide, $Bi_2Te_2S$. An uncommon mineral found in gold-quartz veins and in contact metamorphic deposits associated with tellurides and sulfides.

*Cochise County:* As an exsolution phenomenon in galena of the tungsten veins in the Texas Canyon stock (Cooper and Silver, 1964). Precious Chemical Company property, Little Dragoon Mountains (UA 6323).

*Yavapai County:* Bradshaw Mountains, in small quantities at the Montgomery mine (Guild, 1910). Two miles south of Bradshaw City, as bladed crystals in quartz, associated with pyrite (Genth, 1890).

*Yuma County:* Reported from near Vicksburg, but the exact locality is not known.

## TETRAHEDRITE

Copper iron antimony sulfide, $(Cu,Fe)_{12}Sb_4S_{13}$. A common sulfosalt mineral of widespread occurrence in hydrothermal veins with other sulfosalts and sulfides. Forms a complete solid solution series with the arsenic end-member, tennantite (which see).

*Cochise County:* Tombstone district, notably at the Toughnut (UA 36), Lucky Cuss, and Ingersol mines, but present in most of the ores of the district, and containing silver (Butler et al., 1938; Rasor, 1939; Romslo and Ravitz, 1947). Warren district, Cole mine (S 115256). Pearce Hills, at the Commonwealth mine, with proustite. Cochise district, Johnson Camp area, in small quartz veins at the Republic mine (Cooper and Silver, 1964), and in fissure veins of the Moore ore body (Cooper and Huff, 1951). Chiricahua Mountains, Humboldt mine, where it is argentiferous and associated with alabandite and other sulfides, rhodonite, and rhodochrosite (Hewett and Rove, 1930).

*Gila County:* Globe district, as crystals in cavities in the Old Dominion mine (Lausen, 1923). Payson district, the principal ore mineral at the Silver Butte mine where it reportedly carried considerable amounts of silver (Lausen and Wilson, 1927). Banner district, identified on the dump of the 79 mine (Keith, 1972).

*Graham County:* Aravaipa district, Grand Reef mine.

*Mohave County:* Reported from the Hualpai Mountains, with galena (UA 148).

*Pima County:* In the Santa Rita Mountains, at the Silver Spur and Busterville mines in the Helvetia-Rosemont district, and at the Summit mine, Greaterville district (Schrader and Hill, 1915). Pima district, at the Helmet Peak mine (Ransome, 1922), and the Paymaster mine (Waller, 1960; Nye, 1961). Cerro Colorado Mountains, Cerro Colorado (Heintzelman) mine, with stromeyerite and native silver. Tucson Mountains, very rare as spots on galena (Guild, 1917).

*Pinal County:* Pioneer district, at the Silver King mine (Guild, 1917); assays showed up to 3,000 ounces of silver per ton, but much of the value was thought to have been due to undetected stromeyerite. Abundant in the Magma mine below the 900-foot level (Short et al., 1943). Galiuro Mountains, at the Blue Bird and Childs-Aldwinkle mines, Copper Creek district, associated with stromeyerite (Kuhn, 1938; Simons, 1964).

*Santa Cruz County:* Santa Rita Mountains, Tyndall district, at the Alta, Treasure Vault, and other mines; Wrightson district, as fine crystals at American Boy mine (Schrader and Hill, 1915). Also at several localities in the Patagonia Mountains, including the World's Fair mine where it is argentiferous (UA 152, 7340). Ruby district, at the Montana and Idaho mines (Warren and Loofburrow, 1932); analyses by F. E. Gregory and R. W. Loofburrow on 100-mg samples yielded the following results:

Idaho

| | | | | | |
|---|---|---|---|---|---|
| S | 24.90 | Zn | 8.52 | Sb | 22.08 |
| Ag | 1.21 | As | 4.50 | Insol. | 3.50 |
| Cu | 32.40 | | | TOTAL | 97.11% |

Montana

| | | | | | |
|---|---|---|---|---|---|
| S | 24.63 | Cu | 30.87 | Fe | 1.45 |
| Ag | 4.00 | Zn | 7.60 | As | 2.52 |
| Pb | 2.48 | | | Sb | 25.83 |
| | | | | TOTAL | 99.38% |

Also at the Warsaw property (S 118120).

*Yavapai County:* Wickenburg Mountains, with enargite, argentiferous tennantite, nickeline, and native silver at the Monte Cristo mine (Bastin, 1922). Black Hills, at the United Verde, Shea, Yeager, and Shylock mines (Anderson and Creasey, 1958). Bradshaw Mountains, at several properties in the Walker, Hassayampa, Agua Fria, and other districts of the range (Lindgren, 1926). A partial analysis of tetrahedrite from the Shea mine, Verde district, follows (Anderson and Creasey, 1958):

| | | | | | |
|---|---|---|---|---|---|
| Cu | 31.7 | As | 2.1 | Ag (oz. | |
| Fe | 3.3 | S | 22.8 | per ton) | 583.57 |
| Sb | 23.4 | | | Au (oz. | |
| | | | | per ton) | 0.21 |

Genth (1868) reported a "Fahlerz-like" mineral associated with quartz at the Goodwin mine, near Prescott.

# THALENITE

Yttrium silicate hydroxide, $Y_3Si_3O_{10}(OH)$. A rare mineral known to occur in granite pegmatites.

*Mohave County:* From a prospect on claims formerly known as the Guy Hazen group, located about 5 miles southwest of Kingman in $NW\frac{1}{4}$ sec. 15, T.20N, R.17W (A. Pabst, pers. comm., 1974); associated with microcline and cyrtolite (metamict zircon) in a quartz outcrop which is part of a weathered pegmatite in coarse granitic rock, as rough grayish to flesh-colored crystals up to several centimeters in length that are bleached white to a depth of up to 3 mm. The specific gravity is 4.41 (Pabst and Woodhouse, 1964).

# THAUMASITE

Calcium silicate hydroxide carbonate sulfate hydrate, $Ca_3Si(OH)_6(CO_3)(SO_4) \cdot 12H_2O$.

*Cochise County:* Tombstone district, Lucky Cuss mine, in small fissures and replacing limestone (Butler et al., 1938; Schaller, 1939) (UA 23, 615).

# THENARDITE

Sodium sulfate, $Na_2SO_4$. Of widespread occurrence in the arid southwest where it may be prevalent in dry lakes and playas; also formed as white incrustations on lavas. An ephemeral mineral which is quite common as white crusts on dark rocks, especially in the more arid parts of the state. It is more abundant in Arizona than the few localities noted would suggest.

*Cochise County:* San Simon Basin, about 7 miles northeast of Bowie, in tuff-bearing bedded lake deposits, associated with a variety of zeolite minerals (Regis and Sand, 1967).

*Coconino County:* As tufts in ice caves and as coatings on basaltic lavas near Sunset Crater.

*Pima County:* As ephemeral, thin, spotty crusts on volcanic rocks of the Tucson Mountains.

*Pinal County:* Reported from near Maricopa.

*Yavapai County:* In salt deposits of the Verde Valley lake beds; 3 miles southwest of Camp Verde, with halite, mirabilite, glauberite (Silliman, 1881; Blake, 1890; Guild, 1910; Peirce, 1969). An analysis by George M. Dunham gave:

| | | | | | |
|---|---|---|---|---|---|
| Cl | 0.095 | CaO | 0.12 | Na₂O | (42.96) |
| SO₃ | 56.41 | MgO | 0.12 | Insol. | 0.39 |
| | | | | TOTAL | 100.10% |

Specific gravity: 2.681.

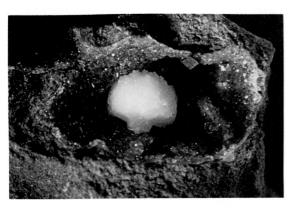

**Thomsonite.**
Superior, Pinal County. Julius Weber.

## THOMSONITE

Sodium calcium aluminum silicate hydrate, $NaCa_2(Al_5Si_5)O_{20} \cdot 6H_2O$. A member of the zeolite group, thomsonite typically occurs in cavities and vesicles in mafic igneous rocks, especially in amygdaloidal basalts.

*Cochise County:* Warren district.
*Mohave County:* San Francisco district, near Oatman, in amygdules in vesicular basalt (UA 8515, 8516).
*Pima County:* Mount Fagan, Santa Rita Mountains, with pennine, pumpellyite, prehnite, epidote, and copper (Bideaux et al., 1960).
*Pinal County:* As radiating fibrous material in amygdules in basalts, just east of the highway about midway between Tucson and Florence.

## *THORITE

Thorium silicate, $ThSiO_4$. Commonly containing substantial amounts of uranium in substitution for thorium. A primary mineral most commonly found in pegmatites.

*Mohave County:* Uranium Basin claims, as the variety uranothorite; erratically distributed in a vein at the contact between granite and pegmatite (Adams and Staatz, 1969).
*Yavapai County:* Fairview claims, south side of the Verde River, in small amounts in veinlets cutting stockwork in green metavolcanic rocks (Olson and Adams, 1962).
*Yuma County:* Scott Lode, near Quartzsite, in small amounts in a quartz vein cutting a biotite schist (Adams and Staatz, 1969).

## Thulite (a variety of ZOISITE)

## *THURINGITE

Iron magnesium aluminum silicate hydroxide, $(Fe^{2+},Fe^{3+},Mg)_6(Al,Si)_4O_{10}(OH)_8$. A member of the chlorite group.

*Yavapai County:* Verde district, United Verde mine area, where it is present as a constituent of the "black schist," a chlorite-rich aggregate which has replaced rocks in the region (Anderson and Creasey, 1958).

## *TILASITE

Calcium magnesium arsenate fluoride, $CaMg(AsO_4)F$. A rare secondary mineral which occurs in veins in limestone and dolomitic limestones; associated with braunite or hausmannite. The Bisbee discovery constitutes the third world occurrence.

*Cochise County:* Warren district, in outcrop near the White Tail Deer mine, as complex crystals up to 6 mm in size, with braunite, conichalcite, and calcite in veinlets in clean crystalline limestone. (Williams, 1970; Bladh et al., 1972) (UA 9679).

## *TINZENITE

Calcium manganese iron aluminum borosilicate hydroxide, $(Ca,Mn,Fe)_3Al_2(BO_3)(SiO_3)_3(OH)$. A member of the axinite group. Most common occurrence is in contact metamorphic aureoles formed where intrusive rocks have invaded sediments, especially limestones; commonly associated with other calcium silicate minerals.

*Cochise County:* Huachuca Mountains, as yellow cleavage plates in massive quartz (H 95332).

## Titanandalusite
## (a variety of ANDALUSITE)

## TITANITE (sphene)

Calcium titanium silicate, $CaTiSiO_5$. A common accessory of igneous rocks; also found in pegmatites and skarn.

*Cochise County:* Tombstone district, as microscopic crystals in granodiorite and porphyritic rocks (Butler et al., 1938). With wollastonite, grossular, epidote, and other contact silicate minerals in the calcareous rocks near Johnson Camp, Little Dragoon Mountains (Cooper, 1957). Warren district, where it is a common accessory mineral in the Juniper Flat Granite, sometimes as well-formed crystals in vugs with chlorite and quartz.

*Gila County:* Globe-Miami district, locally abundant in the Willow Spring Granodiorite (Peterson, 1962); Castle Dome mine, as a minor accessory mineral in quartz monzonite (Peterson et al., 1946).

*Graham County:* Aravaipa district, Mount Turnbull, in micro-pegmatite at the Fisher prospect.

*Greenlee County:* Morenci district, an accessory mineral in quartz monzonite, associated with zircon (Schwartz, 1947).

*Mohave County:* Aquarius Mountains, as small euhedral and long anhedral grains in pegmatite in a granite dike, associated with monazite, apatite, cronstedtite, chevkinite, and quartz; cross-cutting and apparently replacing chevkinite (Kauffman and Jaffe, 1946). Cerbat Mountains, Wallapai district, of common occurrence as an alteration product in wall rocks associated with sulfide vein deposits, as long stringers and lenses replacing muscovite (Thomas, 1949). Near Wright Creek ranch, Blue Bird (Bountiful Beryl) prospect, in several irregular pegmatite dikes in schist and granite (Schaller et al., 1962).

*Pima County:* Linda Lee claim, east flank of the Quijotoa Mountains, as tiny grains in spessartine-actinolite rock and in quartz monzonite (Williams, 1960). Ajo district, a primary constituent of all the igneous rocks of the mining district (Gilluly, 1937). An accessory mineral in the rocks of the Silver Bell area (Kerr, 1951).

*Yavapai County:* Bradshaw Mountains, in schist at the Butternut mine, Big Bug district; as large crystals in granodiorite on the Springfield group of claims, Pine Grove district.

## *TODOROKITE

Manganese calcium magnesium manganese oxide hydrate, $(Mn,Ca,Mg)Mn_3^{4+}O_7 \cdot H_2O$. A rare secondary mineral formed from the alteration of other manganese-bearing minerals.

*Cochise County:* Turquoise district, from prospect pits north of Courtland, intimately mixed with birnessite; the mixtures form films on fractures in altered quartz monzonite.

## TOPAZ

Aluminum silicate fluoride hydroxide, $Al_2SiO_4(F,OH)_2$. A mineral of widespread occurrence, topaz is typically associated with granites, pegmatites, and rhyolites, and also occurs in greisen and pneumatolytic bodies.

*Cochise County:* Dos Cabezas Mountains, as masses weighing up to several hundred pounds, with quartz and muscovite in coarse-grained granite, 6 miles east-southeast of Dos Cabezas and ¾ mile southeast of the Cottonwood mine, on the William DeBorde property.

*Mohave County:* Aquarius Range, south of the Rare Metals mine, in pegmatites as euhedral crystals up to 2½ inches long (Heinrich, 1960).

*Pima County:* Ajo district, a rare mineral found in the New Cornelia Quartz Monzonite west of the Gibson Arroyo fault (Gilluly, 1937).

*Pinal County:* Near Saddle Mountain, Winkelman area, in rhyolite with pseudobrookite, spessartine, and bixbyite (Richard L. Jones, pers. comm.).

## TORBERNITE

Copper uranyl phosphate hydrate, $Cu(UO_2)_2(PO_4)_2 \cdot 8\text{-}12H_2O$. A member of the autunite group of layer-structure minerals. The water content is variable between 8 and 12; desiccation will produce the lower hydrated form, metatorbernite, and the inversion may or may not be reversible. A secondary mineral which occurs in the oxidized portions of uranium deposits; commonly associated with metatorbernite, metazeunerite, autunite, and other secondary uranium minerals.

*Coconino County:* Cameron district, at a number of properties including the Huskon claims; Grand Canyon National Park, west of Maricopa Point, in a shear zone in Coconino Sandstone.

*Gila County:* Sierra Ancha Mountains, with bassetite and coffinite in the Dripping Spring Quartzite.

*Mohave County:* Hack Canyon mine, associated with metatorbernite and tyuyamunite.

*Navajo County:* Monument Valley, Monument No. 1 mine (Witkind and Thaden, 1963); Mitten No. 2 mine (Witkind, 1961).

*Pima County:* Reported to occur on native copper in the New Cornelia open pit mine at Ajo. Linda Lee claims, with gummite and hematite in a vein in arkose (Robinson, 1955). Silver Bell district, in the Oxide Pit of the Silver Bell mine as transparent, emerald-green, euhedral tabular crystals to 3 mm; readily alters to apple-green, opaque metatorbernite on exposure to air. Associated with turquoise(?) crystals and cacoxenite, both of which grow on torbernite (Kenneth W. Bladh, pers. comm., 1974).

*Pinal County:* Mineral Creek district, from a locality one mile south of Kelvin, where it is associated with turquoise (Charles L. Fair, pers. comm., 1975).

*Yavapai County:* Eureka district, in small amounts with bermanite and triplite in a pegmatite on the 7U7 ranch (Hurlbut, 1936).

*Yuma County:* Castle Dome Mountains (UA 6162).

## TOURMALINE

A group of minerals having complex compositions which may be represented generally by the formula $(Na,Ca)(Mg,Fe^{2+},Fe^{3+},Al,Li)_3$-$Al_6(BO_3)_3Si_6O_{18}(OH)_4$. A number of species and varieties of tourmaline are known which usually differ in color. A mineral of widespread occurrence, tourmaline is found in granite pegmatites, pneumatolytic veins, and granites. It also occurs as a product of metasomatism involving boron. Its resistance to mechanical and chemical weathering results in its accumulation in sands and in some sedimentary rocks. By far the most abundant tourmaline variety is the black, iron-bearing schorl. Unless indicated to the contrary, the listings below refer to that variety.

*Apache and Navajo Counties:* Common as a heavy mineral in the basal member of the Monitor Butte Sandstone, associated with apatite, leucoxene, zircon (Young, 1964).

*Cochise County:* Warren district, as nests of small, prismatic crystals in muscovite, northwest of Bisbee. Johnny Lyon Hills, as a common mineral at the contact zone of the Johnny Lyon Granodiorite (Cooper and Silver, 1964).

*Coconino County:* Grand Canyon, as black crystals in pegmatites at Hermit Creek.

*Gila County:* Globe-Miami district, a widespread constituent of the Pinal Schist, as a bluish black variety (Peterson, 1962).

*Maricopa County:* Mazatzal Mountains, in cinnabar veins (Lausen, 1926). In the pegmatites of the White Picacho district, as schorl and elbaite, some of the latter being of the colorful "watermelon" variety (Jahns, 1952). Mummy Mountain, northeast of Phoenix.

*Mohave County:* Cottonwood district, near the Wright Creek ranch, in several irregular pegmatite dikes in schist and granite, as large, podlike masses enclosing crystals of microcline, quartz, albite, fluorite, and sphene; also as abundant well-formed, prismatic crystals in the cores of pegmatites (Schaller et al., 1962). In small amounts with beryl in the pegmatites of the Cerbat Mountains (Thomas, 1953).

*Pima County:* Santa Rita Mountains, with magnetite and siderite at the Iron Mask mine, Old Baldy district (Schrader, 1917). Sierrita Mountains, as veinlike masses with quartz in granite, Papago district. Widely distributed in the Little Ajo Mountains, as brilliant black prisms coating joint planes in the Cardigan Gneiss and other rocks (Gilluly, 1937). Sierra Blanca Mountains (UA 2465). North end of the Baboquivari Mountains (UA 2470).

*Pinal County:* Santa Catalina Mountains, in pegmatites in the Oracle granite (Guild, 1910; Ward, 1931). In vein quartz in the Pinal Schist near post-Cambrian granite contacts (Ransome, 1919). Galiuro Mountains, Copper Creek district, as radiating groups of slender prismatic crystals, notably in the breccia pipe of the American Eagle mine (Simons, 1964); Childs-Aldwinkle mine, as a gangue mineral in the breccia pipe deposit (Kuhn, 1941), and at Copper Creek Canyon, as elbaite and gray to green crystals in masses and small fan-shaped aggregates enclosed in quartz crystals, some of which are twinned after the Japanese law (William and Mildred Schupp, pers. comm.; Gary Edson, pers. comm., 1972).

*Santa Cruz County:* Present in considerable amounts in contact metamorphic deposits in the Patagonia district in the Duquesne and Washington Camp areas; Tyndall district (Schrader, 1917).

*Yavapai County:* Bradshaw Mountains, in pegmatites of the Bradshaw granite, in lenses in schist and scattered through the schist near granite contacts. In veins of the Prescott district and as blue-gray prisms in quartz and dolomite at the Iron King mine, Big Bug district (Lindgren, 1926; Creasey, 1952). Eureka district, in pegmatites of the Bagdad area (Anderson, 1950).

*Yuma County:* Dome Rock Mountains, with magnetite and siderite in gangue in cinnabar veins. Plomosa Mountains, as black, massive material associated with scheelite and quartz at the Night Hawk and White Dike mines, 5.7 miles south of Quartzsite (Dale, 1959).

## TREMOLITE

Calcium magnesium silicate hydroxide, $Ca_2Mg_5Si_8O_{22}(OH)_2$. A member of the amphibole group; typically occurs in contact metamorphosed limestones and dolomites with other calcium silicate minerals. In Arizona, a common gangue mineral associated with contact metamorphic base metal deposits.

*Cochise County:* Warren district, as an abundant gangue mineral of the unoxidized pyritic ores. Tombstone district, as long, fibrous masses at the Toughnut mine (Butler et al., 1938). Little Dragoon Mountains, in the metamorphosed dolomites (Cooper and Huff, 1951; Cooper, 1957; Cooper and Silver, 1964).

*Gila County:* Banner district, Christmas mine, associated with diopside, a prominent gangue mineral formed by contact metamorphism of dolomitic rocks (Peterson and Swanson, 1956; Perry, 1969); 79 mine, with andradite, diopside, and epidote in contact metamorphosed limestones (Keith, 1972).

*Greenlee County:* Clifton-Morenci district, in contact metamorphosed limestones with other calc-silicate minerals (Lindgren, 1905; Moolick and Durek, 1966).

*Maricopa County:* Incipiently replacing limestone near Aguila (UA 5907).

*Mohave County:* At the Copper World mine, near Yucca, as a gangue mineral in a lenticular copper-zinc sulfide ore body (Rasor, 1946). Hualpai Mountains, Antler mine (Romslo, 1948).

*Pima County:* Pima district, Pima mine, abundant in limestone hornfels with diopside and grossular (Journeay, 1959); in the tactites and skarns of the Twin Buttes mine (Stanley B. Keith, pers. comm., 1973); Mission mine, in the hornfels host rock (Richard and Courtright, 1959); Mineral Hill mine and other properties, associated with mineral deposits of contact metamorphic origin (Ransome, 1922). Helvetia district, in contact metamorphosed sedimentary rocks at several properties including the Copper World, Leader, Narragansett, and Isle Royal mines (Creasey and Quick, 1955).

*Santa Cruz County:* Patagonia Mountains, as gangue at the Westinghouse property (Schrader and Hill, 1915); Washington Camp and Duquesne areas, in contact metamorphic deposits in limestone (Schrader, 1917).

*Yuma County:* Dome Rock Mountains, as an asbestiform variety in marbleized limestones of the northern part of the range.

## TRIDYMITE

Silicon dioxide, $SiO_2$. A high temperature polymorph of silica which typically occurs in felsic volcanic rocks, commonly with sanidine and hornblende or augite.

*Mohave County:* Cerbat Mountains, in a tuffaceous breccia found among stratiform volcanic rocks along the western edge of the Chloride district (Thomas, 1953).

*Pima County:* Roskruge Mountains, with cristobalite and clay in cavities in andesite.

## TRIPHYLITE

Lithium iron manganese phosphate, $Li(Fe^{2+},Mn^{2+})PO_4$. A primary mineral which occurs in granite pegmatites. A complete substitutional series probably extends between triphylite and lithiophilite, $Li(Mn^{2+},Fe^{2+})PO_4$.

*Yavapai County:* White Picacho district, with lithiophilite in the Midnight Owl and other pegmatites, as equant to stubby prismatic crystals up to six inches long with rough faces (Jahns, 1952).

## TRIPLITE

Manganese iron magnesium calcium phosphate fluoride hydroxide, $(Mn,Fe,Mg,Ca)_2(PO_4)(F,OH)$. A primary mineral which occurs in phosphate-rich granite pegmatites.

*Maricopa and Yavapai Counties:* In the pegmatites of the White Picacho district, especially well crystallized in the Midnight Owl pegmatite, as rough-faced tabular crystals up to 7 inches in length (Jahns, 1952).

*Yavapai County:* Eureka district, as a spherical aggregation about 2 feet in diameter in a small pegmatite knot on the 7U7 ranch (Hurlbut, 1936; Leavens, 1967). An analysis by F. A. Gonyer is as follows:

| | | | | | |
|---|---|---|---|---|---|
| FeO | 11.68 | $Na_2O$ | 0.52 | $P_2O_5$ | 33.32 |
| MnO | 34.55 | $H_2O$ | 0.75 | F | 8.02 |
| CaO | 2.48 | | | $(-)O = F$ | 3.38 |
| MgO | 11.87 | | | TOTAL | 99.81% |

Also found at other localities in the same general area.

## TROILITE

Iron sulfide, FeS. Closely related to pyrrhotite. Known to occur in Arizona only in metallic meteorites.

*Coconino County:* Meteor Crater, as a minor constituent of metallic spheroids which were probably formed upon impact of the Canyon Diablo meteorite; intergrown with schreibersite, occurring interstitially along grain boundaries of the kamacite cores. Also associated with mag-

**Turquoise.**
Bisbee, Cochise County. University of Arizona collection. Jeff Kurtzeman.

hemite and a goethite-like iron oxide, both produced by weathering (Mead et al., 1965).

## *TSUMEBITE

Lead copper phosphate sulfate hydroxide, $Pb_2Cu(PO_4)(SO_4)(OH)$. A very rare secondary mineral formed in the oxidized portion of base metal deposits. The Arizona tsumebite probably constitutes the second and third reported world occurrences.

*Greenlee County:* Morenci, with wulfenite

**Tsumebite.**
Morenci, Greenlee County. R. A. Bideaux.

in the oxidized portion of the Clay ore body (Bideaux et al., 1960) (UA 2393, 2394, 5687).
*Pinal County:* Mammoth district, Mammoth–St. Anthony mine, very rare as yellow-green spherules, with mimetite and wulfenite.

## TUNGSTITE

Tungstic oxide hydrate, $WO_3 \cdot H_2O$. An uncommon mineral formed by the alteration of primary tungsten minerals such as wolframite and scheelite, with which it is usually associated. Reported as occurring in small amounts at a number of the tungsten districts of the state.

## TURQUOISE

Copper aluminum phosphate hydroxide hydrate, $CuAl_6(PO_4)_4(OH)_8 \cdot 5H_2O$. Turquoise is usually iron bearing and is considered to be an end-member of a series which terminates with the iron analogue, chalcosiderite (which see). A widely distributed mineral often thought to be of secondary origin; found associated with limonite and clay minerals in the upper, oxidized portions of some base metal deposits in the southwestern United States. A few representative occurrences are listed.

**Tyuyamunite.**
Monument Valley, Navajo County. Julius Weber.

*Cochise County:* Turquoise district, near Gleeson, mined at Turquoise Mountain where it occurs as stringers up to a few inches wide, and as small nugget-like masses in granite and in Bolsa Quartzite (Crawford and Johnson, 1937) (UA 9688). Warren district, as minute stringers in massive pyrite on the 1200-foot level of the Cole shaft (UA 1243), and also as fine masses of large size in the Lavender open pit mine (UA 6669). Reported from the Pearce area (Guild, 1910).

*Gila County:* Globe-Miami district, in small amounts with copper ores in a number of deposits; as gem material of great beauty at the Sleeping Beauty mine near Miami (Jackson, 1955) (H 106506; UA 8944); Castle Dome and Copper Cities mines, in substantial quantities (Peterson et al., 1946, 1951; Peterson, 1962; Simmons and Fowells, 1966) (UA 1162, 1163). Also reported from Canyon Creek.

*Graham County:* Lone Star district, Gila Mountains, intimately associated with jarosite and alunite in the oxidized zone of the Safford porphyry copper deposit (Robinson and Cook, 1966).

*Greenlee County:* Clifton-Morenci district, as thin plates and nodules in close association with a diabase dike system which crosses the Morenci ore body (Moolick and Durek, 1966).

*Maricopa County:* Reported 12 miles east of Morristown. Several mines worked near Mineral Peak are said to have yielded large quantities of gem quality material.

*Mohave County:* Cerbat Mountains, Wallapai district, Ithaca Peak (Sterrett, 1908; Crawford and Johnson, 1937; Thomas, 1949; Eidel et al., 1968), as gem material in porphyry cutting schist and gneiss (Schrader, 1907; Guild, 1910) (UA 1161, 6675; BM 86815). Turquoise Mountain, as gem quality material in veins in gold-bearing quartz (Frenzel, 1898).

*Pima County:* Pima district, Esperanza mine (Schmidt et al., 1959; Loghry, 1972). Silver Bell district, in the Oxide Pit of the Silver Bell mine, as transparent, tabular green crystals to 1 mm, that occur in subparallel aggregates; associated with torbernite and cacoxenite at the contact of an andesite dike (Kenneth W. Bladh, pers. comm., 1974).

*Pinal County:* Mineral Creek district, from a locality one mile south of Kelvin, where it is associated with torbernite (Charles L. Fair, pers. comm., 1975).

*Yavapai County:* Reported to occur in the county, but the specific locality is not known.

## *TYROLITE

Copper calcium arsenate carbonate hydroxide hydrate, $Cu_5Ca(AsO_4)_2(CO_3)(OH)_4 \cdot 6H_2O$. A rare secondary mineral found in oxidized portions of copper deposits.

*Pinal County:* Superstition Mountains, as films on mercurian tetrahedrite (variety schwatzite).

## TYUYAMUNITE

Calcium uranyl vanadate hydrate, $Ca(UO_2)_2$-$(VO_4)_2 \cdot 5\text{-}8H_2O$. A widespread secondary mineral associated with carnotite, uranophane, and other uranium and vanadium minerals. Resembles carnotite but is less common in occurrence.

*Apache County:* Lukachukai Mountains, Mesa No. 1 and No. 5 mines; Carrizo Mountains, Cove Mesa, Chinle area. Black Mesa Basin at numerous claims (Garrels and Larsen, 1959). Monument Valley, Monument No. 2 mine (Rosenzweig et al., 1954; Weeks et al., 1955; Mitcham and Evensen, 1955; Finnell, 1957; Evensen and Gray, 1958; Young, 1964). At Garnet Ridge, near Dinnehotso (Witkind and Thaden, 1963; Gavasci and Kerr, 1968).

*Cochise County:* Warren district, Cole mine, lining cracks and fissures in limestone country rock adjacent to massive chalcocite-covellite ore bodies (Hutton, 1957) (UA 6158, 6239; S 96427); also at the Campbell mine (AM 21108).

*Coconino County:* Cameron area, Black Point-Murphy mine (Austin, 1964).

*Mohave County:* Hack Canyon.

*Navajo County:* Monument Valley, at several mines and claims, notably at the Monument No. 1 mine (Holland et al., 1958; Evensen and Gray, 1958; Witkind, 1961; Witkind and Thaden, 1963).

# U

## *UMOHOITE

Uranyl molybdate hydrate, $(UO_2)MoO_4 \cdot 4H_2O$. A secondary mineral formed during the early stages of oxidation of primary uranium minerals.

*Coconino County:* Cameron district, Alyce Tolino mine, as a blue-black, opaque, isotropic mineral contained in sooty masses and carbonaceous trash replacements (Hamilton and Kerr, 1959).

## *UNNAMED SPECIES
### (Sodium analogue of ZIPPEITE)

Sodium uranyl sulfate hydrate. The sodium analogue of zippeite, $(UO_2)_2(SO_4)(OH)_2 \cdot 4H_2O$. A secondary mineral reported, to our knowledge, only from the locality noted here.

*Gila County:* Sierra Ancha Mountains, Cherry Creek district, at the Sue mine, where it was found on only one specimen as a pulverulent yellow coating on bassetite and saleeite; faintly fluorescent in yellowish green (Granger and Raup, 1969).

## URANINITE

Uranium oxide, $UO_2$. Natural uranium dioxide is usually partially oxidized. The principal occurrence of uraninite is in hydrothermal veins. In Arizona, it is most notable in the sedimentary rocks of the Colorado Plateau deposits where it may be associated with coffinite and a wide variety of secondary uranium, vanadium, and copper minerals. The unit cell parameters of the cubic mineral have been collected by Frondel (1958) for a number of Arizona occurrences. These include: Workman Creek, Globe, 5.430 Å; Monument No. 2 mine, Apache County, 5.427 Å, 5.421 Å, 5.415 Å, 5.412 Å; Cato Sells mine, Apache County, 5.40 Å; Monument No. 2 mine, 5.40Å; Huskon mine, Cameron, 5.39 Å.

*Apache County:* Lukachukai Mountains, at several claims and mines including Luki, Mesa 1¾, Mesa 2, Cove School, and Mesa 4½ (Joralemon, 1952). Monument Valley, Monument No. 2 mine (Rosenzweig et al., 1954; Mitcham and Evensen, 1955; Jensen, 1958; Witkind and Thaden, 1963; Young, 1964); Bootjack mine; Garnet Ridge, near Dinnehotso (Witkind and Thaden, 1963).

*Cochise County:* Warren district, reported in very small amounts in some of the underground workings of the district (Richard Graeme, pers. comm., 1974).

*Coconino County:* Hosteen Nez claim, near Tuba City. Cameron district (Austin, 1964), Henry Sloan No. 1 mine, with marcasite in calcite cement in sandstone bordering carbonaceous fossil wood; Ramco No. 22 mine; Huskon No. 7 mine; Sun Valley mine, as replacements of rounded detrital grains with pyrite and sphalerite (Petersen et al., 1959; Petersen, 1960); Arrowhead mine (Rosenzweig et al., 1954; Holland et al., 1958); Boyd Teise No. 2 mine, in flattened nodules with pyrite in sandstone (Atomic Energy Comm., Grand Junction office mineral collection). Orphan mine, situated near park headquarters on the south rim of the Grand Canyon, as the fairly pure mineral in disseminations and as veins and lenses up to several inches thick in a nearly vertical, circular or ellipsoidal pipe-like body of collapse breccia developed primarily in the Coconino Sandstone of Permian Age. Associated are pyrite and other sulfides and sulfosalts of copper, lead, zinc, cobalt, nickel, and molybdenum. Secondary minerals of uranium are common in the workings (Granger and Raup, 1962).

*Gila County:* At a number of localities in the Dripping Spring Quartzite adjacent to diabase (Granger, 1956); Turquoise mines at T.5N, R.15E, associated with pyrite, marcasite, and chalcopyrite (Granger and Raup, 1959, 1962). Northern Gila County, in Dripping Spring Quartzite, at the Black Bush, Hope, Little Joe, Lucky Stop, Red Bluff, Rock Canyon, Suckerite, Sue, Tomato Juice, and Workman deposits, as fissure and open-space fillings or as lenses and blebs in host rock (Granger and Raup, 1969). Globe district, Old Dominion vein; Iron Cap mine, near Miami, with iron oxides (S 112722).

*Mohave County:* At Hack Canyon (UA 6115, 6116, 6118).

*Navajo County:* Found at a number of localities including the Monument No. 1 mine, Monument Valley (Evensen and Gray, 1958; Witkind and Thaden, 1963). Petrified Forest, near Holbrook, Ruth group of claims; Stinking Spring, near Hunt (Rosenzweig et al., 1954).

*Pima County:* Rincon Mountains, Black Rock claims.

*Santa Cruz County:* Oro Blanco district, at the Annie Laurie prospect near Ruby (Granger and Raup, 1962; Anderson and Kurtz, 1955). Also at Alamo Spring with uranophane and autunite. Happy Jack mine, Santa Rita Mountains, Wrightson district (Butler and Allen, 1921).

*Yavapai County:* Hillside mine (Granger and Raup, 1962).

## *URANOCIRCITE

Barium uranyl phosphate hydrate, $Ba(UO_2)_2(PO_4)_2 \cdot 8(?)H_2O$. An uncommon member of the autunite group; of secondary origin.

*Coconino County:* Cameron area, associated with meta-uranocircite in some of the uranium deposits of the area (Austin, 1964).

## URANOPHANE

Calcium uranyl silicate hydrate, $Ca(UO_2)_2$-$Si_2O_7 \cdot 6H_2O$. One of the commonest of the secondary uranium minerals; occasionally occurs as fine crystals.

*Apache County:* Monument Valley, Monument No. 2 and Cato Sells mines (Witkind and Thaden, 1963).

*Coconino County:* Cameron area, present in minor amounts in uranium ore in the Shinarump Sandstone, in paleo stream channels, and in the Chinle Formation in sandy portions of mounds, associated with uraninite, metatorbernite, and meta-autunite (Holland et al., 1958); at the Black Point-Murphy mine where it is intimately mixed with beta-uranophane in Pleistocene gravels (Austin, 1964).

*Gila County:* Northwest Gila County, in Dripping Spring Quartzite at the Fairview, Little Joe, and Red Bluff deposits (Granger and Raup, 1969).

*Pima County:* At the Linda Lee claims, on the east flank of the Quijotoa Mountains, as fracture coatings in a contact metamorphic assemblage (Williams, 1960).

*Santa Cruz County:* In a vein near Alamo Springs, associated with autunite and uraninite. Santa Rita Mountains, Duranium claims, in arkosic sandstone with kasolite and autunite (Robinson, 1954). Nogales district, White Oak property, with kasolite and oxidized lead ore in shear zones in rhyolite (Granger and Raup, 1962).

*Yavapai County:* Weaver Mountains, Peeples Valley mine (Granger and Raup, 1962).

## *URANOSPINITE

Calcium uranyl arsenate hydrate, $Ca(UO_2)_2$-$(AsO_4)_2 \cdot 10H_2O$. A member of the autunite group. Of secondary origin.

*Coconino County:* Grand Canyon National Park, the Orphan mine (AM 33454).

## *UREYITE

Sodium chromium silicate, $NaCrSi_2O_6$. A rare member of the pyroxene group. Known only from iron meteorites.

*Coconino County:* Canyon Diablo area, occurs associated with krinovite in graphite nodules in the octahedrite meteorite; also associated with roedderite, high albite, richterite, and chromite (Olsen and Fuchs, 1968).

## UVANITE

Uranium vanadium oxide hydrate, $U_2V_6O_{21} \cdot 15H_2O$ (?). An uncommon mineral associated with the uranium-vanadium ores of the Colorado Plateau where it probably forms by the alteration of uraninite and, possibly, of tyuyamunite.

*Apache County:* Monument Valley, Monument No. 2 mine (Frondel, 1958).

## *UVAROVITE

Calcium chromium silicate, $Ca_3Cr_2(SiO_4)_3$. An uncommon member of the garnet group; known to occur with chromite in serpentines and granular limestones.

*Mohave County:* Cerbat Range, Wallapai district, an emerald-green garnet found in migmatite just east of the Tennessee mine staff house is believed to be this mineral (Thomas, 1953).

# V

## *VAESITE

Nickel sulfide, $NiS_2$. A rare mineral of primary origin; a member of the pyrite group.

*Coconino County:* Orphan mine, Grand Canyon, as brilliant tin-white crystals up to 2 mm in diameter perched on coarse yellow barite crystals; $a = 5.632$ Å.

## *VALLERIITE

Iron copper sulfide magnesium aluminum hydroxide, $2(Fe,Cu)_2S_2 \cdot 3(Mg,Al)(OH)_2$. An uncommon mineral thought to be of high temperature origin.

*Gila County:* Banner district, Christmas mine, as rod-shaped grains in chalcopyrite in the pyrrhotite-chalcopyrite outer ore zone of ore bodies; replacing dolomite (Perry, 1969).

*Pima County:* Pima district, associated with magnetite in the Pima mine.

## VANADINITE

Lead vanadate chloride, $Pb_5(VO_4)_3Cl$. A comparatively rare mineral which occurs in some oxidized lead deposits, typically associated with wulfenite, cerussite, and descloizite. The discovery of vanadinite in the nineteenth century in Arizona at localities such as the Vulture and

Red Cloud mines aroused considerable interest among eastern mineralogists and led to a number of reports in the *American Journal of Science*. A sparse but widely distributed mineral in the state.

*Cochise County:* Tombstone district, on calcite at the Tribute and Tombstone Extension mines (Butler et al., 1938); as masses of brownish crystals at the Gallagher Vanadium property near Charleston (S 15016; UA 6903). Huachuca Mountains, at the Davis property (UA 8172). Near Elfrida, in the Swisshelm Mountains (UA 5668, 9349).

*Coconino County:* Bright Angel Creek, as hollow encrustation pseudomorphs of brown vanadinite after calcite scalenohedrons, in part only encrusting calcite (Frondel, 1935). At Supai, as brown crusts on twinned calcite crystals (S 94280).

*Gila County:* Dripping Spring Mountains, Banner district, 79 mine (Keith, 1972), also at the McHur, Premier, and C & B (International) properties (Allen and Butler, 1921). Castle Dome district, in lead-bearing veins with wulfenite, cerussite, galena, and fluorite; Mabel mine (H 101521); Silver Dollar mine (H 101518). Tonto Basin (UA 8603). With calcite and siderite from near Hayden (S 115636; UA 134). Globe-Miami district, Lockwood claims, and at the Clark and Stewart claims near the Old Dominion mine; Defiance and Albert Lea deposits (Peterson, 1962). Apache mine, as fine specimens of deep red, well-crystallized, stocky prismatic vanadinite, characteristically having a background of dark mottramite on mottled quartzite of the Precambrian Pioneer Formation. Crystals are commonly 1–2 mm in length, although the best range up to 6–7 mm in length. Crystals up to 2–3 cm have been reported (Guild, 1910; Wilson, 1971) (UA 6228, 7092).

*Greenlee County:* As doubly-terminated single crystals in gravels of the Haggin placers along the Coronado Trail, north of Metcalf.

*Maricopa County:* Observed by Silliman (1881) in the Hieroglyphic Mountains, with wulfenite in veins. Vulture Mountains, at the Collateral, Phoenix, Montezuma, and Frenchman mines. Painted Rock Mountains, at the Rowley mine, near Theba (Wilson and Miller, 1974); White Picacho district (Jahns, 1952).

*Mohave County:* Cerbat Mountains, El Dorado and Climax mines; Gold Basin district (Schrader, 1909). Western mine, as sheaf-like

**Vanadinite.**
Apache mine, Gila County. Tom Trebisky.

bundles and single, doubly-terminated crystals of the variety endlichite.

*Pima County:* Tucson Mountains, in the Old Yuma mine as exceptionally beautiful specimens (Jenkins and Wilson, 1920), and as pseudomorphs after wulfenite (Frondel, 1935) (AM 18716; UA 6294; BM 1968, 1097, and others). The indices of refraction of a ground crystal from the Old Yuma mine were determined for red light by Bowman (1903) using the method of minimum deviation with the following result: E = 2.299, O = 2.354. Empire Mountains, Total Wreck mine (Schrader and Hill, 1915; Schrader, 1917). South Comobabi Mountains, Cababi district, Mildren and Steppe claims (Williams, 1963). Pima district, Twin Buttes mine (Stanley B. Keith, pers. comm., 1973). Helvetia-Rosemont district, as a secondary mineral in replacement deposits in metamorphosed limestones, associated with wulfenite, jarosite, and iron and manganese oxides (Schrader, 1917).

*Pinal County:* Mammoth district, Mammoth–St. Anthony mine, where its occurrence was first noted in the literature by Genth (1887), with descloizite; Royal Dane property, 7 miles southeast of Oracle. Dripping Spring Mountains, 4 miles east of Kelvin (Newhouse, 1934). As the arsenian variety endlichite, with wulfenite, Table Mountain mine, Galiuro Mountains. Pioneer district, Black Prince mine, as doubly-terminated crystals (Blake, 1881; Penfield, 1886) (AM 15129). Slate Mountains, Jackrabbit mine, as coarse crystalline masses with manganese oxides in vugs in limestone (Hammer, 1961). Bear Cat claims, northeast side of the Santa Catalina Mountains (Dale, 1959).

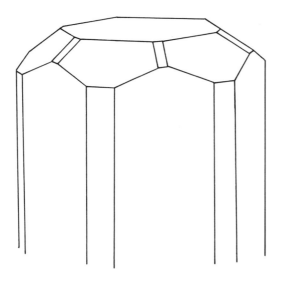

**Vanadinite.**
Silver-Lead claim, Mildren and Steppe area
(Williams, 1962).

*Santa Cruz County:* Patagonia Mountains, Patagonia district, Mowry mine, and on the J. C. Holmes claims near Patagonia, on fracture planes in quartz vein filling (Pellegrin, 1911); Harshaw district, Flux mine; Tyndall district, Cottonwood Canyon, Glove mine, with wulfenite and smithsonite (Olson, 1966).

*Yavapai County:* As fine specimens from the property of the Kirkland gold mines; near the Silver Belt mine, Big Bug district. Humboldt, as masses of ¼ inch, yellow-brown crystals (H 103031).

*Yuma County:* Trigo Mountains, Silver district, as fine crystal aggregates, Silver Clip and Princess mines, and as deep red, brilliant crystals at the Red Cloud mine where they were observed by Silliman (1881) who noted the great beauty and extraordinary color of the crystals (see also Hills, 1890); Hamburg mine (H 64115); crystals of outstanding beauty occurred at Ronaldo Pacheco's mine (S 48793). Castle Dome district, in channels and vugs (Wilson, 1933). Also reported from the Chocolate Mountains. Muggins Mountains, at the Red Knob mine where it is associated with wulfenite, cuprite, chalcedony, limonite, weeksite, opal, and carnotite (Honea, 1959; Outerbridge, 1960).

## *VANDENDRIESSCHEITE

Lead uranium oxide hydrate, $PbU_7O_{22} \cdot 12H_2O$. A secondary mineral closely associated with uraninite of which it is an oxidation alteration product.

*Apache County:* Monument Valley, Monument No. 2 mine (specimen collected by Joseph Urban).

## VANOXITE

Vanadium oxide hydrate, $V_4^{4+}V_2^{5+}O_{13} \cdot 8H_2O$. An associate of the uranium-vanadium ores of the Colorado Plateau where it occurs massively, as cement in sandstone.

*Apache County:* Lukachukai Mountains, also at King Tut Mesa, and on the Rattlesnake anticline (all north of the Lukachukai Mountains), impregnating certain beds of the Salt Wash Sandstone; associated with carnotite (Masters, 1955).

## *VARISCITE

Aluminum phosphate hydrate, $AlPO_4 \cdot 2H_2O$. Formed under near-surface conditions by the action of phosphatic waters on aluminous rocks.

*Cochise County:* Warren district, Cole mine, as unusual pale green, massive, ferrian material (Sinkankas, 1964).

## VAUQUELINITE

Lead copper chromate phosphate hydroxide, $Pb_2Cu(CrO_4)(PO_4)(OH)$. A rare mineral of secondary origin found in the oxidized portions of metalliferous deposits typically associated with other chromates including phoenicochroite and crocoite.

*Cochise County:* About 8 miles north of Benson, in quartz veins cutting gneiss, in minor amounts with pyromorphite, chrysocolla, and perite (Phelps Dodge Corp., pers. comm., 1972).

*Maricopa County:* Vulture region east of Trilby Wash, at the Collateral, Chromate, Blue Jay, and Phoenix properties (Silliman, 1881); south of Wickenburg at several localities including the Moon Anchor mine, Rat Tail claim, and Potter-Cramer property, as a minor mineral in the oxidized zone of lead-zinc ore, associated with wickenburgite, willemite, mimetite, phoenicochroite, and hemihedrite (Williams, 1968; Williams et al., 1970; Williams and Anthony, 1970).

*Pinal County:* Tortilla Mountains, Florence Lead-Silver mine, in an oxide mineral assemblage which replaced galena, sphalerite, pyrite, and tennantite along a sheared and mineralized fault zone separating limestone and quartzite, associated with a variety of other oxidized minerals including wulfenite, hemihedrite, willemite, cerussite, minium, and mimetite (Williams and Anthony, 1970).

## VERMICULITE

Magnesium iron aluminum silicate hydroxide hydrate, $(Mg,Fe,Al)_3(Al,Si)_4O_{10}(OH)_2 \cdot 4H_2O$. A member of the group of mica minerals having chemical properties analogous to those of the smectite (montmorillonite) minerals. The vermiculites have the property of expanding remarkably normal to the micaceous layers when rapidly heated.

*Cochise County:* Willcox Playa, where it is, after illite and montmorillonite, the most abundant clay mineral in the younger playa sediments (Pipkin, 1968).

*Maricopa County:* Reported from near Aguila and from a locality between Wickenburg and the Vulture Mountains (North and Jensen, 1958) (UA 8753). On the Bar FX ranch, southwest of Wickenburg (Wilson and Roseveare, 1949).

*Mohave County:* In a deposit 15 miles southwest of Kingman in the Hualpai Mountains (Wilson, 1944; *Engr. and Min. Jour.,* 1940).

*Pima County:* Tucson Mountains, where the variety jefferisite was recognized in rocks metamorphosed by igneous intrusions (Brown, 1939).

*Pinal County:* Near Oracle (Moore, 1969).

*Yuma County:* As the variety jefferisite, formed by the alteration of biotite (UA 8651); near Bouse (UA 8646).

## *VESIGNIEITE

Barium copper vanadate hydroxide, $BaCu_3(VO_4)_2(OH)_2$. A rare secondary mineral associated with other oxidized uranium minerals.

*Apache County:* Monument Valley, Monument No. 2 mine (Joseph Urban, pers. comm., 1967) (UA 9501; BM 1967, 148).

*Pinal County:* Copper Creek district, in a prospect pit 2 miles south of Mercer Ranch, as crude, rounded crystals up to 1.5 mm in length of a rich pistachio-green color; occurs in albitized quartz monzonite with cuprite and chrysocolla. A chemical analysis by the Schwartzkopf Laboratory follows:

| | | |
|---|---|---|
| CuO 36.97 | V₂O₅ 32.78 | SrO <0.005 |
| BaO 23.69 | | H₂O 5.95 % |

Specific gravity: Observed, 4.604; calculated, 4.682. Monoclinic, 2/m. Unit cell parameters: $a = 7.134$, $b = 5.910$, $c = 5.122$ Å, beta = $103°33.7'$.

## VESUVIANITE

Calcium magnesium aluminum silicate hydroxide, $Ca_{10}Mg_2Al_4(SiO_4)_5(Si_2O_7)_2(OH)_2$. Occurs typically in contact metamorphosed limestones where it is associated with other calcium silicate minerals.

*Cochise County:* Tombstone district, in the contact zone of Comstock Hill; Lucky Cuss mine, with monticellite, hillebrandite, and thaumasite (Butler et al., 1938). Little Dragoon Mountains, as pale green square, vertically-striated prisms up to 2 mm long, Johnson district (Romslo, 1949; Cooper and Huff, 1951; Cooper, 1957; Cooper and Silver, 1964).

*Gila County:* Dripping Spring Mountains, Christmas mine, in fibrous masses showing anomalous birefringence and color zoning; moderately common in skarn (Ross, 1925b; David Perry, pers. comm., 1967).

*Greenlee County:* Clifton-Morenci district, where it is of local occurrence with other calc-silicate minerals including garnet, tremolite, diopside, and epidote in a large contact metamorphosed zone in the northern part of Morenci (Moolick and Durek, 1966).

*Pima County:* From an unspecified locality at the northern end of the Baboquivari Mountains, and in schist at the east end of the Coyote Mountains (UA 8095). In contact metamorphosed rocks of the Tucson Mountains (Brown, 1939). Pima district, present in the skarns and tactites of the Twin Buttes mine (Stanley B. Keith, pers. comm., 1973).

*Santa Cruz County:* Patagonia Mountains, Washington Camp, as lime-green crystal fragments in a wash uphill behind the local store (Bideaux et al., 1960).

*Yuma County:* Locally abundant in metamorphosed limestones. Gila Mountains, in a contact zone (Wilson, 1933).

## VIVIANITE

Iron phosphate hydrate, $Fe_3(PO_4)_2 \cdot 8H_2O$. Occurs in the weathered portion of base metal deposits and as an alteration product of primary iron-manganese phosphates in pegmatites.

*Maricopa and Yavapai Counties:* White Picacho district, as finely crystalline, bluish gray films in triplite (Jahns, 1952).

## VOGLITE

Calcium copper uranyl carbonate hydrate, $Ca_2Cu(UO_2)(CO_3)_4 \cdot 6H_2O$ (?). A very rare secondary mineral formed as an alteration product of uraninite. At the Elias mine, Joachimsthal, Bohemia, where it occurs with liebigite (Frondel, 1958).

*Navajo County:* Red Mesa district, near the Red Mesa trading post. This is thought to be the only known occurrence other than the original locality.

## VOLBORTHITE

Copper vanadate hydrate, $Cu_3(VO_4)_2 \cdot 3H_2O$. A rare secondary mineral formed in the oxidized zones of base metal deposits; also associated with certain of the uranium deposits of the Colorado Plateau region.

*Apache County:* Monument Valley, Monument No. 2 mine, as a barian variety (Witkind and Thaden, 1963) (UA 1569, 9797).

*Gila County:* From an undisclosed locality north of Globe (Robert O'Haire, pers. comm., 1972).

*Navajo County:* Monument Valley, Mitten No. 2 mine (Witkind and Thaden, 1963); Monument No. 1 mine (Holland et al., 1958; Witkind, 1961).

## VOLTAITE

Potassium iron sulfate hydrate, $K_2Fe_5^{2+}Fe_4^{3+}(SO_4)_{12} \cdot 18H_2O$ (?). Found at a number of localities throughout the world, commonly associated with other sulfates under conditions which suggest that it may be a low temperature, late-stage, primary mineral; also of secondary origin.

*Cochise County:* Warren district, in irregular, porous crusts several inches thick in the Copper Queen mine, associated with coquimbite, kornelite, copiapite, and rhomboclase (Merwin and Posnjak, 1937).

*Pinal County:* Pioneer district, Magma mine, as crystals up to ¼ inch, associated with copiapite.

*Yavapai County:* Black Hills, at the United Verde mine as black, resinous cubo-octahedral crystals up to 5 mm, formed as the result of burning pyritic ore (Anderson, 1927; Lausen, 1928; Hutton, 1959) (UA 64, 3074, and others; H 85680, 90536). An analysis by C. A. Anderson gave the following result:

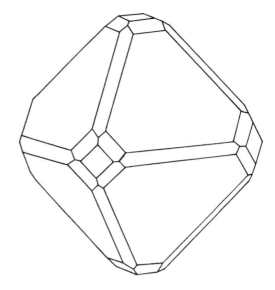

**Voltaite.**
United Verde mine (Lausen, 1928).

| | | | | | |
|---|---|---|---|---|---|
| $SO_3$ | 47.83 | FeO | 8.82 | $K_2O$ | 4.52 |
| $Al_2O_3$ | 6.06 | MgO | 1.55 | $H_2O$ | 16.13 |
| $Fe_2O_3$ | 14.34 | | | TOTAL | 99.25% |

Specific gravity: 2.75.

The aluminum content apparently is the highest reported for the species.

# W

## Wad

Manganese oxides. Wad is a generic or field term for aggregates of manganese oxides whose mineralogy has not been defined in detail. It stands in relation to the manganese oxide minerals in much the same manner that "limonite" and "bauxite" stand with respect to the undifferentiated iron and aluminum oxide and hydroxide minerals respectively. Many occurrences of black manganese oxides in Arizona must be referred to by this appellation until they are more carefully examined. Where knowledge of their mineralogies permits, reference is made to specific species in this compilation (Jones and Ransome, 1920); Wilson and Butler, 1930; Farnham and Stewart, 1958).

*Cochise County:* At Tombstone it occurs as veins and replacements which were mined for their silver content (UA 6956). Warren district, as bodies of considerable size. Cochise district, at Johnson Camp (UA 9359).

*Coconino County:* Long Valley region, as nodules and masses in the Kaibab Limestone, Iron Mine Canyon; from near Williams (Farnham and Stewart, 1958) (UA 7370, 7751).

*Gila County:* Globe district, as earthy material along many faults. Roosevelt Lake area (UA 7423). Medicine Butte area, in the Apache and Accord manganese deposit, as fracture fillings and as cement in conglomerate of Cenozoic age (Moore, 1968). Banner district, where it is common in many of the fault zones and solution channels, associated with oxide copper and iron minerals, often as bands alternating with goethite (Keith, 1972).

*Graham County:* Arivaca district.

*Greenlee County:* Ash Peak district, in shear zones with manganite and pyrolusite.

*Maricopa County:* Big Horn Mountains, Aguila district, as extensive bedded deposits (Lasky and Webber, 1944, 1949) (UA 6999). Little Chemehuevis Valley, Topock district. Aguila district, at the Sambo Aguila, Black Crow, Pumice, Valley View and Yarnell, Purple Pansy, Kat Head, Black Rock, Black Bart and many other groups of claims (Farnham and Stewart, 1958).

*Mohave County:* As extensive low-grade deposit in the Artillery Peak Mountains (Lasky and Webber, 1949; Farnham and Stewart, 1958).

*Pinal County:* Pioneer district, abundant in fractures and fault planes; as veins and irregular masses on the east 1,600-foot level, Magma mine. Globe district, in surface gash veins at the Old Dominion mine. Black Hills, Camp Grant Wash, at the Tarr and Harper mine. Galiuro Mountains, Florence district, Chamberlain mine; Blue Bird mine, Copper Creek district.

*Santa Cruz County:* Santa Rita Mountains, Rosario group. Patagonia Mountains, Mowry, La Plata, Hermosa, Jarrilla, and Isabella mines.

*Yavapai County:* Castle Creek district, 23 miles northeast of Morristown. Reported from near Mayer. At a number of properties extending across the southern part of the county, including the Harris, Box Canyon, Black Rock, Blind Child, Black Dome, Black Buck, Cummings, Fiscus and Mitchell, Burmeister, Black Duke, La McCoy and other properties (Farnham and Stewart, 1958).

*Yuma County:* In the Granite Wash Hills, east of Bouse. Also about 33 miles west of Congress Junction. Planet mine, with hematite, malachite, azurite, and chrysocolla (Bancroft, 1911). Plomosa and Trigo districts (Farnham and Stewart, 1958).

## WAVELLITE

Aluminum phosphate hydroxide hydrate, $Al_3(PO_4)_2(OH)_3 \cdot 5H_2O$. An uncommon but widespread mineral formed by low-grade metamorphism and as a product of hydrothermal alteration in mineral deposits.

*Gila County:* Globe-Miami district, Castle Dome mine, where it is localized along fractures crossing the trend of ore veins; probably of hypogene origin (Peterson, 1947; Peterson et al., 1951).

## *WEEKSITE

Potassium uranyl silicate hydrate, $K_2(UO_2)_2$-$(Si_2O_5)_3 \cdot 4H_2O$. A rare secondary mineral intimately associated with other oxidized uranium minerals; closely resembles uranophane.

*Yuma County:* Muggins Mountains, Red Knob claims, as radial aggregates of fibrous to acicular crystals coating and intergrown with chalcedony, wulfenite, carnotite, vanadinite, cuprite, azurite, calcite, gypsum, and limonite (Honea, 1959; Outerbridge et al., 1960).

## *WEISSITE

Copper telluride, $Cu_5Te_3$ or, perhaps, $Cu_{2-x}$-Te with $x = 0$ to $0.33$. Found associated with pyrite, petzite, rickardite, and tellurium at a few localities.

*Yuma County:* Dome Rock Mountains, where it occurs in trace amounts with blebs of bornite in massive djurleite which is in a quartz-tourmaline gangue (Williams and Matter, 1975).

## WHERRYITE

Lead copper carbonate sulfate chloride hydroxide oxide, $Pb_4Cu(CO_3)(SO_4)_2(Cl,OH)_2O$. A rare secondary mineral found in the oxidized portion of some lead deposits. The occurrence noted below is the type and only known locality. The species has been further substantiated by McLean (1970).

*Pinal County:* Mammoth district, Mammoth–St. Anthony mine, where it occurs in small vugs on the 760-foot level with matlockite, leadhillite, hydrocerussite, paralaurionite, diaboleite, phosgenite, chrysocolla, anglesite, and cerussite (Fahey et al., 1950). An analysis by Fahey gave:

| PbO | 72.9 | $CO_2$ | 3.1 | Insol. | 2.2 |
|-----|------|--------|-----|--------|-----|
| CuO | 7.3 | Cl | 0.9 | Oxygen | |
| $SO_3$ | 13.0 | $H_2O$ | 1.2 | Corr. | −0.2 |

TOTAL 100.4%

## *WHEWELLITE

Calcium oxalate hydrate, $Ca(C_2O_4) \cdot H_2O$. Most commonly of organic origin; found in the remains of dead plants. Also known to occur in hydrothermal veins and in coal seams. Observed as very abundant minute needle-like, colorless crystals in the cup-like depressions in the bases of dead agave in Arizona's southwestern desert. Locally, the material is so abundant as to sparkle in sunlight.

## *WICKENBURGITE

Lead aluminum calcium silicate hydroxide, $Pb_3Al_2CaSi_{10}O_{24}(OH)_6$. A rare secondary lead mineral formed from the oxidation of lead ores; locally it is relatively abundant in the Wickenburg area. The Potter-Cramer property is the type locality.

*Maricopa County:* At several localities south of Wickenburg including the Moon Anchor mine and the Potter-Cramer property, as transparent, colorless to rarely salmon-pink crystals; also massive or dull white granular masses. Forms in the oxidized portion of galena-sphalerite veins; associated with a large number of secondary minerals, some of which are quite exotic; for example, phoenicochroite, crocoite, $\beta$-duftite, ajoite, vauquelinite, shattuckite, minium, and hemihedrite (Williams, 1968). A chemical analysis gave:

| PbO | 44.0 | $Al_2O_3$ | 7.6 | CaO | 3.80 |
|-----|------|-----------|-----|-----|------|
| $SiO_2$ | 42.1 | | | $H_2O+$ | 3.77 |

TOTAL 101.27%

## WILLEMITE

Zinc silicate, $Zn_2SiO_4$. An uncommon mineral found in crystalline limestones, possibly formed by metamorphism of other zinc minerals; far more common as a secondary or mesogene mineral in the oxidized portions of zinc deposits. Most willemite does not fluoresce.

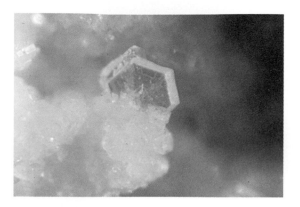

**Wickenburgite.**
Wickenburg, Maricopa County. Julius Weber.

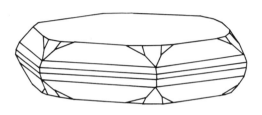

**Wickenburgite.**
Potter-Cramer mine (Williams, 1968).

*Cochise County:* Warren district, as a fluorescent variety, on the 1500 foot level of the Campbell mine (UA 6793). Chiricahua Mountains, Hill Top mine, as small white- to rose-colored prisms in cavernous rock (Pough, 1941) (AM 21263). Gunnison Hills, as minute, glassy prisms which fluoresce straw-yellow, in metamorphosed limestones at the Texas Arizona mine (Cooper, 1957); also in a quartz vein near the Little Fanny mine (Cooper and Silver, 1964).

*Gila County:* Abundant at the Apache mine, north of Globe, with vanadinite; also at the Defiance mine, with vanadinite and descloizite (Petersen, 1960).

*Greenlee County:* Clifton-Morenci district, as small grayish crystals in garnet rock in the Modoc open cut on Modoc Mountain (Lindgren, 1905).

*Maricopa County:* South of Wickenburg, at several localities including the Moon Anchor mine, Rat Tail claim, and Potter-Cramer property, as a secondary mineral in the oxidized zone of lead-zinc veins (Williams, 1968); Tonopah

Belmont mine (Robert O'Haire, pers. comm., 1970). Painted Rock Mountains, at the Rowley mine near Theba (Wilson and Miller, 1974).

*Pima County:* Tucson Mountains, as crystals on the dump of a prospect 1 mile south of the Old Yuma mine (UA 5691) and at the Old Yuma mine. Waterman Mountains, as small, barrel-shaped crystals at the Silver Hill mine near the head of the inclined surface tram. South Comobabi Mountains, Cababi district, Mildren and Steppe claims (Williams, 1963). Cimarron Mountains, at the Paul Hinshaw property, T.11S, R.2E, as fluorescent material (Wilson and Roseveare, 1949; Funnell and Wolfe, 1964).

*Pinal County:* Mammoth district, Mammoth–St. Anthony mine, as small, colorless rhombs and bluish barrel-shaped crystals on wulfenite and vanadinite (Fahey, 1955) (H 101143). In veinlets with calcite from the McCarthy-Hinshaw property south of Casa Grande; highly fluorescent. Galiuro Mountains, Table Mountain mine, in vugs with conichalcite, plancheite, and malachite (Bideaux et al., 1960) (UA 6421; S 113802); Tortilla Mountains, Florence Lead-Silver mine, with hemihedrite, wulfenite, cerussite, and other oxidized secondary minerals (Williams and Anthony, 1970); Silver Reef mine, near Casa Grande (UA 6855). Slate Mountains, Jackrabbit mine, as the most abundant zinc ore mineral (Hammer, 1961).

*Yuma County:* Reported from an unspecified locality near Wenden; Plomosa Mountains, Black Mesa mine, at Sec. 16, T.3N, R.17W, as

brilliant, elongate, hexagonal crystals with aurichalcite and malachite in limonite; willemite is non-fluorescing (David Shannon, pers. comm., 1971); Trigo Mountains, Red Cloud mine, Silver district, as minute prismatic crystals, often associated with wulfenite (Gary Edson, pers. comm., 1972).

### Withamite
(a variety of CLINOZOISITE)

### WITHERITE
Barium carbonate, $BaCO_3$. An uncommon mineral found in low-temperature hydrothermal veins with barite and galena.

*Yuma County:* Castle Dome Mountains, as gangue in lead ores at the De Luce mine, Castle Dome district.

### *WITTICHENITE
Copper bismuth sulfide, $Cu_3BiS_3$. An uncommon mineral of primary origin in vein deposits; less commonly formed during secondary enrichment of copper ores. Usually associated with chalcocite.

*Cochise County:* Warren district, as a primary mineral in some of the mines; observed as exsolution blebs in bornite from an unspecified mine in the district (Julie DeAzevado Harlan, unpublished manuscript, 1966).

*Pinal County:* Pioneer district, reported in the hypogene ores of the Magma mine (Hammer and Peterson, 1968).

### WOLFRAMITE
Iron tungstate, $FeWO_4$, to manganese tungstate, $MnWO_4$, forming a continuous substitutional series. The end-members are termed ferberite (ca. 80% or greater Fe) and huebnerite (ca. 80% or greater Mn). Minerals of intermediate or unspecified composition are termed wolframite. In the listing here species not specifically identified as to compositional type are referred to as wolframite. Most abundant in quartz veins in granite and schist associated with pegmatites. Commonly the only metallic mineral present, but locally accompanied by scheelite or sulfides. The species are widespread but sparse in Arizona and only representative occurrences are noted.

*Cochise County:* Whetstone Mountains, in veins and as replacements with scheelite on the

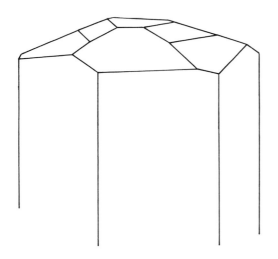

**Willemite.**
Mildren and Steppe area (Williams, 1962).

western slope of the range (Hess, 1909; Dale et al., 1960). Little Dragoon Mountains, in quartz veins and in placer gravels (Kellogg, 1906; Wilson, 1941; Palache, 1941; Dale et al., 1960). Dragoon Mountains, near S.P. Railroad station of Dragoon Summit (Guild, 1910), as crystals of huebnerite 2 inches long. An analysis by Peppel (Hobbs, 1905) gave:

| | | |
|---|---|---|
| SiO 1.10 | MnO 22.87 | FeO 0.81 |
| WO$_3$ 75.10 | | TOTAL 99.88% |

At Russellville, 8 miles north of Dragoon Station, as a highly manganiferous huebnerite in quartz veins cutting porphyritic granite; as prismatic and bladelike masses, associated with small quantities of scheelite and gold (Blake, 1898). Cochise district, Little Dragoon Mountains, in quartz veins and in placer gravels (Rickard, 1904; Kellogg, 1906), and, as huebnerite, the principal ore mineral in the tungsten veins in the Texas Canyon Quartz Monzonite; as platy crystals up to 10 cm long (Wilson, 1941; Palache, 1941a; Cooper and Silver, 1964).

*Gila County:* Pinal Mountains, at several places south of Globe; west of Miami on Spring Creek. Southwest of Young, as huebnerite at the Bobtail mine, with galena and sphalerite (Ransome, 1903; Dale, 1961). Mazatzal Mountains, near Four Peaks (Dale, 1961).

*Maricopa County:* Cave Creek district, as ferberite containing 2.19 percent combined columbium-tantalum as oxides (Wherry, 1915). White Picacho district, as wolframite in pegmatites. As ferberite, associated with tungstite, cuprotungstite, fluorite, molybdenite, pyrite, and chalcopyrite at the Gold Cliff group, Sec. 11, T.6N, R.4E, about 27 miles north of Phoenix (Dale, 1959).

*Mohave County:* Hualpai Mountains, in quartz veins of the Boriana (Hobbs, 1944), Telluride Chief, Laxton, and Moon properties. Aquarius Range (*Engr. Min. Jour.,* 1910; Dale, 1961), in quartz veins in the Williams and other mines in the vicinity of Boner Canyon (Hobbs, 1944); west of Cottonwood Cliffs and west of Greenwood Peak. White Hills, at the O.K. mine, Gold Basin district (Schrader, 1909; Wilson, 1941; Dale, 1961).

*Pima County:* Las Guijas and Baboquivari Mountains (Dale et al., 1960), and Comobabi Mountains (Guild, 1930; Wilson, 1941) (H 89695). In quartz veins cutting granite, Arivaca district, as huebnerite (Blake, 1896). An analy-

sis of huebnerite from the Las Guijas district by Easton and Moss (1966) gave:

| | | |
|---|---|---|
| FeO 1.2 | MoO$_3$ 0.4 | WO$_3$ 74.5 |
| MnO 21.8 | | SiO$_2$ 1.0 |
| | | TOTAL 98.9% |

Cerro Colorado district, in quartz veins (H 89698).

*Pinal County:* Mammoth district, Tarr property, northwest of the Mammoth-St. Anthony mine, with schorl (Dale, 1959). Galiuro Mountains, at the Blue Bird mine (Wilson, 1941). With scheelite in quartz veins in the Antelope Peak area and at a number of properties in the Crook National Forest, northeast of Superior (Dale, 1959).

*Santa Cruz County:* Patagonia Mountains, Mount Benedict and at Reagen Camp, with scheelite in narrow quartz veins in lamprophyre dikes cutting quartz monzonite, Nogales district (Schrader, 1917); Red Mountain property (Hill, 1910). San Cayetano district, southeast of Calabasas (Pellegrin, 1911; Dale et al., 1960).

*Yavapai County:* Bradshaw Mountains, Tip Top mine, Thule Creek area (De Wolfe, 1916; Dale, 1961); Eureka district, in quartz veins at the Black Pearl and Joy properties, south of Camp Wood (Dale, 1961); Silver Mountain district (Wilson, 1941), and in the Camp Verde district (Dale, 1961).

## WOLLASTONITE

Calcium silicate, CaSiO$_3$. A common member of the pyroxenoid group found in contact metamorphosed limestones; a common gangue mineral in mineral deposits of contact metamorphic origin.

*Cochise County:* Tombstone district, as radiating fibrous masses in the Silver Thread and West Side mines (Butler et al., 1938). In garnetite near the Texas Canyon stock, commonly associated with vesuvianite (Cooper and Huff, 1951; Cooper and Silver, 1964). With grossular, epidote, sphene, and other contact silicates in calcareous rocks near Johnson Camp, Little Dragoon Mountains (Cooper, 1957). Dragoon Mountains, abundant in contact metamorphosed limestones at the Abril mine (Perry, 1964). Paradise area, Chiricahua Mountains (Dale et al., 1960).

*Gila County:* Dripping Spring Mountains,

Christmas mine, Banner district, where it is a common mineral, usually associated with chert nodules in marble of the Naco Formation; also locally present in skarn (Ross, 1925b; Perry, 1969).

*Maricopa County:* Sierra Estrella Mountains, in the NW ¼ Sec. 33, T.2S, R.1E, as radiating fibrous masses associated with garnet in metamorphosed limestone (Robert O'Haire, pers. comm., 1971).

*Mohave County:* Hualpai Mountains, in gangue at the Antler mine (Romslo, 1948).

*Pima County:* Santa Rita Mountains, Helvetia-Rosemont district, where large masses of limestone are completely converted to wollastonite (Creasey and Quick, 1955). Sierrita Mountains, Pima district, abundant throughout metamorphosed rocks of the Mineral Hill area; Twin Buttes mine, in skarns and tactites (Stanley B. Keith, pers. comm., 1973). Silver Bell district, where it is locally abundant in contact metamorphosed limestones (Stewart, 1912) (UA 4510; AM 31425); Atlas mine area, in tactites (Agenbroad, 1962).

*Santa Cruz County:* Duquesne-Washington Camp areas, in contact metamorphosed limestones (Schrader and Hill, 1915).

*Yuma County:* Harcuvar Mountains, Cabrolla prospect, where an entire bed of limestone is replaced (Bancroft, 1911).

## WULFENITE

Lead molybdate, $PbMoO_4$. An uncommon mineral generally considered to be of secondary origin. Widely distributed in parts of Arizona in the oxidized portions of lead-bearing deposits some of which have produced wulfenites which in the magnificence of their crystal development or color are the equal of any in the world. Many connoisseurs believe the wulfenite from the Red Cloud mine in Yuma County to be unsurpassed for the beauty of its deep red color.

*Cochise County:* Tombstone district, commonly present in significant amounts with silver ores, as clusters and rosettes of crystals (Butler et al., 1938). Turquoise district, Gleeson, at the Defiance mine as magnificent specimens in substantial amounts in a limestone cavern (Bideaux et al., 1960); also at the Mystery, Silver Bill, and Tom Scott workings. Chiricahua Mountains, Hilltop mine, as groups of deep yellow crystals (H 94678; UA 2626). Pearce district, Pearce mine, lining cavities with embolite (Endlich, 1897). Warren district, as small crystals in the Campbell orebody between the 1,700 and 2,500 levels; associated with copper, malachite, cerussite, smithsonite, azurite, and mimetite (Phelps Dodge Corp., pers. comm., 1974).

*Gila County:* Dripping Spring Mountains, Banner district, 79 mine, as clear, unflawed crystals to two inches on an edge (Keith, 1972); also at the McHur, Premier, and London Range properties. Less abundant in the Globe district, Castle Dome mine area, as small, pointed prisms, rarely over 3 mm long (Peterson, 1947; Peterson et al., 1951). As the variety chillagite at Sleeping Beauty Mountain, 2 miles northwest of the Inspiration mine (UA 8197).

*Graham County:* Aravaipa district, at the Silver Coin and Dogwater mines (Simons, 1964).

*Maricopa County:* Excellent, bright orange to yellow crystals up to 2 cm on an edge, typically associated with mimetite and barite, occur at the Rowley mine in the Painted Rock Mountains (Wilson and Miller, 1974) (BM 1970, 59). As a rare mineral in the pegmatites of the White Picacho district (Jahns, 1952).

*Mohave County:* Rawhide Mountains, near Artillery Peak; Gold Basin district, at the Climax mine.

*Pima County:* Tucson Mountains, Old Yuma mine, as deep orange-red crystal groups with spectacular vanadinite (Guild, 1910, 1911; Newhouse, 1934). Empire Mountains, Total Wreck and Hilton mines (UA 136, etc.). South Comobabi Mountains, Cababi district, Mildren and Steppe claims, with a variety of exotic secondary minerals formed from the oxidation of sulfide ores (Williams, 1963). Silver Bell district as brownish plates, with fluorite (Stewart, 1912), and as crystals showing obvious tetartohedrism (Williams, 1966). Pima district, Twin Buttes open pit mine (Stanley B. Keith, pers. comm., 1973).

*Pinal County:* Mammoth district, Mammoth–St. Anthony mine, as light yellow to bright red crystals containing tungsten (Guild, 1910; Newhouse, 1934; Peterson, 1938; Galbraith and Kuhn, 1940; Palache, 1941b; Fahey, 1955; Petersen et al., 1959; Fleischer, 1959) (H 101755; UA 8208; BM 1961, 539, and others). Black Prince mine, Pioneer district. Dripping Spring Mountains, 4 miles east of Kelvin (Newhouse, 1934). Tortilla Mountains, Florence Lead-Silver mine, with willemite, hemihedrite, vauquelinite, minium, mimetite, and other min-

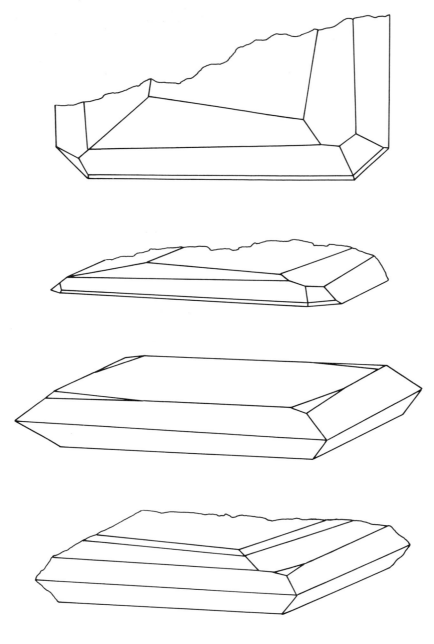

**Wulfenite.**
Old Yuma mine (Guild, 1911).

erals as oxidation products from the alteration of sphalerite and galena; adjacent country rocks contain chromium (Williams and Anthony, 1970). Copper Creek district, Blue Bird mine (Kuhn, 1951). At the Lee Reagan property, as brown plates up to ½ inch on an edge, with orange mimetite (S 97899).

*Santa Cruz County:* Santa Rita Mountains, Wrightson district, Gringo mine, with native gold; Tyndall district, Glove mine, as remarkable crystal aggregates of various colors and habits, some large crystals measuring 4 or more inches along the edge (Olson, 1966) (UA 2760, 2801; BM 1957, 56, etc.); J. C. Holmes claims near

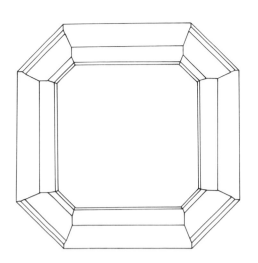

**Wulfenite.**
Mildren mine (Williams, 1962).

**Wulfenite.**
Old Yuma mine, Tucson Mountains, Pima County.
J. R. Thompson collection. Jeff Kurtzeman.

**Wulfenite in place.**
Glove mine, Cottonwood Canyon, Santa Rita
Mountains, Santa Cruz County. H. J. Olson.

by many as the world's premier wulfenite locality
for the remarkable crystals which have found
their way into collections throughout the world.
Red Cloud mine wulfenites are a brilliant orange-
red color and up to 2 inches on an edge (Silliman, 1881; Blake, 1881; Foshag, 1919; Wilson,
1933; Fleischer, 1959) (UA 9949). Specimens
exhibiting unusual forms were taken from the
Melissa mine (UA 4281). Muggins Mountains,
Red Knob mine, with weeksite, vanadinite, and
cuprite (Honea, 1959; Outerbridge et al., 1960).

Patagonia, with vanadinite, descloizite, and cerussite, on fracture planes in quartz vein filling
(Pellegrin, 1911). Patagonia Mountains, Palmetto district, as beautifully crystallized specimens associated with galena, cerussite, and native
silver (Schrader, 1917).

*Yavapai County:* Bradshaw Mountains, locality unspecified.

*Yuma County:* Trigo Mountains, Silver district, at the Hamburg and other properties, and
notably at the Red Cloud mine which is regarded

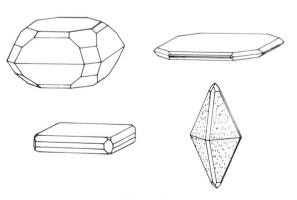

**Wulfenite.**
Florence Lead-Silver mine (Williams, 1966).

## WURTZITE

Zinc sulfide, ZnS. A rare and unstable hexagonal polymorph of zinc sulfide. Wurtzite is known to occur in several polymorphic modifications related to one another by minor structural differences.

*Pinal County:* Mammoth district, Mammoth–St. Anthony mine, where it is reported to occur below the 900 level.

# X

## *XENOTIME

Yttrium phosphate, $YPO_4$. Widely distributed in felsic igneous rocks as an accessory mineral; also in pegmatites and in gneiss.

*Gila County:* At two localities in the Diamond Butte quadrangle (northern part of the county), of authigenic origin in the hematite-rich portion of feldspathic sandstone of the Dripping Spring Formation (Gastil, 1954).

*Mohave County:* Aquarius Range, Columbite prospect, occurs as sparse crystals up to 1 inch across (Heinrich, 1960). Virgin Mountains, with monozite in Precambrian granite gneiss (Young and Sims, 1961). Near Mesquite, on the Nevada state line (Clark County, Nev.), with monazite in granite augen gneiss (Overstreet, 1967).

*Pima County:* North side of the Rincon Mountains, with gadolinite in biotite gneiss (Bideaux et al., 1960).

## *XONOTLITE

Calcium silicate hydrate, $CaSiO_3 \cdot nH_2O$, having uncertain water content. A retrograde low temperature metamorphic mineral usually derived from wollastonite.

*Gila County:* Common at the open pit Christmas mine in wollastonite marbles and derived from wollastonite by alteration. As compact, snow-white masses of chalky appearance associated with laumontite, stilbite, apophyllite, and stringhamite. Partial analysis of this material by M. Duggan (Phelps Dodge Corp.) suggests the formula $CaSiO_3 \cdot H_2O$.

# Y

## *YAVAPAIITE

Potassium iron sulfate, $KFe(SO_4)_2$. A rare secondary mineral formed as a result of a mine

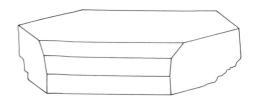

**Yavapaiite.**
United Verde mine (Hutton, 1959).

fire at the United Verde mine at Jerome. This is the type and probably the only known locality.

*Yavapai County:* United Verde mine, of limited occurrence forming cement in rubble exposed in open pit operations, also as rare, short, stumpy crystals. A secondary mineral probably formed as a result of burning sulfide ores, associated with voltaite, sulfur, and jarosite (Hutton, 1959a; Graeber and Rosenzweig, 1971; Anthony et al., 1972).

**Yedlinite.**
Mammoth–St. Anthony mine, Pinal County. Neal Yedlin.

## *YEDLINITE

Hydrous oxychloride of lead and chromium, $Pb_6Cl_6CrX_6Y_2$ with $X = O$ or (OH) and $Y = H_2O$ or (O,OH). A very rare secondary mineral, known only from the Tiger locality noted here. Associated with a complex group of exotic secondary minerals typical of the Mammoth-St. Anthony mine.

*Pinal County:* Mammoth district, found sparingly on a few specimens in the Mammoth-St. Anthony mine where it is associated with diaboleite, quartz, wulfenite, dioptase, phosgenite, and wherryite. As red-violet rhombohedral crystals up to 1 mm long; $a = 12.868$ Å, $c =$

9.821. Density: about 5.85 g/cc. This is the type locality (McLean et al., 1974).

## *YTTROTANTALITE

Yttrium uranium iron tantalum columbium oxide, $(Y,U,Fe)(Ta,Nb)O_4$. A rare mineral which occurs in granite pegmatites.

*Mohave County:* Rare Metals mine, Aquarius Range (Heinrich, 1960).

# Z

## *ZEUNERITE

Copper uranyl arsenate hydrate, $Cu(UO_2)_2$-$(AsO_4)_2 \cdot 10\text{-}16H_2O$. A member of the autunite group. The occurrence noted may be of meta-zeunerite (5 to 8 $H_2O$), a member of the meta-autunite group, because the fully hydrated form dehydrates rapidly upon storage after collection. A secondary mineral associated with other uranium minerals.

*Coconino County:* Grandview mine, as very good quality crystals, up to ¼ inch on an edge, with brochantite, olivenite, scorodite, and chalcoalumite (John S. White, Jr., pers. comm., 1974; collected by Richard L. Jones).

*Mohave County:* Hacks Canyon (AM 26907).

## *ZINCITE

Zinc oxide, $ZnO$. A rare mineral, except in the remarkable deposits at Franklin and Sterling Hill, New Jersey.

*Maricopa County:* Tonopah-Belmont mine, as dull red-brown earthy crusts; formed from sphalerite-bearing ore by a mine fire. (Collected by William Hunt.)

## ZINNWALDITE

Potassium lithium iron aluminum silicate hydroxide fluoride, $K(Li,Fe,Al)_3(Si,Al)_4O_{10}$-$(OH,F)_2$. An uncommon member of the mica group, commonly associated with other lithium-bearing minerals, topaz, cleavelandite, beryl, tourmaline, fluorite, and monazite in tin veins and pegmatites.

*Maricopa County:* Rare in the pegmatites of the White Picacho district, as very dark cleavable crystals having golden brown cleavage flakes; associated with spodumene and amblygonite (Jahns, 1952).

*Santa Cruz County:* At the Line Boy mine, Duquesne district, Patagonia Mountains (Wilson and Roseveare, 1949).

## ZIPPEITE

Uranyl sulfate hydroxide hydrate, $(UO_2)_2$-$(SO_4)(OH)_2 \cdot 4H_2O$. A secondary mineral sometimes found on the walls of mine workings and dumps; also commonly associated with other secondary uranium minerals and gypsum.

*Coconino County:* Cameron district, present in minor amounts in uranium ore in Shinarump paleo stream channels and in the Chinle Formation in sandy patches of mounds, associated with uraninite, metatorbernite, and meta-autunite; at the Huskon Nos. 7 and 8 claims (Holland et al., 1958). At the Sun Valley mine, associated with a zippeite-like mineral, an unnamed uranyl phosphate, and ilsemannite (Petersen et al., 1959).

*Gila County:* Cherry Creek area, Sue mine, as the sodium analogue of zippeite (Granger and Raup, 1969).

*Navajo County:* Holbrook district, on the Ruth claims; Monument Valley at the Monument No. 1 and Mitten No. 2 mines (Witkind, 1961; Witkind and Thaden, 1963).

## ZIRCON

Zirconium silicate, $ZrSiO_4$. A common accessory mineral in igneous rocks. Because of its resistant nature it is present in many sedimentary and metamorphic rocks. Usually occurs as very small crystals but which may assume large sizes in pegmatites. The variety cyrtolite is glassy and non-crystalline owing to structural damage caused by alpha particle bombardment from included radioactive elements.

*Apache County:* An accessory mineral found in boulders of garnet-bearing gneiss, associated with apatite and muscovite; also as an accessory in the Navajo Sandstone, associated with magnetite and tourmaline (Gavasci and Kerr, 1968).

*Apache and Navajo Counties:* Monument Valley, one of the most common heavy minerals in uranium- and vanadium-bearing Shinarump Conglomerate; also common in the basal Monitor Butte Sandstone above the Shinarump, associated with apatite, barite, leucoxene, and tourmaline (Young, 1964).

*Cochise County:* Warren district, as small crystals in Pinal Schist with tourmaline, and in granite northwest of Bisbee. Tombstone district, as microscopic grains in light-colored intrusive rocks. Zircons from the Johnny Lyon Granodiorite have yielded an age of 1655 ± 20 million years for the intrusive rock unit (Silver and Deutsch, 1963).

*Gila County:* Found in the Pinal Schist, Madera Diorite, and Ruin Granite (Ransome, 1919). Globe district, Castle Dome mine, in quartz monzonite.

*Graham County:* Santa Teresa Mountains, as the variety cyrtolite, with monzonite in pegmatite (Robert O'Haire, pers. comm., 1972).

*Greenlee County:* Clifton-Morenci district, in granite (Reber, 1916; Peterson et al., 1946).

*Mohave County:* The variety cyrtolite has been reported from north of Kingman.

*Pima County:* Ajo district, as small stout crystals in the New Cornelia Quartz Monzonite (Gilluly, 1937; Hutton and Vlisidis, 1960). Santa Rita Mountains, in granite porphyry and schist (Catanzaro and Kulp, 1964). An accessory mineral in rocks of the Silver Bell area (Kerr, 1951).

*Santa Cruz County:* Patagonia Mountains, in granite (Schrader, 1913, 1917).

*Yavapai County:* Bradshaw Mountains, sparingly in the Bradshaw Granite (Anderson, 1950). Reported in possible commercial quantities in the black sands of the Kirkland-Copper Basin placers (Wilson, 1961).

## ZOISITE

Calcium aluminum silicate hydroxide, $Ca_2Al_3(SiO_4)_3(OH)$. Common in argillaceous calcareous sandstones as a product of medium grade thermal or contact metamorphism. Much of what has been called zoisite is doubtless clinozoisite, the latter species being by far the more abundant of the two.

*Cochise County:* Cochise district, Johnson Camp area, sparse in contact metamorphosed rocks containing other calc-silicate minerals, at the Republic mine; locally as the variety thulite, in vugs and as coatings on joint surfaces (Cooper, 1957; Cooper and Silver, 1964). Tombstone district, microscopically in igneous rocks (Butler et al., 1938). Warren district (Schwartz, 1956). Paradise area, Chiricahua Mountains (Dale et al., 1960).

*Gila County:* Banner district, Christmas mine, as the variety thulite; a rare mineral found in metamorphosed diorite and in skarn (Perry, 1969); at the 79 mine, as lovely pink crystals of thulite up to 1 inch long on the *b* axis; as coatings on fracture surfaces in metamorphosed limestones.

*Greenlee County:* Clifton-Morenci district, a product of contact metamorphism, found in highly altered granite (Reber, 1916).

*Mohave County:* Cerbat Mountains, Wallapai district, as an alteration product in wall rock of sulfide-bearing veins (Thomas, 1949).

*Pima County:* Sierrita Mountains, Sierrita open pit mine, as an alteration product of dioritic rocks (Roger Lainé, written comm., 1972).

*Yavapai County:* Bradshaw Mountains, in scattered lenses in schist. Eureka district, as an accessory mineral in bodies of titaniferous magnetite in the gangue of the copper deposits at Bagdad.

*Yuma County:* Dome Rock Mountains, in the wall rocks of cinnabar veins.

## ZUNYITE

Aluminum silicate hydroxide fluoride chloride, $Al_{13}Si_5O_{20}(OH,F)_{18}Cl$. A rare mineral which probably forms as a result of metamorphic or intense hydrothermal activity.

*Pinal County:* Mammoth district, San Manuel mine, in very small quantities in highly altered monzonite porphyry (Schwartz, 1953).

*Yuma County:* Dome Rock Mountains, as small, transparent, buff-colored crystals from the Big Bertha Extension mine, about 5 miles southwest of Quartzsite (UA 5455).

PART FOUR

# BIBLIOGRAPHY

# Bibliography

Adams, J. W., and M. H. Staatz (1969) Rare earths and thorium, *in* Mineral and water resources of Arizona. *Ariz. Bur. Mines Bull.* 180: 245–251.

Adams, S. F. (1920) A microscopic study of vein quartz. *Econ. Geology,* 15: 623–664.

Adler, I., and E. J. Dwornik (1961) Electroprobe analysis of schreibersite (rhabdite) in the Canyon Diablo meteorite. *U.S. Geol. Survey Prof. Paper* 424-B: 263–265.

Agenbroad, L. D. (1962) The geology of the Atlas mine area, Pima County, Arizona. Univ. Arizona M.S. thesis, 39 p.

Allen, E. T. (1916) The composition of natural bornite. *Am. Jour. Sci.* 41: 409–413.

Allen, J. E. and R. Balk (1954) Mineral resources of Fort Defiance and Tohatchi quadrangles, Arizona and New Mexico. *New Mexico Bur. Mines Mineral Res. Bull.* 36.

Allen, M. A. (1922) Arizona gold placers. *Ariz. Bur. Mines Bull.* 118, *Econ. Series* 18, 24 p.

————, and G. M. Butler (1918–19) Manganese. *Ariz. Bur. Mines Bull.* 91, *Mineral Tech. Series* 19, 32 p.

————, and ———— (1919–20) Barytes. *Ariz. Bur. Mines Bull.* 99, *Mineral Tech. Series* 22, 18 p.

————, and ———— (1921) Asbestos. *Ariz. Bur. Mines Bull.* 113, *Mineral Tech. Series* 24, 31 p.

————, and ———— (1921) Fluorspar. *Ariz. Bur. Mines Bull.* 114, *Mineral Tech. Series* 19.

————, and ———— (1921) Vanadium. *Ariz. Bur. Mines Bull.* 115, *Mineral Tech. Series* 26.

Anders, E., and M. E. Lipschutz (1966) Critique of paper by N. L. Carter and G. C. Kennedy, "Origin of diamonds in Canyon Diablo and Novo-Urei meteorites." *Jour. Geophys. Res.* 71: 663–672.

Anderson, C. A. (1927) Voltaite from Jerome, Arizona. *Amer. Min.* 12: 287–290.

———— (1950) Alteration and metallization in the Bagdad porphyry copper deposit, Arizona. *Econ. Geol.* 45: 609–628.

———— (1969) Copper, *in* Mineral and water resources of Arizona. *Ariz. Bur. Mines Bull.* 180: 117–156.

————, and S. C. Creasey (1958) Geology and ore deposits of the Jerome area, Yavapai County, Arizona. *U.S. Geol. Survey Prof. Paper* 308, 185 p.

————, and J. T. Nash (1972) Geology of the massive sulfide deposits at Jerome, Arizona — a reinterpretation. *Econ. Geol.* 67: 845–863.

————, E. H. Scholz, and J. D. Strobell, Jr. (1955) Geology and ore deposits of the Bagdad area, Yavapai County, Arizona. *U.S. Geol. Survey Prof. Paper* 278, 103 p.

Anderson, R. Y., and E. B. Kurtz, Jr. (1955) Biochemical reconnaissance of the Annie Laurie Uranium Prospect, Santa Cruz County, Arizona. *Econ. Geol.* 50: 227–232.

Anonymous (1892a) General mining news. *Engr. Mining Jour.* 53: No. 16.

———— (1892b) General mining news. *Engr. Mining Jour.* 53: No. 20.

———— (1892c) General mining news. *Engr. Mining Jour.* 54: No. 14.

———— (1892d) General mining news. *Engr. Mining Jour.* 54: No. 19.

———— (1896) *Engr. Mining Jour.* 62: No. 26.

———— (1897a) General mining news. *Engr. Mining Jour.* 64: No. 2.

———— (1897b) General mining news. *Engr. Mining Jour.* 64: No. 6.

———— (1900) White gold. *Engr. Mining Jour.* 69: 354.

———— (1904) The Silverbell Camp, Arizona. *Engr. Mining Jour.* 77: 639.

———— (1910) Mining news. *Engr. Mining Jour.* 89: No. 26.

———— (1910) A tungsten deposit in western Arizona. *Eng. Mining Jour.* 90: 1103.

———— (1911) Tungsten in Arizona. *Engr. Mining Jour.* 93: 39.

———— (1912) The mining news. *Engr. Mining Jour.* 97: 395.

———— (1915) The mining news. *Engr. Mining Jour.* 100: 862.

———— (1915–16) Directory of Arizona minerals. *Ariz. Bur. Mines Bull.* 3, *Mineral Tech. Series* 1, 14 p.

[ 213 ]

Anonymous *(cont'd)*

———— (1916–17) Mineralogy of useful minerals in Arizona. *Ariz. Bur. Mines Bull.* 41, *Econ. Series* 11, 70 p.

———— (1918) Editorial correspondence. *Engr. Mining Jour.* 106: 726.

———— (1922) Onyx in Arizona. *Engr. Mining Jour.* 114: 408.

———— (1926) Large marble beds reported in Arizona. *Engr. Mining Jour.* 121: 416.

———— (1940) News of the industry. *Engr. Mining Jour.* 141: 76.

Anthony, J. W. (1951) Geology of the Montosa-Cottonwood Canyons area, Santa Cruz County, Arizona. Univ. Arizona M.S. thesis, 84 p.

————, R. L. DuBois, and H. E. Krumlauf (1955) Gypsum. *In* Mineral Resources of the Navajo-Hopi Indian Reservations, Arizona-Utah, 2, 78–83. Univ. of Arizona, Tucson.

————, and R. B. Laughon (1970) Kinoite, a new hydrous copper calcium silicate mineral from Arizona. *Amer. Min.* 55: 709–713.

————, and W. J. McLean (1976) Jurbanite, a new post-mine aluminum sulfate mineral from San Manuel, Arizona. *Amer. Min.* 61.

————, W. J. McLean, and R. B. Laughon (1972) The crystal structure of yavapaiite: a discussion. *Amer. Min.* 57.

Arnold, L. C. (1964) Supergene mineralogy and processes in the San Xavier mine area, Pima County, Arizona. Univ. Arizona M.S. thesis, 63 p.

Austin, S. R. (1957) Recent uranium redistribution in the Cameron, Arizona deposits. *Nuclear Eng. & Sci. Cong. — Proc. 2nd Conf.* 2: 332–338.

———— (1964) Mineralogy of the Cameron area, Coconino County, Arizona. *U.S. Atomic Energy Comm. RME-99, Tech. Inf. Serv., Oak Ridge,* 99.

Axelrod, J. M., F. S. Grimaldi, C. Milton, and K. J. Murata (1951) The uranium minerals from the Hillside mine, Yavapai County, Arizona. *Amer. Min.* 36: 1–22.

Ayres, V. L. (1924) Pyrite from Tucson, Arizona. *Amer. Min.* 9: 91–92.

Bain, G. W. (1952) The age of the lower cretaceous from Bisbee, Arizona, uraninite. *Econ. Geol.* 47: 305–315.

Baker, A., III (1960) Chalcopyrite blebs in sphalerite at Johnson Camp, Arizona. *Econ. Geol.* 55: 387–398.

Baldwin, E. J. (1971) Environments of deposition of the Moenkopi Formation in north-central Arizona. Univ. Arizona Ph.D. dissertation, 208 p.

Balk, R. (1954) Petrology section *in* Mineral resources of Fort Defiance and Tohatchi quad-

rangles, Arizona and New Mexico. *N. Mex. Bur. Mines, Mineral Res. Bull.* 36: 192.

Ball, S. H., and T. M. Broderick (1919) Magmatic iron ore in Arizona. *Engr. Mining Jour.* 107: 353.

Balla, J. C. (1962) The geology and geochemistry of beryllium in southern Arizona. Univ. Arizona M.S. thesis, 76 p.

Bancroft, H. (1909) Notes on the occurrence of cinnabar in central western Arizona. *U.S. Geol. Survey Bull.* 430: 151–153.

———— (1911) Reconnaissance of the ore deposits in northern Yuma County, Arizona. *U.S. Geol. Survey Bull.* 451: 130 p.

Bannerjee, A. K. (1957) Structural and petrological study of the Oracle granite. Univ. Arizona Ph.D. dissertation, 112 p.

Barney, W. G. (1904) The Silverbell mountains, Arizona. *Eng. Min. Jour.* 78: 755.

Barrington, J., and P. F. Kerr (1961) Breccia pipe near Cameron, Arizona. *Geol. Soc. Amer. Bull.* 72: 1661–1674.

————, and ———— (1962) Alteration effects at Tuba dike, Cameron, Arizona. *Geol. Soc. Amer. Bull.* 73: 101–112.

————, and ———— (1963) Collapse features and silica plugs near Cameron, Arizona. *Geol. Soc. Amer. Bull.* 74: 1237–1258.

Barter, C. F. (1962) Geology of the Owl Head mining district, Pinal County, Arizona. Univ. Arizona M.S. thesis, 73 p.

Bastin, E. S. (1922) Primary native-silver ores near Wickenburg, Arizona, and their bearing on the genesis of the silver ores of Cobalt, Ontario. *U.S. Geol. Survey Bull.* 735: 131–155.

———— (1925) Origin of certain rich silver ores near Chloride and Kingman, Arizona. *U.S. Geol. Survey Bull.* 750-B: 17–39.

Bateman, A. M. (1923) An Arizona asbestos deposit. *Econ. Geol.* 18: 663–684.

———— (1929) Some covellite-chalcocite relationships. *Econ. Geol.* 24: 424–439.

Batty, J. U., H. D. Snedden, G. M. Potter, and B. K. Shibler (1947) Concentration of fluorite ores from Arizona, California, Idaho, Montana, Nevada, and Wyoming. *U.S. Bur. Mines Rept. Inv.* 4133.

Beaumont, E. C., and G. H. Dixon (1965) Geology of the Kayenta and Chilchinbito quadrangles, Navajo County, Arizona. *U.S. Geol. Survey Bull.* 1020-A: A1–A28.

Beckman, R. T., and W. H. Kerns (1965) Mercury in Arizona, *in* Mercury Potential of the United States. *U.S. Bur. Mines Inf. Circ.* 8252.

Bejnar, W. (1952) Geology of the Ruin Basin area, Gila County, Arizona. Univ. Arizona Ph.D. dissertation, 103 p.

Bennett, P. J. (1957) Geology and mineralization

of the Sedimentary Hills area. Univ. Arizona M.S. thesis, 40 p.

Berry, L. G. (1951) Observations on conichalcite, cornwallite, euchroite, liroconite and olivenite. *Amer. Min.* 36: 484–503.

———, and R. M. Thompson (1962) X-ray powder data for ore minerals: The Peacock Atlas. *Geol. Soc. Amer. Mem.* 85: 281 p.

Best, M. G. (1970) Kaersutite-peridotite inclusions and kindred megacrysts in basanic lavas, Grand Canyon, Arizona. *Contr. Min. Petr.* 27: 25–44.

Bickerman, M. (1962) A geologic-geochemical study of the Cat Mountain rhyolite. Univ. Arizona M.S. thesis, 42 p.

——— (1966) Geological and geochemical studies of the Roskruge Range, Pima County, Arizona. Univ. Arizona Ph.D. dissertation, 112 p.

——— (1967) Isotopic studies in the Roskruge Mountains, Pima County, Arizona. *Geol. Soc. Amer. Bull.* 78: 1029–1036.

Bideaux, R. A. (1970) A multiple Japan law quartz twin. *Mineral Record,* 1: 33.

———, S. A. Williams, and R. W. Thomssen (1960) Some new occurrences of minerals of Arizona. *Ariz. Geol. Soc. Digest,* 3: 53–56.

Bird, A. T. (1916–17) Resources of Santa Cruz County. *Ariz. Bur. Mines Bull.* 29, *County Resources Series* 1: 27 p.

Bishop, O. M. (1935) Geology and ore deposits of the Richmond Basin area, Gila County, Arizona. Univ. Arizona M.S. thesis, 50 p.

Bissett, D. H. (1958) A survey of hydrothermal uranium occurrences in southeastern Arizona. Univ. Arizona M.S. thesis, 88 p.

Bladh, K. L. (1972) Petrology of O'Leary Peak volcanics, Coconino County, Arizona. Univ. Arizona M.S. thesis, 129 p.

Bladh, K. W. (1973) The clay mineralogy of selected fault gouges. Univ. Arizona M.S. thesis, 70 p.

———, R. K. Corbett, and W. J. McLean (1972) The crystal structure of tilasite. *Amer. Mineral.* 57: 1880–1884.

Blake, D. W. (1971) Geology, alteration, and mineralization of the San Juan mine area, Graham County, Arizona. Univ. Arizona M.S. thesis, 85 p.

Blake, F. H. (1884) Vanadinite in Pinal County, Arizona. *Amer. Jour. Sci.* 28, *3rd Series:* 145.

Blake, W. P. (1855) Observations on the extent of the gold region of California and Oregon with notices of mineral localities in California, and of some remarkable specimens of crystalline gold. *Amer. Jour. Sci. Series 2,* 20: 72–85.

——— (1865) Iron regions of Arizona. *Amer. Jour. Sci. Series 2,* 40: 388.

——— (1881a) Vanadinite in Arizona. *Amer. Jour. Sci.* 22, *3rd Series:* 235.

——— (1881b) On the occurrence of vanadinites of lead at the Castle Dome mines in Arizona. *Amer. Jour. Sci.* 22: 410–411.

——— (1890) Mineralogical notes. *Amer. Jour. Sci.* 39, *3rd Series:* 43–45.

——— (1896) Gypsum in Arizona. *Amer. Geol.* 18: 394.

——— (1897) The Fortuna gold mine. *Engr. Mining Jour.* 63: 26.

——— (1898) Wolframite in Arizona. *Engr. Mining Jour.* 65: 21.

——— (1899) Mining in Arizona. *Engr. and Mining Jour.* 67: 5.

——— (1902) The geology of the Galiuro Mountains, Ariz., and of the gold-bearing ledge known as Gold Mountain. *Engr. Mining Jour.* 73: 546.

——— (1903) Tombstone and its mines. *Transactions, AIME,* 34: 668–670.

——— (1904) Mining in the southwest. *Engr. Mining Jour.* 77: 35.

——— (1905) Iodobromite in Arizona. *Amer. Jour. Sci.* 19, *4th Series:* 230.

——— (1909) Minerals of Arizona: their occurrence and associations, with notes on their composition. *A report to the Governor of Arizona, Tucson,* 64 p.

——— (1910) Manganese ore in an unusual form. *Transactions, AIME,* 41: 647–649.

Blanchard, R. (1944) Chemical and mineralogical composition of twenty typical "limonites." *Amer. Min.* 29: 111–114.

———, and P. F. Boswell (1925) Notes on the oxidation products derived from chalcopyrite. *Econ. Geol.* 20: 613–638.

———, and ——— (1930) Limonite types derived from bornite and tetrahedrite. *Econ. Geol.* 25: 557–580.

———, and ——— (1935) Limonite of molybdenite derivation. *Econ. Geol.* 30: 313–319.

Bock, C. M. (1962) The distribution of some selected alkali metals and alkaline earths in the Stronghold Granite, Cochise County, Arizona. Univ. Arizona Ph.D. dissertation, 95 p.

Bohn, E., and W. Stöber (1966) Coesit und Stishovit als isolierte natürliche Mineralien. *Neues Jahrb. Min.* 89–96.

Bolfa, J., R. Chevallier, H. de la Roche, and R. Kern (1961) Contributions à l'étude des 'ilménites' du sud-est du Madagascar et du Senegal. Relations avec la nature de l'arizonite? *Bull. Soc. Franc. Min. Crist.* 84: 33–39.

Bollin, E. M., and P. F. Kerr (1958) Uranium mineralization near Cameron, Arizona, *in* New Mexico Geol. Soc. Guidebook 9th Field Conf., Black Mesa Basin, northeastern Arizona: 164–168.

Boswell, P. J., and R. Blanchard (1927) Oxidation products derived from sphalerite and galena. *Econ. Geol.* 22: 419–453.

Bowen, N. L., and R. W. G. Wyckoff (1926) A petrologic and X-ray study of the thermal dissociation of dumortierite (from Clip, Arizona). *Jour. Wash. Acad. Sci.* 16: 178–189.

Bowles, O. (1940) Onyx marble and travertine. *U.S. Bur. Mines Inf. Circ.* 6751R.

Bowman, H. L. (1903) Note on the refractive indices of pyromorphite, mimetite, and vanadinite. *Min. Mag.* 13: 324–329.

Brady, L. F. (1929) Pre-historic Arizona meteorite. *Pan-Amer. Geol.* 51: 287–288.

——— (1931) A suspected meteoric specimen from northern Arizona. *Amer. Jour. Sci.* 21, *5th Series:* 173–177.

Bramlette, M. N., and E. Posnjak (1933) Zeolitic alteration of pyroclastics. *Amer. Min.* 18: 167–171.

Braun, E. R. (1969) Geology and ore deposits of the Marble Peak area, Santa Catalina Mountains, Pima County, Arizona. Univ. Arizona M.S. thesis, 75 p.

Brett, R. (1967) Metallic spherules in impactite and tektite glasses. *Amer. Min.* 52: 721–733.

———, and G. T. Higgins (1967) Cliftonite in meteorites: a proposed origin. *Science,* 156: 819–820.

———, and R. A. Yund (1964) Sulfur-rich bornites. *Amer. Min.* 49: 1084–1098.

Brinsmade, R. B. (1907) Copper in northern Arizona. *Engr. Min. Jour.* 84: 962.

Brittain, R. L. (1954) Geology and ore deposits of the western portion of the Hilltop mine area, Cochise County, Arizona. Univ. Arizona M.S. thesis, 97 p.

Bromfield, C. S., and A. F. Schride (1956) Mineral resources of the San Carlos Indian Reservation, Arizona. *U.S. Geol. Survey Bull.* 1027-N: 613–690.

Bromfield, J. P. (1950) Geology of the Maudina mine area, northern Santa Catalina Mountains, Pinal County, Arizona. Univ. Arizona M.S. thesis, 63 p.

Brophy, G. P., and M. F. Sheridan (1965) Sulfate studies IV: The jarosite-natrojarosite-hydronium jarosite solid solution series. *Amer. Min.* 50: 1595–1607.

Brown, R. L. (1926) Geology and ore deposits of the Twin Buttes district. Univ. Arizona M.S. thesis, 40 p.

Brown, W. H. (1939) Tucson Mountains, an Arizona basin and range type. *Geol. Soc. Amer. Bull.* 50: 697–760.

Browne, J. F. (1958) The geology of the Cuprite mine area, Pima County. Univ. Arizona M.S. thesis, 39 p.

Brownell, G. M. (1959) A beryllium detector for field exploration. *Econ. Geol.* 54: 1103–1114.

Brush, G. J. (1873) On a compact anglesite from Arizona. *Amer. Jour. Sci.* 5, *3rd Series:* 421–422.

Bryant, D. G. (1968) Intrusive breccias associated with ore, Warren (Bisbee) mining district, Arizona. *Econ. Geol.* 63: 1–12.

———, and H. E. Metz (1966) Geology and ore deposits of the Warren mining district, *in Geology of the porphyry copper deposits, southwestern North America,* S. R. Titley and C. L. Hicks (eds.); Univ. Ariz. Press, Tucson: 189–203.

Bryner, L. (1959) Geology of the South Comobabi Mountains and Ko Vaya Hills, Pima County, Arizona. Univ. Arizona Ph.D. dissertation, 156 p.

Buddhue, J. D. (1940) An analysis of lawrencite in the Mount Elden, Arizona, meteorite. *Pop. Astron.* 48: 561.

Buerger, M. S. (1942) The unit cell and space group of claudetite, $As_2O_2$ (Abstr.) *Amer. Min.* 27: 216.

Buerger, N. W. (1942) X-ray evidence of the existence of the mineral digenite, $Cu_9S_5$. *Amer. Min.* 27: 712–716.

Bunch, J. E., and L. H. Fuchs (1969) A new mineral: brezinaite, $Cr_3S_4$, and the Tucson meteorite. *Amer. Min.* 54: 1509–1518.

Burchard, E. F. (1930) Iron ore on Canyon Creek, Fort Apache Indian Reservation. *U.S. Geol. Survey Bull.* 821-C: 51–75.

——— (1943) Results of exploration for iron ore in far western states. *Econ. Geol.* 38: 85–86.

Buseck, P. R., and C. B. Moore (1966) A coarse octahedrite from Bloody Basin, Arizona. *Jour. Ariz. Acad. Sci.* 4: 67–70.

Butler, B. S. (1929) Strontium deposits near Aguila, Arizona. *Dept. Interior Press Mem.* 31445.

———, E. D. Wilson, et al. (1938) Some Arizona ore deposits. *Ariz. Bur. Mines Bull.* 145, *Geol. Series* 12: 136 p.

———, E. D. Wilson, and C. A. Rasor (1938) Geology and ore deposits of the Tombstone district, Arizona. *Ariz. Bur. Mines Bull.* 143, *Geol. Series* 10: 114 p.

Butler, G. M. (1928) Corrections to Volume 13. *Amer. Min., 13,* 594.

———, and M. A. Allen (1921) Uranium and vanadium. *Ariz. Bur. Mines Bull.* 117, *Mineral Tech. Series* 27.

Campbell, I. (1936) On the occurrence of sillimanite and staurolite in Grand Canyon. *Grand Canyon Nat. Hist. Assoc. Bull.* 5: 17–22.

———, and E. T. Schenk (1950) Camptonite dikes near Boulder Dam, Arizona. *Amer. Min.* 35, 9–10: 671–692.

Canney, F. C., W. L. Lehmbeck, and F. E. Williams (1967) Mineral resources of the Pine Mountain primitive area, Arizona. *U.S. Geol. Survey Bull.* 1230-J: 1–45.

Cannon, R. S., A. P. Pierce, J. C. Antweiler, and K. L. Buck (1961) The data of lead isotope geology related to problems of ore genesis. *Econ. Geol.* 56: 1–38.

Carter, N. L., and G. C. Kennedy (1964) Origin of diamonds in the Canyon Diablo and Novo Urei meteorites. *Jour. Geophys. Res.* 69: 2403–2421.

Carter, T. L. (1911) Gold placers in Arizona. *Engr. Min. Jour.* 90: 561.

Catanzaro, E. J., and J. L. Kulp (1964) Discordant zircons from the Little Belt (Montana), Beartooth (Montana) and Santa Catalina (Arizona) mountains. *Geochim. Cosmochim. Acta* 28: 87–124.

Cederstrom, D. J. (1946) Geology of the central Dragoon Mountains, Arizona. Univ. Arizona Ph.D. dissertation, 93 p.

Cesbron, F. (1964) Contribution à la minéralogie des sulfates de fer hydraté. *Bull. Soc. Franc. Min. Crist.* 87: 125–143.

Chaffee, M. A. (1964) Dispersion patterns as a possible guide to ore deposits in the Cerro Colorado district, Pima County, Arizona. Univ. Arizona M.S. thesis, 69 p.

Chao, E. C. T., J. J. Fahey, and J. Littler (1962) Stishovite, a very high pressure new mineral from Meteor Crater, Arizona. *Jour. Geophys. Res.* 67: 419–421.

——, E. M. Shoemaker, and B. M. Madsen (1960) First natural occurrence of coesite. *Science* 132: 220–222.

Chapman, T. L. (1947) San Manuel copper deposit, Pinal County, Arizona. *U.S. Bur. Mines, Rept. Inv.* 4108.

Chenoweth, W. L. (1955) The geology and the uranium deposits of the northwest Carrizo area, Apache County, Arizona, *in* Four Corners Geol. Soc. (1st) Field Conf., *Guidebook of parts of Paradox, Black Mesa, and San Juan Basins:* 177–185.

—— (1967) The uranium deposits of the Lukachukai Mountains, *in* New Mexico Geol. Soc. Guidebook 18th Field Conf., *Guidebook of Defiance-Zuni-Mt. Taylor region, Arizona and New Mexico:* 78–85.

Christ, C. L., and J. R. Clark (1955) The crystal structure of murdochite. *Amer. Min.* 40: 907–916.

Clark, J. L. (1956) Structure and petrology pertaining to a beryl deposit, Baboquivari Mountains, Arizona. Univ. Arizona M.S. thesis, 49 p.

Clarke, O. M., Jr. (1953) Geochemical prospecting for copper at Ray, Arizona. *Econ. Geol.* 48: 39–45.

Cochran, H. M. (1914) The petrography of the Twin Hills, Pima County, Arizona. Univ. Arizona M.S. thesis, 19 p.

Coleman, R. G., and M. Delavaux (1957) Occurrence of selenium in sulphides from some elementary rocks of the western United States. *Econ. Geol.* 5: 499–527.

Cooper, J. R. (1957) Metamorphism and volume losses in carbonate rocks near Johnson Camp, Cochise County, Arizona. *Geol. Soc. Amer. Bull.* 68: 577–610.

—— (1960) Some geologic features of the Pima mining district, Pima County, Arizona. *U.S. Geol. Survey Bull.* 1112-C: 63–103.

—— (1961) Turkey-track porphyry — a possible guide for correlation of Miocene rocks in southeastern Arizona. *Ariz. Geol. Soc. Digest,* 4: 17–35.

——, and L. C. Huff (1951) Geological investigations and geochemical prospecting experiment at Johnson, Arizona. *Econ. Geol.* 46: 731–756.

——, and L. T. Silver (1964) Geology and ore deposits of the Dragoon quadrangle, Cochise County, Arizona. *U.S. Geol. Survey Prof. Paper* 416, 196 p.

Crawford, W. P. (1930) Notes on rickardite, a new occurrence. *Amer. Min.* 15: 272–273.

——, and F. Johnson (1937) Turquoise deposits of Courtland, Arizona. *Econ. Geol.* 32: 511–523.

Creasey, S. C. (1952) Geology of the Iron King mine, Yavapai County, Arizona. *Econ. Geol.* 47: 24–56.

—— (1959) Some phase relations in the hydrothermally altered rocks of porphyry copper deposits. *Econ. Geol.* 54: 351–373.

—— (1966) Hydrothermal alteration, *in Geology of the porphyry copper deposits, southwestern North America,* S. R. Titley and C. L. Hicks (eds.); Univ. Ariz. Press, Tucson: 51–74.

——, and J. D. Pelletier (1965) Geology of the San Manuel area, Pinal County, Arizona. *U.S. Geol. Survey Prof. Paper* 471, 64 p.

——, and G. L. Quick (1955) Copper deposits of part of Helvetia mining district, Pima County, Arizona. *U.S. Geol. Survey Bull.* 1027-F: 301–323.

Cress, S. M., and C. Feldman (1943) Platinum minerals identified in western alunite. *Engr. Mining Jour.* 144: 106.

Crittenden, M. D., F. Cuttitta, H. J. Rose, Jr., and M. Fleischer (1962) Studies of manganese oxide minerals VI. Thallium in some manganese oxides. *Amer. Min.* 47: 1461–1467.

Culin, L., Jr. (1915–16) Gypsum. *Ariz. Bur. Mines Bull.* 19, *Mineral Tech. Series* 10: 8 p.

—— (1916) Magnesite. *Ariz. Bur. Mines Bull.* 14, *Mineral Tech. Series* 7: 8 p.

—— (1916–17) Celestite and strontianite. *Ariz. Bur. Mines Bull.* 35, *Mineral Tech. Series* 13: 4 p.

—— (1916–17) Gems and precious stones of Arizona. *Ariz. Bur. Mines Bull.* 48, *Mineral Tech. Series* 17: 7 p.

Culin, L., Jr. *(cont'd)*
—— (1917–18) Mica. *Ariz. Bur. Mines Bull.* 16, *Mineral Tech. Series* 8: 10 p.

Cummings, J. B. (1946a) Exploration of the Packard fluorspar property, Gila County, Arizona. *U.S. Bur. Mines Rept. Inv.* 3880.

—— (1946b) Exploration of New Planet iron deposit, Yuma County, Arizona. *U.S. Bur. Mines Rept. Inv.* 3982.

Cunningham, J. E. (1964) Geology of the north Tumacacori foothills, Santa Cruz County, Arizona. Univ. Arizona Ph.D. dissertation, 139 p.

Dale, U. B. (1959) Tungsten deposits of Yuma, Maricopa, Pinal, and Graham Counties, Arizona. *U.S. Bur. Mines Rept. Inv.* 5516.

—— (1961) Tungsten deposits of Gila, Yavapai, and Mohave Counties, Arizona. *U.S. Bur. Mines Inf. Circ.* 8078.

——, L. A. Stewart, and W. A. McKinney (1960) Tungsten deposits of Cochise, Pima, and Santa Cruz Counties, Arizona. *U.S. Bur. Mines Rept. Inv.* 5650.

Davis, D. G., and W. Breed (1968) Rock fulgurites on the San Francisco Peaks, Arizona. *Plateau (Quarterly of the Mus. of Northern Arizona)* 41: 34.

Davis, J. D. (1971) The distribution and zoning of the radio elements potassium, uranium, and thorium in selected porphyry copper deposits. Univ. Arizona M.S. thesis, 130 p.

Davis, R. E. (1955) Geology of the Mary G mine area, Pima County, Arizona. Univ. Arizona M.S. thesis, 51 p.

Davis, S. R. (1975) The Hardshell silver deposit, Harshaw mining district, Santa Cruz County, Arizona (abstr.) Sympos. New Mex. Geol. Soc. and Ariz. Geol. Soc. *Abstr. Progr.* 6.

Day, D. T., and R. H. Richards (1906) Investigation of black sands from placer mines. *U.S. Geol. Survey Bull.* 285: 150–164.

Dean, K. C., H. D. Snedden, and W. W. Agey (1952) Concentration of oxide manganese ores from vicinity of Winkelman, Pinal County, Arizona. *U.S. Bur. Mines Rept. Inv.* 4848.

Deere, W. A., R. A. Howie, and J. Zussman (1962) *Rock forming minerals, vol. 3, Sheet silicates.* Longmans, Green and Co., Ltd., London.

DeKalb, C. (1895) Onyx marbles. *Transactions AIME* 25: 562–563.

Dell'Anna, L., and C. C. Garavelli (1967) Plancheite di Capo Calamita (Isola d'Elba). *Periodico Min., Roma* 36: 125–146.

Denton, T. C. (1947a) Old Reliable copper mine, Pinal County, Arizona. *U.S. Bur. Mines Rept. Inv.* 4006.

—— (1947b) Aravaipa lead-zinc deposits, Graham County, Arizona. *U.S. Bur. Mines Rept. Inv.* 4007.

——, and P. S. Haury (1946) Exploration of the Reward (Vekol) zinc deposit, Pinal County, Arizona. *U.S. Bur. Mines Rept. Inv.* 3975.

——, and C. A. Kumke (1949) Investigation of Snowball fluorite deposit, Maricopa County, Arizona. *U.S. Bur. Mines Rept. Inv.* 4540.

DeWolf, W. P. (1916) Tungsten in Arizona. *Engr. and Mining Jour.* 101: 680.

Diery, H. D. (1964) Petrography and petrogenetic history of a quartz monzonite intrusive, Swisshelm Mountains, Cochise County, Arizona. Univ. Arizona M.S. thesis, 100 p.

Diller, J. S. and J. E. Whitfield (1889) Dumortierite from Harlem, N.Y., and Clip, Arizona. *Amer. Jour. Sci., 3rd Series,* 37: 216–220.

Dimick, A. (1957) Arizona's Peloncillo agate. *Gems and Minerals,* 242: 24–27.

Dings, M. G. (1951) The Wallapai mining district, Cerbat Mountains, Mohave County, Arizona. *U.S. Geol. Survey Bull.* 978-E: 123–163.

Douglas, J. (1899) The Copper Queen mine, Arizona. *Transactions, AIME,* 29: 511–546.

Dreyer, R. M. (1939) Darkening of cinnabar in sunlight. *Amer. Min.* 24: 457–460.

Duke, A. (1960) *Arizona gem fields. 2nd ed.* Southwest Printers, Yuma, 132 p.

Dunham, M. S. (1937) The geology of the Blue Jay mine area, Helvetia, Arizona. Univ. Arizona M.S. thesis, 43 p.

Earley, J. W., B. B. Osthaus, and I. H. Milne (1953) Purification and properties of montmorillonite. *Amer. Min.* 38: 707–724.

Easton, A. J., and A. A. Moss (1966) The analysis of molybdates and tungstates. *Min. Mag.* 35: 995–1002.

Eastwood, R. L. (1970) A geochemical-petrological study of mid-Tertiary volcanism in parts of Pima and Pinal Counties, Arizona. Univ. Arizona Ph.D. dissertation, 212 p.

Eckel, E. B. (1930) Geology and ore deposits of the Mineral Hill area, Pima County, Arizona. Univ. Arizona M.S. thesis, 51 p.

—— (1930) Boxwork siderite: an analogous occurrence of silica and siderite. *Econ. Geol.* 25: 290–292.

Eidel, J. J. (1966) The crystallization and mineralization of a porphyry copper stock, Ithaca Peak, Mohave County, Arizona. *Econ. Geol.* 61: 1305–1306.

——, J. E. Frost, and D. M. Clippinger (1968) Copper-Molybdenum mineralization at Mineral Park, Mohave County, Arizona. *In* J. D. Ridge, Ed., Ore deposits of the United States, 1933–1967. *Am. Inst. Min. Met. Eng.* 1258–1281.

El Goresy, A. (1965) Mineralbestand and Strukturen der Graphit — und Sulfidenschlüsse in Eisenmeteoriten. *Geochim. et Cosmochim. Acta* 29: 1131–1151.

Elsing, M. J., and R. E. S. Heineman (1936) Arizona metal production. *Ariz. Bur. Mines Bull.* 140: 112 p.

Endlich, F. M. (1897) The Pearce mining district. *Engr. Mining Jour.* 63: No. 23.

Enlows, H. E. (1939) Geology and ore deposits of the Little Dragoon Mountains. Univ. Arizona Ph.D. dissertation, 52 p.

——— (1955) Welded tuffs of Chiricahua National Monument, Arizona. *Geol. Soc. Amer. Bull.* 66: 1215–1246.

Erickson, R. C. (1969) Petrology and geochemistry of the Dos Cabezas Mountains, Cochise County, Arizona. Univ. Arizona Ph.D. dissertation, 441 p.

Ernst, T. (1943) Über Schmelzglekgewict im System $Fe_2O_3$-FeO-$TiO_2$ und Bemerkungen über die Minerale Pseudobrookit und Arizonit. *Zeits. Angew. Min.* 4: 394–409.

Evensen, C. G., and I. B. Gray (1958) Evaluation of uranium ore guides, Monument Valley, Arizona and Utah. *Econ. Geol.* 53: 639-662.

Fahey, J. J. (1955) Murdochite, a new copper lead oxide mineral. *Amer. Min.* 40: 905–906.

——— (1964) Recovery of coesite and stishovite from Coconino Sandstone of Meteor Crater, Arizona. *Amer. Min.* 49: 1643–1647.

———, E. B. Daggett, and S. G. Gordon (1950) Wherryite, a new mineral from the Mammoth mine, Arizona. *Amer. Min.* 35: 93–98.

Faick, J. N. (1958) Geology of the Ord mine, Mazatzal Mountains quicksilver district, Arizona. *U.S. Geol. Survey Bull.* 1042-R: 685–698.

———, and F. A. Hildebrand (1958) An occurrence of molybdenian stolzite in Arizona. *Amer. Min.* 43: 156–159.

Fairbanks, E. E. (1923) Mineragraphic notes on manganese minerals. *Amer. Min.* 8: 209.

Fanfani, L., A. Nunzi, and P. F. Fanazzi (1971) The crystal structure of butlerite. *Amer. Min.* 56: 751–757.

Farnham, L. L., and R. Havens (1957) Pikes Peak iron deposits, Maricopa County, Arizona. *U.S. Bur. Mines Rept. Inv.* 5319.

———, and L. A. Stewart (1958) Manganese deposits of western Arizona. *U.S. Bur. Mines Inf. Circ.* 7843.

———, ———, and C. W. DeLong (1961) Manganese deposits of eastern Arizona. *U.S. Bur. Mines Inf. Circ.* 7990.

Farrington, O. C. (1891) On crystallized azurite from Arizona. *Amer. Jour. Sci.* 41, *3rd Series*: 300–307.

Fearing, J. L., Jr. (1926) Some notes on the geology of the Jerome district, Arizona. *Econ. Geol.* 21: 757–773.

Feiss, J. W. (1929) The geology and ore deposits of Hiltano Camp, Arizona. Univ. Arizona M.S. thesis, 41 p.

Ferraresso, C. (1967) Thermoluminescence of clay minerals. *Amer. Min.* 52: 1288–1296.

Feth, J. H., and J. W. Anthony (1949) Spheroidal structures in Arizona volcanics. *Amer. Jour. Sci.* 247: 791–801.

Field, W. (1966) Sulfur isotope method for discriminating between sulfates of hypogene and supergene origin. *Econ. Geol.* 61: 1428–1435.

Finch, W. I. (1959) Geology of uranium deposits in Triassic rocks of the Colorado Plateau region. *U.S. Geol. Survey Bull.* 1074-D: 125–164.

——— (1967) Geology of the epigenetic uranium deposits in the United States. *U.S. Geol. Survey Prof. Paper* 538, 121 p.

Finnell, T. L. (1957) Structural control of uranium ore at the Monument No. 2 mine, Apache County, Arizona. *Econ. Geol.* 52: 25–35.

———, C. G. Bowles, and J. H. Soulé (1967) Mineral resources of the Mount Baldy primitive area, Arizona. *U.S. Geol. Survey Bull.* 1230-H: H1–H14.

Fischer, R. P. (1937) Sedimentary deposits of copper, vanadium-uranium and silver in Southwestern United States. *Econ. Geol.* 32: 906–951.

Flagg, A. L. (1958) *Mineralogical journeys in Arizona.* Fred H. Bitner, Scottsdale, 93 p.

Fleischer, M. (1959) The geochemistry of rhenium, with special reference to its occurrence in molybdenite. *Econ. Geol.* 54: 1406–1413.

——— (1966) Index of new mineral names, discredited minerals, and changes of mineralogical nomenclature in volumes 1–50 of The American Mineralogist. *Amer. Min.* 51: 1247–1357.

———. (1975) *1975 glossary of mineral species.* Mineral. Record, Inc., Bowie, Maryland, 145 p.

———, and W. E. Richmond (1943) The manganese oxide minerals: a preliminary report. *Econ. Geol.* 38: 269–286.

Fletcher, L. (1890) The meteoric iron of Tucson. *Mineral Mag.* 9: 16–36.

Flinter, B. H. (1959) The alteration of Maylayan ilmenite grains and the question of "arizonite." *Econ. Geol.* 54: 720–729.

Foote, A. E. (1891) A new locality for meteoric iron with a preliminary notice of the discovery of diamonds in the iron. *Amer. Jour. Sci.* 42: 413–417.

——— (1891) A new locality for meteoric iron with a preliminary notice of the discovery of diamonds in the iron. *Amer. Assoc. for the Advanc. Sci. Proc.* 40: 279–283.

Foote, W. M. (1912) Preliminary note on the shower of meteoric stones near Holbrook, Navajo County, Arizona, July 19, 1912, including a reference to the Perseid swarm of meteors visible from July 11th to August 22d. *Amer. Jour. Sci.* 34: 437–456.

Ford, W. E. (1902) On the chemical composition of dumortierite (from Clip, Arizona). *Amer. Jour. Sci., 4th Series,* 14: 426–430.

——— (1914) New occurrences of spangolite. *Amer. Jour. Sci., 4th Series,* 38: 503–504.

———, and W. M. Beadley (1915) On the identity of footeite with connellite together with the description of two occurrences of the mineral. *Amer. Jour. Sci., 4th Series,* 39: 570–676.

Foshag, W. F. (1919) Famous mineral localities: Yuma County, Arizona. *Amer. Min.* 4: 149–150.

Frank, T. R. (1970) The geology and mineralization of the Saginaw mine area, Pima County, Arizona. Univ. Arizona M.S. thesis, 136 p.

Frenzel, A. B. (1898) A turquoise deposit in Mohave County, Arizona. *Engr. Mining Jour.* 66: 24.

Frondel, C. (1935) Catalog of mineral pseudomorphs in the American Museum. *Bull. Amer. Mus. Nat. Hist. LXVII.*

—— (1941) Paramelaconite: a tetragonal oxide of copper. *Amer. Min.* 26: 657–672.

—— (1943) Mineralogy of the oxides and carbonates of bismuth. *Amer. Min.* 28: 521–535.

—— (1949) Crystallography of spangolite. *Amer. Min.* 34: 181–187.

—— (1956) Mineral composition of gummite. *Amer. Min.* 41: 539–568.

—— (1958) Systematic mineralogy of uranium and thorium. *U.S. Geol. Survey Bull.* 1064, 400 p.

—— (1962) *The system of mineralogy, vol. III, Silica minerals.* John Wiley and Sons: 334.

——, and E. W. Heinrich (1942) New data on hetaerolite, hydrohetaerolite, coronadite and hollandite. *Amer. Min.* 27: 48–56.

——, and U. B. Marvin (1967) Lonsdaleite, a new hexagonal polymorph of diamond. *Nature* 214: 587–589.

——, and —— (1967) Lonsdaleite, a hexagonal polymorph of diamond. *Amer. Min.* 52: 1579.

——, and F. H. Pough (1944) Two new tellurites of iron: mackayite and blakeite, with new data on emmonsite and durdenite. *Amer. Min.* 29: 211–225.

Frondel, J. W. (1964) Variation of some rare earths in allanite. *Amer. Min.* 49: 1157–1177.

——, and F. E. Wickman (1970) Molybdenite polytypes in theory and occurrence. II. Some naturally-occurring polytypes of molybdenite. *Amer. Min.* 55: 1857–1875.

Funnell, J. E., and E. J. Wolfe (1964) *Compendium on nonmetallic minerals of Arizona.* Southwest Research Inst., San Antonio, 374 p.

Galbraith, F. W. (1935) Geology of the Silver King area, Superior, Arizona. Univ. Arizona Ph.D. dissertation, 154 p.

—— (1941) Minerals of Arizona. *Ariz. Bur. Mines Bull.* 149, 82 p.

—— (1947) Minerals of Arizona. 2nd ed., revised. *Ariz. Bur. Mines Bull.* 153, 97 p.

——, and D. J. Brennan (1959) Minerals of Arizona. 3rd. ed., revised. *Univ. Ariz. Phys. Sci. Bull.* 4: 116 p.

——, and T. H. Kuhn (1940) A new occurrence of dioptase in Arizona. *Amer. Min.* 25: 708–710.

Gardner, E. D. (1936) Gold mining and milling in the Black Mountains, western Mohave County, Arizona. *U.S. Bur. Mines Inf. Circ.* 6901.

Garrels, R. M., and E. S. Larsen, 3d. (1959) Geochemistry and mineralogy of the Colorado Plateau uranium ores. *U.S. Geol. Survey Prof. Paper* 320, 236 p.

Garrison, F. L. (1907) Notes on minerals. *Proc. of the Acad. of Nat. Sci. of Phil.* 59: 445.

Gastil, G. R. (1954) An occurrence of authigenic xenotime. *Jour. Sed. Petrol.* 24: 280–281.

—— (1958) Older Pre-Cambrian rocks of the Diamond Butte quadrangle, Gila County, Arizona. *Geol. Soc. Amer. Bull.* 69: 1495–1514.

Gavasci, A. T., and P. F. Kerr (1968) Uranium emplacement at Garnet Ridge. *Econ. Geol.* 63: 859–876.

Gebhardt, R. C. (1931) The geology and mineral resources of the Quijotoa Mountains. Univ. Arizona M.S. thesis, 63 p.

Genth, F. A. (1855) Analysis of the meteoric iron from Tuczon, Province of Sonora, Mexico. *Amer. Jour. of Sci., series 2,* 20: 119–120.

—— (1868) Contributions to mineralogy. *Amer. Jour. Sci., 3rd Series,* 40: 114.

—— (1887) Contributions to mineralogy. *Proc. Amer. Phil. Soc.* 24: 1–44.

—— (1890) Contributions to mineralogy. *Amer. Jour. Sci. 3rd Series,* 40: 114.

Gerrard, T. A. (1964) Environmental studies of the Fort Apache Member, Supai Formation (Permian), East-Central Arizona. Univ. Arizona Ph.D. dissertation, 187 p.

Gibson, E. K. (1970) Discovery of another meteorite specimen from the 1912 Holbrook, Arizona, fall site. *Meteorites* 5: 57–60.

Gigl, P. D., and F. Dachille (1968) Effect of pressure and temperature on the reversal transitions of stishovite. *Meteorites* 4: 123–136.

Gilluly, J. (1937) Geology and ore deposits of the Ajo quadrangle, Arizona. *Arizona Bur. Mines Bull., Geol. Series* 9, 83 p.

—— (1942) Mineralization of the Ajo copper district, Arizona. *Amer. Min.* 27: 222–223.

—— (1942) The mineralization of the Ajo copper district, Arizona. *Econ. Geol.* 37: 297–309.

—— (1946) The Ajo mining district, Arizona. *U.S. Geol. Survey Prof. Paper* 209, 112 p.

—— (1956) General geology of central Cochise County, Arizona. *U.S. Geol. Survey Prof. Paper* 281, 169 p.

Gordon, S. G. (1923) Recently described "bisbeeite" from the Grand Canyon is cyanotrichite. *Amer. Min.* 8: 92–93.

Graeber, E. J., and A. Rosenzweig (1971) The crystal structures of yavapaiite, $KFe(SO_4)_2$ and goldichite, $KFe(SO_4)_2 \cdot 4H_2O$. *Amer. Min.* 56: 1917–1933.

Granger, H. C. (1955) Dripping Spring quartzite, Arizona, *in* Geologic Investigations of radioactive deposits — Semi annual progress report. *U.S. Geol. Survey TEI* 540.

——— (1956) Dripping Spring quartzite, Geologic Investigations of radioactive deposits. *U.S. Dept. of the Interior TEI* 620: 204–209.

———, and R. B. Raup (1959) Uranium deposits in the Dripping Spring quartzite, Gila County, Arizona. *U.S. Geol. Survey Bull.* 1046-P: 415–486.

———, and ——— (1962) A reconnaissance study of uranium deposits in Arizona. *U.S. Geol. Survey Bull.* 1147-A: 1–54.

———, and ——— (1969) Geology of uranium deposits in the Dripping Spring quartzite, Gila County, Arizona. *U.S. Geol. Survey Prof. Paper* 595, 108 p.

Graybeal, F. T. (1962) The geology and gypsum deposits of the southern Whetstone Mountains, Cochise County, Arizona. Univ. Arizona M.S. thesis, 80 p.

——— (1972) The partition of trace elements among coexisting minerals in some Laramide intrusive rocks in Arizona. Univ. Arizona Ph.D. dissertation, 220 p.

Gregory, H. E. (1916) Garnet deposits on the Navajo Reservation, Arizona and Utah. *Econ. Geol.* 11: 223–230.

——— (1917) Geology of the Navajo Country — a reconnaissance of parts of Arizona, New Mexico, and Utah. *U.S. Geol. Survey Prof. Paper* 93, 161 p.

Grim, R. E. (1968) *Clay mineralogy*. McGraw-Hill, New York, 596 p.

Gross, M. P. (1969) Mineralization and alteration in the Greaterville district, Pima County, Arizona. Univ. Arizona M.S. thesis, 82 p.

Grout, F. E. (1946) Microscopic characters of vein carbonates. *Econ. Geol.* 41: 475–502.

Grundy, W. D. (1953) Geology and uranium deposits of the Shinarump conglomerate of Nokai Mesa, Arizona and Utah. Univ. Arizona M.S. thesis, 88 p.

Gruner, J. W., and L. Gardiner (1952) Mineral association in the uranium deposits of the Colorado Plateau and adjacent regions with special emphasis on those in the Shinarump formation, pt. 3, *Ann. Rept. July 1, 1951–June 30, 1952;* RMO-566, U.S. Atomic Energy Comm. Tech. Inf. Service, Oak Ridge, Tenn.

Guild, F. N. (1905) Petrography of the Tucson Mountains. *Amer. Jour. Sci., 4th Series,* 20: 313.

——— (1907) The composition of molybdite from Arizona. *Amer. Jour. Sci.* 23: 455–456.

——— (1910) *The mineralogy of Arizona.* The Chemical Publishing Co., Easton, Pa., 103 p.

——— (1911) Mineralogische Notizen. *Zeit. Krystal. und Mineral.* 49: 321–331.

——— (1917) A microscopic study of the silver ores and their associated minerals. *Econ. Geol.* 12: 297–353.

——— (1920) Flagstaffite, a new mineral from Arizona. *Amer. Min.* 5: 169–172.

——— (1921) The identity of flagstaffite and terpin hydrate. *Amer. Min.* 6: 133–135.

——— (1922) Flagstaffite, a new Arizona mineral, and its identity with terpin hydrate (abs.), *Science* 55: 543.

——— (1929) Copper pitch ore. *Amer. Min.* 14: 313–318.

——— (1930) The relation of pyrite to wolframite. *Amer. Min.* 15: 451–452.

——— (1934) Microscopic relations of magnetite, hematite, pyrite and chalcopyrite. *Econ. Geol.* 29: 107–120.

——— (1935) Piedmontite in Arizona. *Amer. Min.* 20: 679–692.

Guillebert, C., and M.-T. Le Bihan (1965) Contribution à l'étude structurale des silicates de cuivre: structure atomique de la papagoite. *Bull. Soc. Franc. Min. Crist.* 88: 119–121.

Guillemin, C. (1956) Contribution à la mineralogie des arséniates, phosphates, et vanadates de cuivre. *Bull. Soc. Franc. MM* 79: 7–95; 219–275.

Guiteras, J. R. (1936) Gold mining and milling in the Black Canyon area, Yavapai County, Arizona. *U.S. Bur. Mines Inf. Circ.* 6905.

Hagner, A. E. (1939) Absorptive clays of the Texas Gulf coast. *Amer. Min.* 24: 67–108.

Halva, C. J. (1961) A geochemical investigation of "basalts" in southern Arizona. Univ. Arizona M.S. thesis, 88 p.

Hamilton, P. (1884) *The resources of Arizona, 3rd ed.* A. L. Bancroft & Co., San Francisco.

Hamilton, P. and P. F. Kerr (1959) Umohoite from Cameron, Arizona. *Amer. Min.* 44: 1248–1260.

Hammer, D. F. (1961) Geology and ore deposits of the Jackrabbit area, Pinal County, Arizona. Univ. Arizona M.S. thesis, 156 p.

———, and D. W. Peterson (1968) Geology of the Magma mine area, Arizona. *In* J. D. Ridge, Ed., Ore deposits of the United States, 1933–1967. *Am. Inst. Min. Met. Eng.:* 1282–1310.

Handverger, P. H. (1963) Geology of the Three R mine, Palmetto mining district, Santa Cruz County, Arizona. Univ. Arizona M.S. thesis, 70 p.

Hanson, H. S. (1966) Petrography and structure of the Leatherwood quartz diorite, Santa Catalina Mountains, Pima County, Arizona. Univ. Arizona Ph.D. dissertation, 104 p.

Harcourt, G. A. (1937) The distinction between enargite and famatinite (luzonite). *Amer. Min.* 22: 517–525.

——— (1942) Tables for the identification of ore minerals by x-ray powder patterns. *Amer. Min.* 27: 63–113.

Hardas, A. V. (1966) Stratigraphy of gypsum deposits, south of Winkelman, Pinal County, Arizona. Univ. Arizona M.S. thesis, 45 p.

Harness, C. L. (1942) Strontium minerals. *U.S. Bur. Mines Inf. Circ.* 7200.

Harrer, C. M. (1964) Reconnaissance of iron resources in Arizona. *U.S. Bur. Mines Inf. Circ.* 8236.

Harshman, E. N. (1940) Geology of the Belmont-Queen Creek area, Superior, Arizona. Univ. Arizona Ph.D. dissertation, 167 p.

Haury, P. S. (1947) Examination of lead-zinc mines in the Wallapai mining district, Mohave County, Arizona. *U.S. Bur. Mines Rept. Inv.* 4101.

Havens, R., S. J. Hussey, J. A. McAllister, and K. C. Dean (1954) Beneficiation of oxide manganese and manganese-silver ores from southern Arizona. *U.S. Bur. Mines Rept. Inv.* 5024.

———, G. M. Potter, W. W. Agey, and R. R. Wells (1947) Concentration of oxide manganese ores from the Lake Havasu district, California and Arizona. *U.S. Bur. Mines Rept. Inv.* 4147.

Head, R. E. (1941) Physical characteristics of some low-grade manganese ores. *U.S. Bur. Mines Rept. Inv.* 3560.

Headden, W. P. (1903) Mineralogical notes — Cuprodescloizite, Arizona. *Colo. Sci. Soc. Proc.* 7: 149–150.

Heatwole, D. A. (1966) Geology of the Box Canyon area, Santa Rita Mountains, Pima County, Arizona. Univ. Arizona M.S. thesis, 70 p.

Hedge, C. E. (1961) Sodium potassium ratios in muscovites as a geothermometer. Univ. Arizona M.S. thesis, 29 p.

Heineman, R. E. S. (1927) The geology and ore deposits of the Johnson mining district, Arizona. Univ. Arizona M.S. thesis, 45 p.

——— (1930) A note on the occurrence of monazite in western Arizona. *Amer. Min.* 15: 536–537.

——— (1931) An Arizona gold nugget of unusual size. *Amer. Min.* 12: 267–269.

——— (1935) Sugarloaf Butte alunite. *Eng. Mining Jour.* 136: 138–139.

———, and L. F. Brady (1929) The Winona meteorite. *Amer. Jour. Sci.* 18: 477–486.

Heinrich, E. W. (1960) Some rare-earth mineral deposits in Mohave County, Arizona. *Ariz. Bur. Mines Bull.* 167, *Min. Tech. Series* 51, 22 p.

Henderson, E. P., and S. H. Perry (1949) The Pima County (Arizona) meteorite. *Proc. U.S. Nat. Mus.* 99: 353–355.

———, and ——— (1951) A reinvestigation of the Weaver Mountains, Arizona, meteorite. *Pop. Astron.* 59: 263–266.

Hernon, R. M. (1932) Pegmatitic rocks of the Catalina-Rincon mountains, Arizona. Univ. Arizona M.S. thesis, 65 p.

Hess, F. L. (1909) Note on a wolframite deposit in the Whetstone Mountains, Arizona. *U.S. Geol. Survey Bull.* 380-D: 164–165.

Hewett, D. F. (1925) Carnotite discovered near Aguila, Arizona. *Engr. Mining Jour.* 120: 19.

——— (1964) Veins of hypogene manganese oxide minerals in the southwestern United States. *Econ. Geol.* 59: 1429–1472.

——— (1972) Manganite, hausmannite, braunite: features, modes of origin. *Econ. Geol.* 67: 83–102.

———, E. Callaghan, B. N. Moore, T. B. Nolan, W. W. Ruby, and W. T. Schaller (1936) Mineral resources of the region around Boulder Dam. *U.S. Geol. Survey Bull.* 871: 197.

———, and M. Fleischer (1960) Deposits of the manganese oxides. *Econ. Geol.* 55: 1–55.

———, ———, and N. Conklin (1963) Deposits of the manganese oxides: supplement. *Econ. Geol.* 58: 1–51.

———, and A. S. Radtke (1967) Silver-bearing black calcite in western mining districts. *Econ. Geol.* 62: 1–21.

———, and O. N. Rove (1930) Occurrence and relations of alabandite. *Econ. Geol.* 25: 36–56.

Hey, M. H. (1966) Catalogue of meteorites, 3rd ed. *British Museum (Nat. Hist.), London,* 637 p.

———, and E. E. Fejer (1962) The identity of erionite and offretite. *Mineral. Mag.* 33: 66–67.

Heyman, A. M. (1958) Geology of the Peach-Elgin copper deposit, Helvetia district, Arizona. Univ. Arizona M.S. thesis, 66 p.

Higdon, C. E. (1935) Geology and ore deposits of the Sunshine area, Pima County, Arizona. Univ. Arizona M.S. thesis, 25 p.

Hill, J. M. (1910) Notes on the placer deposits of Greaterville, Arizona. *U.S. Geol. Survey Bull.* 430: 11–22.

——— (1910) Note on the occurrence of tungsten minerals near Calabasas, Arizona. *U.S. Geol. Survey Bull.* 430: 164–166.

——— (1914) Copper deposits of the White Mesa district, Arizona. *U.S. Geol. Survey Bull.* 540-D: 159–163.

——— (1914) The Grand Gulch mining region, Mohave County, Arizona. *U.S. Geol. Survey Bull.* 580: 39–58.

Hillebrand, J. R. (1953) Geology and ore deposits in the vicinity of Pitman Wash, Pinal County, Arizona. Univ. Arizona M.S. thesis, 94 p.

Hillebrand, W. F. (1885) Emmonsite, a ferric tellurite. *Proc. Colo. Sci. Soc.* 2: 20–23.

——— (1886) Brief notice of some recently described minerals. *Amer. Jour. Sci.* 31: 476.

——— (1889a) Antlerite, a basic cupric sulfate. *U.S. Geol. Survey Bull.* 55: 54–55.

——— (1889b) Analyses of three descloizites from new localities. *Amer. Jour. Sci.* 37: 434–439.

Hills, R. C. (1882) Dioptase from Arizona. *Amer. Jour. Sci., 3rd Series,* 23: 325.

——— (1890) Informal communication. *Colo. Sci. Soc. Proc.* 3: 257–258.

Himes, M. D. (1972) Geology of the Pima mine, Pima County, Arizona. Univ. Arizona M.S. thesis, 92 p.

Hobbs, W. W. (1905) Contributions from the Mineralogical Laboratory of the University of Wisconsin. *Amer. Geol.* 36: 179–184.

——— (1944) Tungsten deposits in the Boriana district and the Aquarius Range, Mohave County,

Arizona. *U.S. Geol. Survey Bull.* 940-I: 247–264.

Hoffman, V. J. (1963) Heavy mineral distribution in sands of the Tortolita Mountain pediment, southern Arizona. Univ. Arizona M.S. thesis, 72 p.

Holden, E. F. (1922) Ceruleofibrite, a new mineral. *Amer. Min.* 17: 80–83.

———— (1924) Ceruleofibrite in connellite. *Amer. Min.* 9: 55–56.

Holland, H. D., G. G. Witter, Jr., W. B. Head, and R. Petti (1958) The use of leachable uranium in geochemical prospecting on the Colorado Plateau — II, The distribution of leachable uranium in surface samples in the vicinity of ore bodies. *Econ. Geol.* 53: 190–209.

Honea, R. M. (1959) New data on gastunite, an alkali uranyl silicate. *Amer. Min.* 44: 1047–1056.

———— (1961) New data on boltwoodite, an alkali uranyl silicate. *Amer. Min.* 46: 24.

Houser, F. N. (1949) The geology of the Contention mine area, Twin Buttes, Arizona. Univ. Arizona M.S. thesis, 61 p.

Hovey, E. D. (1900) Note on a calcite group from Bisbee, Arizona. *Bull. of the Amer. Museum of Nat. Hist.* 12: 189–190.

Huff, L. C. (1952) Abnormal copper, lead, and zinc content of soil near metalliferous veins. *Econ. Geol.* 47: 517–542.

————, E. Santos, and R. G. Raabe (1966) Mineral resources of the Sycamore Canyon primitive area, Arizona. *U.S. Geol. Survey Bull.* 1230-F: 1–19.

Hunt, W. F., and E. H. Krause (1916) Note on the variable composition of melanochalcite. *Amer. Jour. Sci.* 41, *4th Series:* 211–214.

Hurlbut, C. S., Jr. (1936) A new phosphate, bermanite, occurring with triplite in Arizona. *Amer. Min.* 21: 656–661.

————, and L. F. Aristarian (1968) Bermanite and its occurrence in Córdoba, Argentina. *Amer. Min.* 53, pt. 1: 416–431.

Hutton, C. O. (1957) Sengierite from Bisbee, Arizona. *Amer. Min.* 42: 408–411.

———— (1959a) Yavapaiite, an anhydrous potassium, ferric sulfate from Jerome, Arizona. *Amer. Min.* 44: 1105–1114.

———— (1959b) An occurrence of pseudomalachite at Safford, Arizona. *Amer. Min.* 44: 1298–1301.

————, and A. C. Vlisidis (1960) Papagoite, a new copper-bearing mineral from Ajo, Arizona. *Amer. Min.* 45: 599–611.

Ipsen, A. D., and H. L. Gibbs (1952) Concentration of oxide manganese ore from Doyle-Smith claims, northern Yuma County, Arizona. *U.S. Bur. Mines Rept. Inv.* 4844.

Irvin, G. W. (1959) Pyrometasomatic deposits at San Xavier mine, *in* Southern Arizona Guidebook II, combined with the Ariz. Geol. Soc. Digest: 195–197.

Isachsen, Y. W., T. W. Mitcham, and H. B. Wood (1955) Age and sedimentary environments of uranium host rocks, Colorado Plateau. *Econ. Geol.* 50: 127–134.

Jackson, D. (1955) Turquoise in eastern Arizona. *Gems and Minerals,* 212: 52–55.

Jahns, R. H. (1952) Pegmatite deposits of the White Picacho district, Maricopa and Yavapai counties, Arizona. *Ariz. Bur. Mines Bull.* 162, *Min. Tech. Series* 46, 105 p.

———— (1953) The genesis of pegmatites. *Amer. Min.* 38: 563–598.

Jenkins, O. P. (1923) Verde River lake beds near Clarkdale, Arizona. *Amer. Jour. Sci.* 5, *5th Series:* 65–81.

————, and E. D. Wilson (1920) A geological reconnaissance of the Tucson and Amole Mountains. *Ariz. Bur. Mines Bull.* 106, *Geol. Series* 2: 17.

Jensen, M. L. (1958) Sulphur isotopes and the origin of sandstone-type uranium deposits. *Econ. Geol.* 53: 598–616.

Johnson, H. S., and W. Thordarson (1966) Uranium deposits of the Moab, Monticello, White Canyon, and Monument Valley districts, Utah and Arizona. *U.S. Geol. Survey Bull.* 1222-H: 1–53.

Johnson, V. H. (1941) Geology of the Helvetia mining district, Arizona. Univ. Arizona Ph.D. dissertation, 111 p.

Johnston, W. P., and D. J. Lowell (1961) Geology and origin of mineralized breccia pipes in Copper Basin, Arizona. *Econ. Geol.* 56: 916–940.

Jones, E. L., Jr. (1915) Gold deposits near Quartzsite, Arizona. *U.S. Geol. Survey Bull.* 620: 45–57.

———— (1916) A reconnaissance in the Kofa Mountains, Arizona. *U.S. Geol. Survey Bull.* 620-H: 151–164.

————, and F. L. Ransome (1920) Deposits of manganese ore in Arizona. *U.S. Geol. Survey Bull.* 710-D: 93–184.

Jones, R. D. (1957) Geology of the Cerro Colorado mining district. Univ. Arizona M.S. thesis, 59 p.

Joralemon, I. B. (1914) The Ajo copper district, Arizona. *Engr. Mining Jour.* 98: 663.

———— (1952) Age cannot wither her varieties of geological experience. *Econ. Geol.* 47: 243–259.

Joseph, P. E. (1915–16) Molybdenum. *Ariz. Bur. Mines Bull.* 5, *Mineral Tech. Series* 3, 15 p.

———— (1915–16) Asbestos. *Ariz. Bur. Mines Bull.* 8, *Mineral Tech. Series* 4, 10 p.

———— (1916) Primary or secondary origin of wulfenite. Univ. Arizona M.S. thesis, 23 p.

———— (1916) Mercury-quicksilver. *Ariz. Bur. Mines Bull* 12, *Mineral Tech. Series* 6, 8 p.

———— (1916) Vanadium. *Ariz. Bur. Mines Bull.* 18, *Mineral Tech. Series* 9, 8 p.

———— (1916–17) Copper. *Ariz. Bur. Mines Bull.* 37, *Mineral Tech. Series* 14, 14 p.

Journeay, J. A. (1959) Pyrometasomatic deposits at Pima mine, *in* Southern Arizona Guidebook II, combined with the 2nd Ariz. Geol. Soc. Digest: 198–199.

———, R. E. Thurmond, et al. (1958) Pima: a three-part story — geology, open pit, milling. *Min. Eng.* 10: 453–462.

Kalousek, G. L., and L. E. Muttort (1957) Studies on the chrysotile and antigorite components of serpentine. *Amer. Min.* 42: 1–22.

Karkhanavala, M. D. (1959) The nature of arizonite. *Econ. Geol.* 54: 1302–1308.

Kartchner, W. E. (1944) The geology and ore deposits of the Harshaw district, Patagonia Mountains, Arizona. Univ. Arizona Ph.D. dissertation, 100 p.

Kauffman, A. J., Jr. (1943) Fibrous sepiolite from Yavapai County, Arizona. *Amer. Min.* 28: 512–520.

———, and H. W. Jaffe (1946) Chevkinite (tscheffkinite) from Arizona. *Amer. Min.* 31: 582–588.

Keith, S. B. (1972) Mineralogy and paragenesis of the 79 mine lead-zinc-copper deposit. *Mineral. Record* 3: 247–264.

Keller, W. D. (1962) Clay minerals in the Morrison Formation of the Colorado Plateau. *U.S. Geol. Survey Bull.* 1150, 90 p.

Kellogg, L. O. (1906) Sketch of the geology and ore deposits of the Cochise mining district, Cochise County, Arizona. *Econ. Geol.* 1: 651–659.

Kemp, J. F. (1905) Secondary enrichment in ore deposits of copper. *Econ. Geol.* 1: 11–25.

——— (1907) Ore deposits at the contacts of intrusive rocks and limestones; and their significance as regards the general formation of veins. *Econ. Geol.* 1: 1–13.

Kerr, P. F. (1951) Alteration features at Silverbell, Arizona. *Geol. Soc. Amer. Bull.* 62: 451–480.

———, A. W. Thomas, and A. M. Langer (1963) The nature and synthesis of ferrimolybdite. *Amer. Min.* 48: 14–32.

Khin, B. S. (1970) Cornetite from Saginaw Hill, Arizona. *Mineral. Record,* 1: 117–118.

Kiersch, G. A. (1947) The geology and ore deposits of the Seventy-nine mine area. Gila County, Arizona. Univ. Arizona Ph.D. dissertation, 124 p.

——— (1949) Structural control and mineralization at the Seventy-nine mine, Gila County, Arizona. *Econ. Geol.* 44: 24–39.

——— (1955) Kaolin, *in* Mineral resources of the Navajo-Hopi Indian Reservations, Arizona-Utah, vol. II: 72–76. Univ. of Arizona, Tucson.

———, and W. D. Keller (1955) Bleaching clay deposits, *in* Mineral resources of the Navajo-Hopi Indian Reservations, Arizona-Utah, vol. II: 41–54. Univ. of Arizona, Tucson.

———, and ——— (1955) Bleaching clay deposits, Sanders-Defiance Plateau district, Navajo County, Arizona. *Econ. Geol.* 50: 469–494.

Kinnison, J. E. (1958) Geology and ore deposits of the southern section of the Amole mining district, Tucson Mountains, Pima County, Arizona. Univ. Arizona M.S. thesis, 123 p.

——— (1966) The Mission copper deposit, Arizona, *in Geology of the porphyry copper deposits, southwestern North America,* S. R. Titley and C. L. Hicks (eds.). Univ. Ariz. Press, Tucson: 281–287.

Knight, L. H., Jr. (1970) Structure and mineralization of the Oro Blanco mining district, Santa Cruz County, Arizona. Univ. Arizona Ph.D. dissertation, 172 p.

Knoerr, A., and M. Eigo (1963) Arizona's newest copper producer — the Christmas mine. *Engr. Mining Jour.* 164: 55.

Koch, S. (1882) Ueber den Wulfenit. *Zeit. Krystal. und Mineral.* 6: 389–409.

Koenig, G. A. (1891) On paramelaconite and the associated minerals. *Proc. Acad. Nat. Sci., Philad.* 284–291.

——— (1892) On paramelaconite and the associated minerals. *Amer. Jour. Sci., 3rd Series,* 43: 158.

——— (1902) On the new species melaconite and keweenawite. *Amer. Jour. Sci.* 14: 404.

Kokkoros, P. (1951) New mineral names (ktenasite). *Amer. Mineral.* 36: 381–382.

Krieger, M. H. (1965) Geology of the Prescott and Paulden quadrangles, Arizona. *U.S. Geol. Survey Prof. Paper* 467, 127 p.

Ksanda, C. J., and E. P. Henderson (1939) Identification of diamond in the Canyon Diablo iron. *Amer. Min.* 24: 677–680.

Kuellmer, F. J. (1960) Compositional variation of alkali feldspars in some intrusive rocks near Globe-Miami, Arizona. *Econ. Geol.* 55: 557–562.

Kuhn, T. H. (1938) Childs-Aldwinkle mine, Copper Creek, Arizona, *in* Some Arizona ore deposits. *Ariz. Bur. Mines Bull.* 145, *Geol. Series* 12: 127–130.

——— (1940) Geology and ore deposits of the Copper Creek, Arizona, area. Univ. Arizona Ph.D. dissertation, 147 p.

——— (1941) Pipe deposits of the Copper Creek area, Arizona. *Econ. Geol.* 36: 512–538.

——— (1951) Bunker Hill district, *in* Arizona zinc and lead deposits. *Ariz. Bur. Mines Bull.* 158, *Geol. Series* 19: 56–65.

Kumke, C. A. (1947) Zonia copper mine, Yavapai County, Arizona. *U.S. Bur. Mines Rept. Inv.* 4023.

———, C. K. Rose, F. D. Everett, and S. W. Hazen (1957) Mining investigations of manganese deposits in the Maggie Canyon area, Artillery

Mountains region, Mohave County, Arizona. *U.S. Bur. Mines Rept. Inv.* 5292.

Kunz, G. F. (1885) On remarkable copper minerals from Arizona. *Annals of the N.Y. Acad. Sci.* 3: 275–278.

——— (1887) Crystals of hollow quartz from Arizona. *Amer. Jour. Sci.* 34, *3rd Series:* 479.

——— (1905) Moissanite, a natural silicon carbide. *Amer. Jour. Sci.* 19: 396.

———, and C. W. Huntington (1893) On the diamond in the Cañon Diablo meteoric iron and on the hardness of carborundum. *Amer. Jour. Sci.* 46: 470–473.

Lacy, W. C., and S. R. Titley (1962) Geological developments in the Twin Buttes district. *Mining Congress Jour.* 48: 62–64.

Ladoo, R. B. (1923) Fluorspar mining in the western states. *U.S. Bur. Mines Rept. Inv.* 2480.

Laidley, R. A. (1966) An x-ray fluorescent analysis study of the distribution of selected elements within the Hopi Buttes volcanics, Navajo County, Arizona. Univ. Arizona Ph.D. dissertation, 99 p.

Laney, R. L. (1971) Weathering of the granodioritic rocks in the Rose Canyon Lake area, Santa Catalina Mountains, Arizona. Univ. Arizona Ph.D. dissertation, 201 p.

Larsen, E. S., and H. E. Vassar (1925) Chalcoalumite, a new mineral from Bisbee, Arizona. *Amer. Min.* 10: 79–83.

Lasky, S. G., and B. N. Webber (1944) Manganese deposits in the Artillery Mountains region, Mohave County, Arizona. *U.S. Geol. Survey Bull.* 936-R: 417–448.

———, and ——— (1949) Manganese resources of the Artillery Mountains region, Mohave County, Arizona. *U.S. Geol. Survey Bull.* 961, 86 p.

Laughlin, A. W. (1960) Petrology of the Molino Basin area of the Santa Catalina Mountains, Arizona. Univ. Arizona M.S. thesis, 53 p.

Laughon, R. B. (1970) New data on guildite. *Amer. Min.* 55: 502–505.

——— (1971) The crystal structure of kinoite. *Amer. Min.* 56: 193–200.

Laurent, Y., and R. Pierrot (1962) Nouvelles données sur la bisbeeite. *Bull. Soc. Franc. Min. Crist.* 85: 177–180.

Lausen, C. (1923) Geology of the Old Dominion mine, Globe, Arizona. Univ. Arizona M.S. thesis, 155 p.

——— (1926) Tourmaline-bearing cinnabar veins of the Mazatzal Mountains, Arizona. *Econ. Geol.* 21: 782–791.

——— (1927a) The occurrence of olivine bombs near Globe, Arizona. *Amer. Jour. Sci.* 14, *5th Series:* 293–306.

——— (1927b) Piedmonite from the Sulphur Springs Valley, Arizona. *Amer. Min.* 12: 283–287.

——— (1928) Hydrous sulphates formed under fumerolic conditions at the United Verde mine. *Amer. Min.* 13: 202–229.

——— (1931a) Gold veins of the Oatman and Katherine districts, Arizona. Univ. Arizona Ph.D. dissertation, 91 p.

——— (1931b) Geology and ore deposits of the Oatman and Katherine districts, Arizona. *Ariz. Bur. Mines Bull.* 131, *Geol. Series* 6, 126 p.

——— (1936) The occurrence of minute quantities of mercury in the Chinle shales at Lee's Ferry, Arizona. *Econ. Geol.* 31: 610–617.

———, and E. D. Wilson (1927a) Gold and copper deposits near Payson, Arizona. *Ariz. Bur. Mines Bull.* 120, *Geol. Series* 4, 37 p.

———, and ——— (1927b) Quicksilver (mercury) resources of Arizona. *Ariz. Bur. Mines Bull.* 122, *Mineral Tech. Series* 29, 113 p.

Lawson, A. C. (1913) The gold in the Shinarump (Chinle) at Paria. *Econ. Geol.* 8: 434–448.

Leavens, P. B. (1967) Reexamination of bermanite. *Amer. Min* 52: 1060–1066.

LeConte, J. L. (1852) Notice of meteoric iron in the Mexican province of Sonora. *Amer. Jour. of Sci., Series 2,* 13: 289–290.

Lee, C. A., and G. C. Borland (1935) The geology and ore deposits of the Cuprite mining district. Univ. Arizona M.S. thesis, 54 p.

Lee, L. C. (1967) The economic geology of portions of the Tombstone-Charleston district, Cochise County, Arizona, in light of 1967 silver economics. Univ. Arizona M.S. thesis, 99 p.

Legge, J. A., Jr. (1939) Paragenesis of the ore minerals of the Miami mine, Arizona. Univ. Arizona M.S. thesis, 47 p.

Lehner, R. E. (1958) Geology of the Clarkdale quadrangle, Arizona. *U.S. Geol. Survey Bull.* 1021-N: 511–592.

Leicht, W. C. (1971) Minerals of the Grandview mine. *Mineral. Record,* 2: 214–221.

Lewis, A. S. (1920) Ore deposits of Cave Creek district, Arizona. *Engr. Mining Jour.* 110: 713.

Lewis, D. V. (1955) Relationships of ore bodies to dikes and sills. *Econ. Geol.* 50: 495–516.

Lindgren, W. (1902) Copper deposits at Clifton, Arizona. *U.S. Geol. Survey Bull.* 213: 133–140.

——— (1903) The copper deposits of Clifton, Arizona. *Engr. Min. Jour.* 75: 705.

——— (1904) The genesis of copper deposits. *Engr. Min. Jour.* 78: 987.

——— (1905) The copper deposits of the Clifton-Morenci district, Arizona. *U.S. Geol. Survey Prof. Paper* 43, 375 p.

——— (1911) Some modes of deposition of copper ores in basic rocks. *Econ. Geol.* 6: 687–700.

Lindgren, W. (*cont'd*)

——— (1926) Ore deposits of the Jerome and Bradshaw Mountains quadrangle, Arizona. *U.S. Geol. Survey Bull.* 782, 192 p.

———, and W. F. Hillebrand (1904) Minerals from the Clifton-Morenci district, Arizona. *Amer. Jour. Sci.* 18, *4th Series:* 448–460.

Lipschutz, M. E. (1965) Origin of atypical meteorites from the Arizona meteorite crater. *Nature* 208: 636–638.

Loghry, D. L. (1972) Characteristics of favorable cappings from several southwestern porphyry copper deposits. Univ. Arizona M.S. thesis, 112 p.

Long, W. J., J. V. Batty, and K. C. Dean (1948) Concentration of miscellaneous oxide manganese ores from Yavapai, Yuma, Maricopa, and Mohave Counties, Arizona. *U.S. Bur. Mines Rept. Inv.* 4291.

Lootens, D. J. (1965) Structure and petrography of the east side of the Sierrita Mountains, Pima County, Arizona. Univ. Arizona Ph.D. dissertation, 225 p.

Loring, W. B. (1947) The geology and ore deposits of the Mountain Queen area, northern Swisshelm Mountains, Arizona. Univ. Arizona M.S. thesis, 65 p.

Lovering, T. S. (1948) Geothermal gradients, recent climate changes, and rate of sulphide oxidation in the San Manuel district, Arizona. *Econ. Geol.* 43: 1–20.

——— (1962) The origin of jasperoid in limestone. *Econ. Geol.* 57: 861–889.

———, L. C. Huff, and H. Almond (1950) Dispersion of copper from the San Manuel copper deposit, Pinal County, Arizona. *Econ. Geol.* 45: 493–514.

Lowell, J. D. (1955) Applications of cross-stratification studies to problems of uranium exploration, Chuska Mountains, Arizona. *Econ. Geol.* 50: 177–185.

——— (1956) Occurrence of uranium in Seth-La-Kai diatreme, Hopi Buttes, Arizona. *Amer. Jour. Sci.* 254: 404–412.

——— (1968) Geology of the Kalamazoo orebody, San Manuel district, Arizona. *Econ. Geol.* 63: 645–654.

———, and J. M. Guilbert (1970) Lateral and vertical alteration-mineralization zoning in porphyry ore deposits. *Econ. Geol.* 65: 373–408.

Ludden, R. W. (1950) Geology of the Campo Bonito area, Oracle, Arizona. Univ. Arizona M.S. thesis, 52 p.

Lynch, D. W. (1966) The economic geology of the Esperanza mine and vicinity, *in Geology of the porphyry copper deposits, southwestern North America*, S. R. Titley and C. L. Hicks (eds.). Univ. Arizona Press, Tucson: 267–279.

——— (1967) The geology of the Esperanza mine and vicinity, Pima County, Arizona. Univ. Arizona M.S. thesis, 70 p.

Mallet, J. W. (1906) A stony meteorite from Coon Butte, Arizona. *Amer. Jour. Sci.* 21: 347–355.

Marjaniemi, D. K. (1969) Geologic history of an ashflow sequence and its source area in the Basin and Range Province of southeastern Arizona. Univ. Arizona Ph.D. dissertation, 176 p.

Marshall, R. R., and O. Joensuu (1961) Crystal habit and trace element content of some galenas. *Econ. Geol.* 56: 758–771.

Marvin, T. C. (1942) The geology of the Hilton Ranch area, Pima County, Arizona. Univ. Arizona M.S. thesis, 60 p.

Mason, B. (1968) Kaersutite from San Carlos, Arizona, with comments on the paragenesis of this mineral. *Min. Mag.* 36: 997–1002.

———, and C. J. Vitaliano (1953) The mineralogy of the antimony oxides and antimonates. *Mineral. Mag.* 30: 100–112.

Masters, J. A. (1955) Geology of the uranium deposits of the Lukachukai Mountains area, northeastern Arizona. *Econ. Geol.* 50: 111–126.

Matter, P., III (1969) Petrochemical variations across some Arizona pegmatites and their enclosing rocks. Univ. Arizona Ph.D. dissertation, 173 p.

Maucher and Rehwald (1961) Bildkartei der Erzmikroskopie. *Umschau Verlog. Frankfurt am. Main.*

Mauger, R. L. (1966) A petrographic and geochemical study of Silver Bell and Pima mining districts, Pima County, Arizona. Univ. Arizona Ph.D. dissertation, 140 p.

Mayo, E. B. (1955) Copper, *in* Mineral resources of the Navajo-Hopi Indian Reservations, Arizona-Utah, vol. I, 19–32, Univ. Arizona, Tucson.

——— (1955) Manganese, *in* Mineral resources of the Navajo-Hopi Indian Reservations, Arizona-Utah, vol. I, 39–47, Univ. Arizona, Tucson.

Mayuga, M. N. (1942) The geology and ore deposits of the Helmet Peak area, Pima County, Arizona. Univ. Arizona Ph.D. dissertation, 124 p.

McCarn, H. L. (1904) The Planet copper mine. *Engr. Mining Jour.* 78: 26.

McClure, F. G. (1915–16) Gold placers of Arizona. *Ariz. Bur. Mines Bull.* 10, *Econ. Series* 5, 14 p.

McCurry, W. G. (1971) Mineralogy and paragenesis of the ores, Christmas mine, Gila County, Arizona. M.S. thesis, Arizona State University, 47 p.

McGetchin, T. R., L. J. Silver, and A. A. Chodos (1970) Titanclinohumite: a possible mineralogical site for water in the upper mantle. *Jour. Geophys. Res.* 75: 255–259.

McKee, E. H., and C. A. Anderson (1971) Age and chemistry of Tertiary volcanic rocks in northcentral Arizona and relation of the rocks to the Colorado Plateaus. *Geol. Soc. Amer. Bull.* 82: 2767–2782.

McKelvey, V. E., J. H. Wiese, and V. H. Johnson (1949) (1952) Preliminary report on the bedded manganese of the Lake Mead region, Nevada and Arizona. *U.S. Geol. Survey Bull.* 948-D: 83–101.

McLain, J. P. (1965) Petrographic analysis of lineaments in the San Francisco volcanic field, Coconino and Yavapai Counties, Arizona. Univ. Arizona M.S. thesis, 31 p.

McLean, W. J. (1970) Confirmation of the mineral species wherryite. *Amer. Min.* 55: 505–508.

———, and J. W. Anthony (1970) The crystal structure of hemihedrite. *Amer. Min.* 55: 1103–1114.

———, and ——— (1972) The disordered "zeolite-like" structure of connellite. *Amer. Min.* 57: 426–438.

———, R. A. Bideaux, and R. W. Thomssen (1974) Yedlinite, a new mineral from the Mammoth mine, Tiger, Arizona. *Amer. Mineral.* 59: 1157–1159.

Mead, C., W. J. Littler, and E. C. T. Chao (1965) Metallic spheroids from Meteor Crater, Arizona. *Amer. Min.* 50: 667–681.

Medhi, P. K. (1964) A geologic study of the Pontatoc mine area, Pima County, Arizona. Univ. Arizona M.S. thesis, 44 p.

Meeves, H. C. (1966) Nonpegmatite beryllium occurrences in Arizona, Colorado, New Mexico, Utah, and four adjacent states. *U.S. Bur. Mines Rept. Inv. R.I.* 6828, 68 p.

———, C. M. Harrer, M. H. Salsbury, A. S. Konselman, and S. S. Shannon, Jr. (1966) Reconnaissance of beryllium-bearing pegmatite deposits in six western states. *U.S. Bur. Mines Inf. Circ.* 8298.

Melhase, J. (1925) Asbestos deposits of Arizona. *Engr. Mining Jour.* 120: 805.

Merrill, G. P. (1895) The onyx marbles: their origin, composition and uses, both ancient and modern. *Report of the U.S. Nat. Museum for 1893.* 558: 561–564.

——— (1922) New meteorites. *Amer. Jour. Sci.* 3: 153–154.

——— (1927) On newly discovered meteoric irons from the Wallapai (Hualapai) Indian Reservation, Arizona. *Proc. U.S. Nat. Mus.* 72: 1–4.

Merrin, S. (1954) The Cretaceous stratigraphy and mineral deposits of the east face of Black Mesa, Apache County, Arizona. Univ. Arizona M.S. thesis, 93 p.

Merwin, H. E., and E. Posnjak (1937) Sulphate encrustations in the Copper Queen mine, Bisbee, Arizona. *Amer. Min.* 22: 567–571.

Merz, J. J. (1967) The geology of the Union Hill area, Silver Bell district, Pima County, Arizona. Univ. Arizona M.S. thesis, 58 p.

Metz, R. A., and A. W. Rose (1966) Geology of the Ray copper deposit, Ray, Arizona, *in Geol-*

ogy of the porphyry copper deposits, southwestern North America, S. R. Titley and C. L. Hicks (eds.). Univ. Ariz. Press, Tucson: 177–188.

Metzger, O. H. (1938) Gold mining and milling in the Wickenburg area, Maricopa and Yavapai Counties, Arizona. *U.S. Bur. Mines Inf. Circ.* 6991.

Michel, F. A., Jr. (1959) Geology of the King mine, Helvetia, Arizona. Univ. Arizona M.S. thesis, 59 p.

Mielke, J. E. (1964) Trace elements investigation of the "turkey track" porphyry, southeastern Arizona. Univ. Arizona M.S. thesis, 91 p.

Miles, C. H. (1965) Metamorphism and hydrothermal alteration in the Lecheguilla Peak area of the Rincon Mountains, Cochise County, Arizona. Univ. Arizona Ph.D. dissertation, 96 p.

Mills, J. W. and H. T. Eyrich (1966) The rate of unconformities in the localization of epigenetic mineral deposits in the United States and Canada. *Econ. Geol.* 61: 1232–1257.

Milton, C. (1944) Stones from trees. *Scient. Month.* 59, Nov.

———, and J. Axelrod (1947) Fused woodash stones: fairchildite (n. sp.) $K_2CO_3 \cdot CaCO_3$, buetschliite (n. sp.) $3K_2CO_3 \cdot 2CaCO_3 \cdot 6H_2O$, and calcite, $CaCO_3$, their essential components. *Amer. Min.* 32: 607–624.

Min, M. M. (1965) Petrography and alteration of the Kitt Peak area, Pima County, Arizona. Univ. Arizona M.S. thesis, 90 p.

Mitcham, T. W., and C. G. Evensen (1955) Uranium ore guides, Monument Valley district, Arizona. *Econ. Geol.* 50: 170–176.

Mitchell, G. J. (1920) Vertical extent of copper ore minerals in the Junction mine, Warren district, Arizona. *Engr. Min. Jour.* 109: 1411.

——— (1921) Rate of formation of copper sulfate stalactites. *Mining and Metallurgy* 170: 30.

Moger, S. R. (1969) The geology of the west central portion of the Patagonia Mountains, Santa Cruz County, Arizona. Univ. Arizona M.S. thesis, 60 p.

Montoya, J. (1967) A new occurrence of planchéite in Arizona. *Rocks and Minerals,* April: 277.

Moolick, R. T., and J. J. Durek (1966) The Morenci district, *in Geology of the porphyry copper deposits, southwestern North America,* S. R. Titley and C. L. Hicks (eds.). Univ. Arizona Press, Tucson: 221–231.

Moore, B. N. (1935) Some strontium deposits of southeastern California and western Arizona. *Trans. of the AIME* 115: 362–365.

——— (1936) Celestite and strontianite, *in Mineral resources of the region around Boulder Dam. U.S. Geol. Survey Bull.* 871: 151–154.

Moore, C. B., and S. L. Tackett (1963) The Bagdad Arizona iron meteorite (Abs.). *Jour. Ariz. Acad. Sci.* 2: 191.

Moore, R. T. (1968) Mineral deposits of the Fort Apache Indian Reservation, Arizona. *Ariz. Bur. Mines Bull.* 177, 84 p.

—————— (1969) Beryllium, *in* Mineral and water resources of Arizona. *Ariz. Bur. Mines Bull.* 180: 102–113.

—————— (1969) Vermiculite, *in* Mineral and water resources of Arizona. *Ariz. Bur. Mines Bull.* 180: 462–464.

—————— (1969) Gold, *in* Mineral and water resources of Arizona. *Ariz. Bur. Mines Bull.* 180: 156–167.

——————, and E. D. Wilson (1965) Bibliography of the geology and mineral resources of Arizona, 1848–1964. *Ariz. Bur. Mines Bull.* 173, 321 p.

Moores, R. C., III (1972) The geology and ore deposits of a portion of the Harshaw district, Santa Cruz County, Arizona. Univ. Arizona M.S. thesis, 98 p.

Morgan, W. C., and M. C. Tallmon (1904) A peculiar occurrence of bitumen and evidence as to its origin. *Amer. Jour. Sci.* 18: 363–377.

Morimoto, N., and A. Gyobu (1971) The composition and stability of digenite. *Amer. Min.* 56: 1889–1909.

Morris, D. F. C., and E. L. Short (1966) Minerals of rhenium. *Mineral. Mag.* 35: 871–873.

Moses, A. J. (1893) Mineralogical notes. *Amer. Jour. Sci.* 45: 489–492.

——————, and L. McI. Luquer (1892) Alabandite from Tombstone. *Univ. Columbia School of Mines Quarterly* 13: 236–239.

Moss, H. A. (1957) The nature of carphosiderite and allied basic sulphates of iron. *Mineral. Mag.* 31: 407–412.

Mouat, M. M. (1962) Manganese oxides from the Artillery Mountains area, Arizona. *Amer. Min.* 47: 744–757.

Moxham, R. M., R. S. Foote, and C. M. Bunker (1965) Gamma-ray spectrometer studies of hydrothermally altered rocks. *Econ. Geol.* 60: 653–671.

Mrose, Mary E., H. J. Rose, Jr., and J. W. Maninenko (1966) Synthesis and properties of fairchildite and buetschliite: their relationship in wood-ash stone formation. *Geol. Soc. Amer. Spec. Paper* 101: 146.

Mrose, M. E., and A. C. Vlisidis (1966) Proof of the formula of shattuckite, $Cu_5(SiO_3)_4 (OH)_2$. (Abs.). *Amer. Min.* 51: 266–267.

Mullens, R. L. (1967) Stratigraphy and environment of the Toroweap Formation (Permian) north of Ashfork, Arizona. Univ. Arizona M.S. thesis, 101 p.

Murdoch, J., and R. W. Webb (1966) Minerals of California. *Calif. Div. Mines and Geol. Bull.* 189, 559 p.

Nagy, B., and G. T. Faust (1956) Serpentines: natural mixtures of chrysotile and antigorite. *Amer. Min.* 41: 838.

Needham, A. B., and W. R. Storms (1956) Investigation of Tombstone district manganese deposits, Cochise County, Arizona. *U.S. Bur. Mines Rept. Inv.* 5188.

Nehru, C. E., and M. Prinz (1970) Petrologic study of the Sierra Ancha sill complex, Arizona. *Geol. Soc. Amer. Bull.* 81: 1733–1766.

Nelson, E. (1966) The geology of Picketpost Mountain, Pinal County, Arizona. Univ. Arizona M.S. thesis, 123 p.

Nelson, F. J. (1963) The geology of the Peña Blanca and Walker Canyon areas, Santa Cruz County, Arizona. Univ. Arizona M.S. thesis, 82 p.

Newberg, D. W. (1964) X-ray study of shattuckite. *Amer. Min.* 49: 1234–1239.

—————— (1967) Geochemical implications of chrysocolla-bearing alluvial gravels. *Econ. Geol.* 62: 932–956.

Newhouse, W. H. (1933) Mercury in native silver. *Amer. Min.* 18: 295–299.

—————— (1934) The source of vanadium, molybdenum, tungsten and chromium in oxidized lead deposits. *Amer. Min.* 19: 209–220.

Nininger, A. D. (1940) Third catalog of meteoritic falls. *Pop. Astron.* 48: 555–560.

Nininger, H. H. (1949) A new type of magnetometer survey at Barringer Crater. *Pop. Astron.* 57: 17–21.

—————— (1956) *Arizona's Meteorite Crater.* World Press, Denver, 232 p.

North, O. S., and N. C. Jensen (1958) Vermiculite, *in* Minerals Yearbook, 1954. *U.S. Bur. Mines:* 1307–1316.

Norton, J. J. (1965) Lithium-bearing benitonite deposit, Yavapai County, Arizona. *U.S. Geol. Survey Prof. Paper* 525-D: 163–166.

Nye, T. (1961) Geology of the Paymaster mine, Pima County, Arizona. *Ariz. Geol. Soc. Digest* vol. IV: 161–168.

—————— (1968) The relationship of structure and alteration to some ore bodies in the Bisbee (Warren) district, Cochise County, Arizona. Univ. Arizona Ph.D. dissertation, 212 p.

O'Hara, M. J., and E. L. P. Mercy (1966) Eclogite, peridotite and pyrope from the Navajo country, Arizona and New Mexico. *Amer. Min.* 51: 336–352.

Olmstead, H. W., and D. W. Johnson (1966) Inspiration geology, *in* Geology of the porphyry copper deposits, southwestern North America, S. R. Titley and C. L. Hicks (eds.). Univ. Arizona Press, Tucson: 143–150.

Olsen, E., and L. Fuchs (1968) Krinovite, $NaMg_2CrSi_3O_{10} =$ a new meteorite mineral. *Science* 161: 786–787.

Olson, G. G., and J. T. Long (1957) Diatomaceous earth in Arizona — a field and laboratory report. *Ariz. Devel. Board, Phoenix,* 28 p.

Olson, H. J. (1961) The geology of the Glove mine, Santa Cruz County, Arizona. Univ. Arizona M.S. thesis, 82 p.

—— (1966) Oxidation of a sulfide body, Glove mine, Santa Cruz County, Arizona. *Econ. Geol.* 61: 731–743.

Olson, J. C., and J. W. Adams (1962) Thorium and rare earths in the United States exclusive of Alaska and Hawaii. *U.S. Geol. Survey Mineral Inv. Resources Map* MR-28.

——, and E. N. Hinrichs (1960) Beryl-bearing pegmatites in the Ruby Mountains and other areas in Nevada and northwestern Arizona. *U.S. Geol. Survey Bull.* 1082-D, 200 p.

Omori, K., and P. F. Kerr (1963) Infrared studies of saline sulfate minerals. *Geol. Soc. Amer. Bull.* 74: 709–734.

Oosterwyck-Gastuche, Mme. van (1967) Etude des silicates de cuivre du Katanga. *Mus. Roy. Africa Centrale, Ann. Ser.* 8, 58.

——, and C. Gregoire (1971) Electron microscopy and diffraction identification of some copper silicates. *Proc. Int. Mineral. Assoc., 7th Gen. Meet., Tokyo,* 1970 (*Mineral. Soc. Japan Spec. Pap.* 1: 196–205). (In *Amer. Min.* 57: 1005–1006.)

Outerbridge, W. F., M. H. Staats, R. Meyrowitz, and A. H. Pommer (1960) Weeksite, a new uranium silicate from the Thomas Range, Juab County, Utah. *Amer. Min.* 45: 39–52.

Overholt, J. L., G. Vaux, and J. L. Rodda (1950) The nature of "arizonite." *Amer. Min.* 35: 117–119.

Overstreet, W. C. (1967) The geologic occurrence of monazite. *U.S. Geol. Survey Prof. Paper* 530, 327 p.

Pabst, A. (1938) Crystal structure and density of delafossite (Abstr.). *Amer. Min.* 23: 175–176.

—— (1961) X-ray crystallography of davidite. *Amer. Min.* 46: 700–718.

——, and R. W. Thomssen (1959) Davidite from the Quijotoa Mountains, Pima County, Arizona (Abs.). *Geol. Soc. Amer. Bull.* 70: 1739.

——, and C. D. Woodhouse (1964) Thalenite from Kingman, Arizona (abs.). *Geol. Soc. Amer. Special Paper* 82: 269.

Palache, C. I. (1926) Notes on new or incompletely described meteorites in the mineralogical museum of Harvard University. *Amer. Jour. Sci.* 12: 136–150.

—— (1934) Contributions to crystallography: claudetite, minasragrite, samsonite, native selenium, iridium. *Amer. Min.* 19: 194–205.

—— (1939a) Antlerite. *Amer. Min.* 24: 293–302.

—— (1939b) Brochantite. *Amer. Min.* 24: 463–481.

—— (1941a) Crystallographic notes: cahnite, stolzite, zincite, ultrabasite. *Amer. Min.* 26: 429–436.

—— (1941b) Diaboleite from Mammoth mine, Tiger, Arizona. *Amer. Min.* 26: 605–612.

—— (1950) Paralaurionite. *Mineral. Mag.* 29: 341–345.

——, H. Berman, and C. Frondel (1944) *The system of mineralogy,* 7th ed., vol. 1, John Wiley, New York, 834 p.

——, and F. A. Gonyer (1932) On babingtonite. *Amer. Min.* 17: 295–303.

——, and L. W. Lewis (1927) Crystallography of azurite from Tsumeb, Southwest Africa, and the axial ratio of azurite. *Amer. Min.* 12: 115–141.

——, and H. E. Merwin (1909) On connellite and chalcophyllite from Bisbee, Arizona. *Amer. Jour. Sci.* 28, *4th Series:* 537–540.

——, and E. V. Shannon (1920) Higginsite, a new mineral of the olivenite group. *Amer. Min.* 5: 155–157.

——, and H. E. Vassar (1926) A note on cyanotrichite. *Amer. Min.* 11: 213–214.

Palmer, C. (1909) Arizonite, ferric metatitanate. *Amer. Jour. Sci.* 28, *4th Series:* 353–356.

Papish, J. (1928) New occurrences of germanium: I. *Econ. Geol.* 23: 660–670.

Papke, K. G. (1952) Geology and ore deposits of the eastern portion of the Hilltop mine area, Cochise County, Arizona. *Univ. Arizona M.S. thesis,* 99 p.

Park, C. F., Jr. (1929) The geology of the San Xavier district. *Univ. Arizona M.S. thesis,* 20 p.

Parker, R. L., and M. Fleischer (1968) Geochemistry of niobium and tantalum. *U.S. Geol. Survey Prof. Paper* 612, 43 p.

Pearl, R. M. (1950) New data on lossenite, louderbackite, zepharovichite, peganite, and sphalerite. *Amer. Min.* 35: 1055–1059.

Peirce, F. L. (1958) Structure and petrography of part of the Santa Catalina Mountains. *Univ. Arizona Ph.D. dissertation,* 86 p.

Peirce, H. W. (1969) Salines, *in* Mineral and water resources of Arizona. *Ariz. Bur. Mines Bull.* 180: 417–424.

——, and T. A. Gerrard (1966) Evaporite deposits of the Permanian Holbrook Basin, Arizona. *2nd Symposium on Salt, The Northern Ohio Geol. Soc., Inc.* 1: 1–10.

Pellegrin, A. L. (1911) Rare minerals in southern Arizona. *Mining World* 34: 450.

Penfield, S. L. (1881) Analysis of jarosite from the Vulture mine, Arizona. *Amer. Jour. Sci.* 21: 160.

—— (1886) Crystallized vanadinite from Arizona and New Mexico. *Amer. Jour. Sci.* 32, *3rd Series:* 441–443.

—— (1890) On spangolite, a new copper mineral. *Amer. Jour. Sci.* 39, *3rd Series:* 370–378.

Peng, C. (1948) The Mountain Maid ore body, Bisbee, Arizona. *Univ. Arizona M.S. thesis,* 37 p.

Percious, J. K. (1968) Geochemical investigation of the Del Bac Hills volcanics, Pima County, Arizona. *Univ. Arizona M.S. thesis,* 29 p.

Perry, D. V. (1964) Genesis of the contact rocks at the Abril mine, Cochise County, Arizona. Univ. Arizona M.S. thesis, 97 p.

——— (1968) Genesis of the contact rocks at the Christmas mine, Gila County, Arizona. Univ. Arizona Ph.D. dissertation, 229 p.

——— (1969) Skarn genesis at the Christmas mine, Gila County, Arizona. *Econ. Geol.* 64: 255–270.

Perry, S. H. (1934) The San Francisco Mountains meteorite. *Amer. Jour. Sci.* 28: 202–218.

Petereit, A. H. (1907) Crystallized native copper from Bisbee, Arizona. *Amer. Jour. Sci.* 23, *4th Series:* 232–233.

Petersen, R. G. (1960) Detrital-appearing uraninite in the Shinarump member of the Chinle Formation in northern Arizona. *Econ. Geol.* 55: 138–149.

———, J. C. Hamilton, and A. T. Myers (1959) An occurrence of rhenium associated with uraninite in Coconino County, Arizona. *Econ. Geol.* 54: 254–267.

Peterson, E. C. (1966) Titanium resources of the United States. *U.S. Bur. Mines Inf. Circ.* 8290.

Peterson, N. P. (1938a) Geology and ore deposits of the Mammoth mining camp, Pinal County, Arizona. Univ. Arizona Ph.D. dissertation, 171 p.

——— (1938b) Geology and ore deposits of the Mammoth mining camp area, Pinal County, Arizona. *Ariz. Bur. Mines Bull.* 144, *Geol. Series* 11, 63 p.

——— (1947) Phosphate minerals in the Castle Dome copper deposit, Arizona. *Amer. Min.* 32: 574–582.

——— (1954) Copper Cities copper deposit, Globe-Miami district, Arizona. *Econ. Geol.* 49: 362–377.

——— (1962) Geology and ore deposits of the Globe-Miami district, Arizona. *U.S. Geol. Survey Prof. Paper* 342, 151 p.

——— (1963) Geology of the Pinal Ranch quadrangle, Arizona. *U.S. Geol. Survey Bull.* 1141-H: 1–18.

———, C. M. Gilbert, and G. L. Quick (1946) Hydrothermal alteration in the Castle Dome copper deposit, Arizona. *Econ. Geol.* 41: 820–840.

———, ———, and ——— (1951) Geology and ore deposits of the Castle Dome area, Gila County. *U.S. Geol. Survey Bull.* 971, 134 p.

———, and R. W. Swanson (1956) Geology of the Christmas copper mine, Gila County, Arizona. *U.S. Geol. Survey Bull.* 1027-H: 351–373.

Phalen, W. C. (1914) Celestite deposits in California and Arizona. *U.S. Geol. Survey Bull.* 540-T: 521–533.

Phelps, H. D. (1946) Exploration of the Copper Butte mine, Mineral Creek mining district, Pinal County, Arizona. *U.S. Bur. Mines Rept. Inv.* 3914.

Phillips, C. H., H. R. Cornwall, and M. Rubin (1971) A holocene ore body of copper oxides and carbonates at Ray, Arizona. *Econ. Geol.* 66: 495–498.

Pierce, W. G., and E. I. Rich (1962) Summary of rock salt deposits in the United States as possible storage sites for radioactive waste materials. *U.S. Geol. Survey Bull.* 1148, 91 p.

Pilkington, H. D. (1961) A mineralogic investigation of some garnets from the Catalina Mountains, Pima County, Arizona. *Ariz. Geol. Soc. Digest,* vol. IV: 117–122.

——— (1962) Structure and petrology of a part of the east flank of the Santa Catalina Mountains, Pima County, Arizona. Univ. Arizona Ph.D. dissertation, 120 p.

Pipkin, B. W. (1967) Mineralogy of the 140-foot core from Willcox Playa, Cochise County, Arizona (abs.). *Amer. Assoc. Petrol. Geol. Bull.* 51: 478.

——— (1968) Clay mineralogy of the Willcox Playa and its drainage basin, Cochise County, Arizona. Univ. Arizona Ph.D. dissertation, 160 p.

Pirsson, L. V. (1891) Mineralogical notes. *Amer. Jour. Sci.* 42, *3rd Series:* 405–406.

Pogue, J. E. (1913) On cerussite twins from the Mammoth mine, Pinal County, Arizona. *Amer. Jour. Sci.* 35, *4th Series:* 90–92.

——— (1915) The Turquois. *Memoir, Nat. Acad. Sci.* vol. XII, *3rd memoir,* 162 p.

Popoff, C. (1940) The geology of the Rosemont mining camp, Pima County, Arizona. Univ. Arizona M.S. thesis, 100 p.

Posnjak, E., E. J. Allen, and H. E. Merwin (1915) The sulphides of copper. *Econ. Geol.* 10: 491–535.

Potter, G. M., and R. Havens (1949) Concentration of oxide manganese ores from the Adams and Woody properties, Coconino County, near Peach Springs, Arizona. *U.S. Bur. Mines Rept. Inv.* 4439.

———, A. O. Ipsen, and R. R. Wells (1946) Concentration of manganese ores from Gila, Greenlee, and Graham Counties, Arizona. *U.S. Bur. Mines Rept. Inv.* 3842.

Pough, F. N. (1941) Occurrence of willemite. *Amer. Min.* 26: 92–102.

Puckett, J. C., Jr. (1970) Petrographic study of a quartz diorite stock near Superior, Pinal County, Arizona. Univ. Arizona M.S. thesis, 48 p.

Qurashi, M. M., and W. H. Barnes (1954) The structures of the minerals of the descloizite and adelite groups: I — descloizite and conichalcite (Part I). *Amer. Min.* 39: 416–435.

Raabe, R. G. (1959) Structure and petrography of the Bullock Canyon, Buehman Canyon area, Pima County, Arizona. Univ. Arizona M.S. thesis, 50 p.

Rabbitt, J. C. (1948) A new study of the anthophyllite series. *Amer. Min.* 33: 263–323.

Radcliffe, D., and W. B. Simmons, Jr. (1971) Austinite: chemical and physical properties in relation to conichalcite. *Amer. Min.* 56: 1359–1365.

Ransome, F. L. (1902) Copper deposits of Bisbee, Arizona. *U.S. Geol. Survey Bull.* 213: 149–157.

———— (1903) Geology of the Globe copper district, Arizona. *U.S. Geol. Survey Prof. Paper* 12, 168 p.

———— (1903) The copper deposits of Bisbee, Arizona. *Engr. Min. Jour.* 75: 444.

———— (1904) The geology and ore deposits of the Bisbee quadrangle, Arizona. *U.S. Geol. Survey Prof. Paper* 21, 168 p.

———— (1913) The Turquoise copper-mining district, Arizona. *U.S. Geol. Survey Bull.* 530: 125–134.

———— (1914) Copper deposits near Superior, Arizona. *U.S. Geol. Survey Bull.* 540-D: 139–158.

———— (1916) Quicksilver deposits of the Mazatzal Range, Arizona. *U.S. Geol. Survey Bull.* 620-I: 111–128.

———— (1919) The copper deposits of Ray and Miami, Arizona. *U.S. Geol. Survey Prof. Paper* 115, 192 p.

———— (1922) Ore deposits of the Sierrita Mountains, Pima County, Arizona. *U.S. Geol. Survey Bull.* 725-J: 407–428.

———— (1923) Geology of the Oatman gold district, Arizona, a preliminary report. *U.S. Geol. Survey Bull.* 743, 58 p.

Rasor, C. A. (1937) Mineralogy and petrography of the Tombstone mining district, Arizona. Univ. Arizona Ph.D. dissertation, 115 p.

———— (1938) Bromyrite from Tombstone, Arizona. *Amer. Min.* 23: 157–159.

———— (1939) Manganese mineralization at Tombstone, Arizona. *Econ. Geol.* 34: 790–803.

———— (1946) Loellingite from Arizona. *Amer. Min.* 31: 406–408.

Raup, R. B. (1954) Reconnaissance for uranium in the United States, southwest district, *in* Geological investigations of radioactive deposits. *Semiannual progress report, U.S. Geol. Survey* TEI-440.

Reber, L. E., Jr. (1916) The mineralization at Clifton-Morenci. *Econ. Geol.* 11: 528–573.

———— (1922) Geology and ore deposits of Jerome district. *Trans. AIME* 66: 3–26.

Reed, R. K. (1966) Structure and petrography of the Fraquita Peak area, Santa Cruz County, Arizona. Univ. Arizona M.S. thesis, 64 p.

Reeds, C. A. (1937) Catalogue of the meteorites in the American Museum of Natural History as of October 1, 1935. *Bull., Amer. Mus. Nat. Hist.* 73: 517–672.

Regis, A. J., and L. B. Sand (1967) Lateral gradation of chabazite to herschelite in the San Simon basin (abs.), *in* Clays and clay minerals, S. W. Baily, ed. *Proc. 15th Natl. Conf. Clays and Clay Minerals:* 193.

Reid, A. M., and S. H. Jeffrey (1970) Pyrope in kimberlite. *Amer. Min.* 55: 1374–1379.

Reid, J. A. (1906) Sketch of the geology and ore deposits of the Cherry Creek deposit, Arizona. *Econ. Geol.* 1: 417–436.

Richard, K., and J. H. Courtright (1959) Some geologic features of the Mission copper deposit, *in* Southern Arizona Guidebook II, combined with the 2nd Ann. Ariz. Geol. Soc. Digest: 201–204.

————, and ———— (1966) Structure and mineralization at Silver Bell, Arizona, *in Geology of the porphyry copper deposits, southwestern North America,* S. R. Titley and C. L. Hicks, (eds.). Univ. Arizona Press, Tucson: 157–163.

Richards, R. A. (1956) Arizona agate hunting. *Rocks and Minerals,* 31: 279–280.

———— (1956) Arizona obsidian notes. *Rocks and Minerals,* 31: 585.

Richmond, W. E. (1940) Crystal chemistry of the phosphates, arsenates and vanadates of the type $A_2XO_4(Z)$. *Amer. Min.* 25: 441–479.

————, and M. Fleischer (1942) Cryptomelane, a new name for the commonest of the "psilomelane" minerals. *Amer. Min.* 27: 607–610.

Rickard, F. (1904) Notes on tungsten deposits in Arizona. *Engr. Min. Jour.* 78: 263.

Robinson, H. F. (1913) The San Franciscan volcanic field, Arizona. *U.S. Geol. Survey Prof. Paper* 76: 213.

Robinson, R. F., and A. Cook (1966) The Safford copper deposit, Lone Star mining district, Graham County, Arizona, *in Geology of the porphyry copper deposits, southwestern North America,* S. R. Titley and C. L. Hicks (eds.). Univ. Arizona Press, Tucson: 251–266.

Robinson, R. L. (1954) Duranium claims, Santa Cruz County, Arizona. *U.S. Atomic Energy Comm. Prelim. Reconn. Rept.* A-P-285.

———— (1955) Linda Lee claims, Pima County, Arizona. *U.S. Atomic Energy Comm. Prelim. Reconn. Rept.* A-P-331.

———— (1956) Lucky Find group, Maricopa County, Arizona. *U.S. Atomic Energy Comm. Prelim. Reconn. Rept.* A-96.

Roedder, E., B. Ingram, and W. E. Hall (1963) Studies of the fluid inclusions III: extractions and quantitative analysis of inclusions in the milligram range. *Econ. Geol.* 58: 353–374.

Rogers, A. F. (1910) Notes on some pseudomorphs, petrifactions and alterations. *Proc. Am. Phil. Soc.* 49: 17–23.

——— (1913) Delafossite, a cuprous metaferrite from Bisbee, Arizona. *Amer. Jour. Sci. 35, 4th Series:* 290–294.

——— (1922) The optical properties and morphology of bisbeeite. *Amer. Min.* 7: 153–154.

——— (1930) A unique occurrence of lechatelierite or silica glass. *Amer. Jour. Sci. 19, 5th Series:* 195–202.

Rogers, W. (1958) Saddle Mountain, Arizona. *Gems and Minerals* 254: 26–28.

Romslo, T. M. (1948) Antler copper-zinc deposit, Mohave County, Arizona. *U.S. Bur. Mines Rept. Inv.* 4214.

——— (1949) Investigation of Keystone and St. George copper-zinc deposits, Cochise County, Arizona. *U.S. Bur. Mines Rept. Inv.* 4504.

——— (1950) Investigation of the Lake Shore copper deposits, Pinal County, Arizona. *U.S. Bur. Mines Rept. Inv.* 4706.

———, and S. F. Ravitz (1947) Arizona manganese-silver ores. *U.S. Bur. Mines Rept. Inv.* 4097.

———, and C. S. Robinson (1952) Copper giant deposits, Pima County, Arizona. *U.S. Bur. Mines Rept. Inv.* 4850.

Rose, A. W. (1970) Zonal relations of wallrock alteration and sulfide distribution at porphyry copper deposits. *Econ. Geol.* 65: 920–936.

Roseboom, E. H., Jr. (1966) An investigation of the system Cu-S and some natural copper sulfides between 25° and 700° C. *Econ. Geol.* 61: 641–672.

Rosenzweig, A., J. W. Gruner, and L. Gardiner (1954) Widespread occurrence and character of uraninite in the Triassic and Jurassic sediments of the Colorado Plateau. *Econ. Geol.* 49: 351–361.

Ross, C. P. (1922) Geology of the lower Gila region, Arizona. *U.S. Geol. Survey Prof. Paper 129-H:* 183–197.

——— (1925a) Geology and ore deposits of the Aravaipa and Stanley mining districts, Graham County, Arizona. *U.S. Geol. Survey Bull.* 763, 120 p.

——— (1925b) Geology of the Saddle Mountain and Banner mining districts. *U.S. Geol. Survey Bull.* 771, 72 p.

Ross, C. S. (1928) Sedimentary analcite. *Amer. Min.* 13: 195–197.

——— (1941) Sedimentary analcite. *Amer. Min.* 26: 627–629.

Ross, M. (1959) Mineralogical applications of electron diffraction. II. Studies of some vanadium minerals of the Colorado Plateau. *Amer. Min.* 44: 322–341.

Rubel, A. C. (1917–18) Tungsten. *Ariz. Bur. Mines Bull.* 11, *Mineral Tech. Series* 5, 12 p.

Ruff, A. W. (1951) The geology and ore deposits of the Indiana mine area, Pima County, Arizona. Univ. Arizona M.S. thesis, 64 p.

Sabels, B. E. (1960) Late Cenozoic volcanism in the San Francisco volcanic field and adjacent areas in North-Central Arizona. Univ. Arizona Ph.D. dissertation, 135 p.

Saha, P. (1961) The system $NaAlSiO_4$ (Nepheline)-$NaAlSi_3O_8$ (Albite)-$H_2O$. *Amer. Min.* 46: 859–884.

Sampson, E. (1924) Discussion of "An Arizona asbestos deposit." *Econ. Geol.* 19: 386–388.

Sand, L. B., and A. J. Regis (1966) An unusual zeolite assemblage, Bowie, Arizona (abs.). *Geol. Soc. Amer. Spec. Paper* 87.

Sandell, W. G., and D. T. Holmes (1948) Concentration of oxide manganese ores from the Aguila district, Arizona. *U.S. Bur. Mines* 4330.

Sanders, J. (1911) Hematite in veins of Globe district. *Eng. Min. Jour.* 92: 1191–1192.

Santmyers, R. M. (1929) Development of the gypsum industry by states. *U.S. Bur. Mines Inf. Circ.* 6173.

Schaller, W. T. (1905) Dumortierite. *U.S. Geol. Survey Bull.* 262: 91–120.

——— (1915) Four new minerals. *Jour. Wash. Acad. Sci.* 5: 7.

——— (1932) Chemical composition of cuprotungstite. *Amer. Min.* 17: 234–237.

——— (1934) Mottramite or psittacinite — a question of nomenclature. *Amer. Min.* 19: 180–181.

——— (1939) Corrections and additions. *Amer. Min.* 24: 346–347.

———, R. E. Stevens, and R. H. Jahns (1962) An unusual beryl from Arizona. *Amer. Min.* 47: 672–699.

———, and A. C. Vlisidis (1958) Ajoite, a new hydrous aluminum copper silicate. *Amer. Min.* 43: 1107–1111.

Schmidt, E. A. (1971) A structural investigation of the northern Tortilla Mountains, Pinal County, Arizona. Univ. Arizona Ph.D. dissertation, 248 p.

Schmidt, H. A., D. M. Clippinger, W. J. Roper, and H. Toombs (1959) Disseminated deposits at the Esperanza copper mine, in Southern Arizona Guidebook II, combined with the 2nd Ariz. Geol. Soc. Digest: 205.

Schrader, F. C. (1907) The mineral deposits of the Cerbat Range, Black Mountains, and Grand Wash Cliffs, Mohave County, Arizona. *U.S. Geol. Survey Bull.* 340: 53–83.

——— (1909) Mineral deposits of the Cerbat Range, Black Mountains, and Grand Wash Cliffs, Mohave County, Arizona. *U.S. Geol. Survey Bull.* 397, 226 p.

——— (1913) Alunite in Patagonia, Arizona, and Bovard, Nevada. *Econ. Geol.* 8: 752–767.

———— (1914) Alunite in granite porphyry near Patagonia, Arizona. *U.S. Geol. Survey Bull.* 540: 347–350.

———— (1917) The geologic distribution and genesis of the metals in the Santa Rita-Patagonia Mountains, Arizona. *Econ. Geol.* 12: 237–269.

———— (1919) Quicksilver deposits of the Phoenix Mountains. *U.S. Geol. Survey Bull.* 690-D: 95–109.

————, and J. M. Hill (1909) Some occurrences of molybdenite in the Santa Rita and Patagonia Mountains, Arizona. *U.S. Geol. Survey Bull.* 430: 154–163.

————, and ———— (1915) Mineral deposits of the Santa Rita and Patagonia Mountains, Arizona. *U.S. Geol. Survey Bull.* 582, 373 p.

————, R. W. Stone, and S. Sanford (1917) Useful minerals of the United States. *U.S. Geol. Survey Bull.* 624: 412.

Schultz, L. G. (1963) Clay minerals in Triassic rocks of the Colorado Plateau. *U.S. Geol. Survey Bull.* 1147-C, 71 p.

Schwartz, G. M. (1921) Notes on textures and relationships in the Globe copper ores. *Econ. Geol.* 16: 322–329.

———— (1928) Experiments bearing on bornite-chalcocite intergrowths. *Econ. Geol.* 23: 381–397.

———— (1934) Paragenesis of the oxidized ores of copper. *Econ. Geol.* 29: 55–75.

———— (1938) Oxidized copper ores of the United Verde Extension mine. *Econ. Geol.* 33: 21–33.

———— (1939) Significance of bornite-chalcocite microtextures. *Econ. Geol.* 34: 399–418.

———— (1947) Hydrothermal alteration in the "porphyry copper" deposits. *Econ. Geol.* 42: 319–352.

———— (1949) Oxidation and enrichment in the San Manuel copper deposit, Arizona. *Econ. Geol.* 44: 253–277.

———— (1952) Chlorite-calcite pseudomorphs after orthoclase phenocrysts, Ray, Arizona. *Econ. Geol.* 47: 665–672.

———— (1953) Geology of the San Manuel copper deposit, Arizona. *U.S. Geol. Survey Prof. Paper* 256, 65 p.

———— (1956) Argillic alteration and ore deposits. *Econ. Geol.* 51: 407–414.

———— (1958) Alteration of biotite under mesothermal conditions. *Econ. Geol.* 53: 164–177.

———— (1959) Hydrothermal alteration. *Econ. Geol.* 54: 161–183.

———— (1966) The nature of primary and secondary mineralization in porphyry copper deposits, *in Geology of the porphyry copper deposits, southwestern North America,* S. R. Titley and C. L. Hicks (eds.). Univ. Arizona Press, Tucson: 41–50.

————, and C. F. Park, Jr. (1932) A microscopic study of ores from the Campbell mine, Bisbee, Arizona. *Econ. Geol.* 27: 39–51.

Searls, F., Jr. (1950) The Emerald Isle copper deposit. *Econ. Geol.* 45: 175–176.

Selfridge, G. C., Jr. (1936) An x-ray and optical investigation of the serpentine minerals. *Amer. Min.* 21: 463–503.

Sell, J. D. (1961) Bedding replacement deposit of the Magma mine, Superior, Arizona. Univ. Arizona M.S. thesis, 48 p.

Shannon, E. V., and F. A. Gonyer (1927) Natrojarosite from Kingman, Arizona. *Jour. Wash. Acad. Sci.* 17: 536–537.

Sharp, R. R., Jr. (1962) Some magnetic properties of a part of Pikes Peak iron deposit, Maricopa County, Arizona. Univ. Arizona M.S. thesis, 135 p.

Shaw, V. E. (1959) Extraction of yttrium and rare earth elements from Arizona euxenite concentrate. *U.S. Bur. Mines Rept. Inv.* 5544.

Shawe, D. R. (1966) Arizona-New Mexico and Nevada-Utah beryllium belts. *U.S. Geol. Survey Prof. Paper* 550-C: 206–213.

Sheikh, A. M. (1966) Geology and ore deposits of Las Guijas tungsten district, Pima County, Arizona. Univ. Arizona M.S. thesis, 44 p.

Shepard, C. U. (1854) Notice of three ponderous masses of meteoric iron, Tucson, Sonora. *Amer. Jour. Sci.* 18, series 2: 369–372.

Sheppard, R. A. (1969) Zeolites, *in* Mineral and water resources of Arizona. *Ariz. Bur. Mines Bull.* 180: 464–467.

———— (1971) Clinoptilolite, of possible economic value in sedimentary deposits of the conterminous United States. *U.S. Geol. Survey Bull.* 1332-B: B1–B15.

Sheridan, M. F., and C. A. Royse, Jr. (1970) Alunite: a new occurrence near Wickenburg, Arizona. *Amer. Min.* 55: 2016–2022.

Shields, J. C. II (1940) Geology and ore deposits of the Dives and Gold Ridge groups, Dos Cabezas, Arizona. Univ. Arizona M.S. thesis, 67 p.

Shoemaker, A. H., and G. Sommers (1924) The geology of the El Tiro mine, Silver Bell, Arizona. Univ. Arizona M.S. thesis, 40 p.

Short, M. N., F. W. Galbraith, E. N. Harshman, T. H. Kuhn, and E. D. Wilson (1943) Geology and ore deposits of the Superior mining area, Arizona. *Ariz. Bur. Mines Bull.* 151, Geol. Series 16, 159 p.

Shride, A. F., ed. (1952) Guidebook for field trip excursions in southern Arizona. *Ariz. Geol. Soc.,* 150 p.

Sigalove, J. J. (1969) Carbon-14 content and origin of caliche. Univ. Arizona M.S. thesis, 72 p.

Silliman, B. (1866) On some mining districts of Arizona near Rio Colorado, with remarks on the climate, etc. *Amer. Jour. Sci.* 41, *2nd series:* 289–308.

Silliman, B. (*cont'd*)

———— (1879) Jarosite (with gold). *Amer. Jour. Sci.* 17, *3rd series:* 73.

———— (1881) Mineralogical notes. *Amer. Jour. Sci.* 22, *3rd series:* 198–205.

Silver, C. (1954) Manganese deposits of the Mogollon Rim, Coconino County, Arizona. *Econ. Geol.* 49: 88–91.

Silver, L. T., and S. Deutsch (1963) Uranium-lead isotopic variations in zircons: a case study. *Jour. Geol.* 71: 721–758.

Simmons, W. W. (1938) The geology of the Cleveland mine area, Gila County, Arizona. Univ. Arizona M.S. thesis, 22 p.

————, and J. E. Fowells (1966) Geology of the Copper Cities mine, *in Geology of the porphyry copper deposits, southwestern North America,* S. R. Titley and C. L. Hicks (eds.). Univ. Arizona Press, Tucson: 151–156.

Simons, F. S. (1963) Composite dike of andesite and rhyolite at Klondyke, Arizona. *Geol. Soc. Amer. Bull.* 74: 1049–1056.

———— (1964) Geology of the Klondyke quadrangle, Graham and Pinal Counties, Arizona. *U.S. Geol. Survey Prof. Paper* 461, 173 p.

————, and E. Munson (1963) Johannsenite from the Aravaipa mining district, Arizona. *Amer. Min.* 48: 1154–1158.

Sinkankas, John (1964) *Mineralogy for amateurs.* D. Van Nostrand Co., Inc., Princeton, N.J., 587 p.

———— (1966) *Mineralogy; a first course.* D. Van Nostrand Co., Inc., Princeton, N.J., 587 p.

Smith, D. (1970) Mineralogy and petrology of the diabasic rocks in a differentiated olivine diabase sill complex, Sierra Ancha, Arizona. *Contr. Min. Petr.* 27: 95–113.

Smith, F. G. (1952) Decrepitation characteristics of garnet. *Amer. Min.* 37: 470–491.

Smith, G. E. (1956) The geology and ore deposits of the Mowry mine area, Santa Cruz County, Arizona. Univ. Arizona M.S. thesis, 44 p.

Smith, L. A. (1927) The geology of the Commonwealth mine. Univ. Arizona M.S. thesis, 73 p.

Smith, L. J., M.D. (1955) Memoir on meteorites — a description of five new meteoric irons, with some theoretical considerations on the origin of meteorites based on their physical and chemical characters. *Amer. Jour. of Sci.* 19, *Series 2:* 153–163.

Smith, W. B. (1887) Mineral localities in the western United States. *Amer. Jour. Sci.* 34: 315.

———— (1887) Dioptase from Pinal County, Arizona. *Proc. of the Colo. Sci. Soc.* 2: 159–160.

Snyder, J. E. (1971) Salt mine pseudomorphs. *Gems and Minerals* 405, June.

Sopp, G. P. (1940) Geology of the Montana mine area, Empire Mountains, Arizona. Univ. Arizona M.S. thesis, 63 p.

Staatz, M. H., W. R. Griffitts, and P. R. Barnett (1965) Differences in the minor element composition of beryl in various environments. *Amer. Min.* 50: 1783.

Stephens, J. D., and R. A. Metz (1967) The occurrence of copper-bearing clay minerals in oxidized portions of the disseminated copper deposit at Ray, Arizona (abs.). *Econ. Geol.* 62: 876–877.

Stern, T. W., L. R. Stieff, H. J. Evans, Jr., and A. H. Sherwood (1957) Doloresite, a new vanadium oxide mineral from the Colorado Plateau. *Amer. Min.* 42: 587–593.

Sterrett, D. B. (1908) Turquoise, *in* Min. Res. of the U.S. U.S. Geol. Surv: 847–852.

Stewart, C. A. (1912) The geology and ore deposits of the Silverbell mining district, Arizona. *Trans. AIME* 43: 240–290.

Stewart, J. C. (1971) Geology of the Morningstar mine area, Greaterville mining district, Pima County, Arizona. Univ. Arizona M.S. thesis, 79 p.

Stewart, L. A. (1933) A study of the Ajo copper ore minerals. Univ. Arizona M.S. thesis, 24 p.

———— (1947) Apache iron deposits, Navajo County, Arizona. *U.S. Bur. Mines Rept. Inv.* 4093.

———— (1955) Chrysotile-asbestos deposits of Arizona. *U.S. Bur. Mines Inf. Circ.* 7706, 124 p.

———— (1956) Chrysotile-asbestos deposits of Arizona. *U.S. Bur. Mines Inf. Circ.* 7745.

————, and P. S. Haury (1947) Arizona asbestos deposits, Gila County, Arizona. *U.S. Bur. Mines Rept. Inv.* 4100.

————, and A. J. Pfister (1960) Barite deposits of Arizona. *U.S. Bur. Mines Rept. Invest.* 5651, 89 p.

Stokes, W. L. (1951) Carnotite deposits in the Carrizo Mountains area, Navajo Indian Reservation, Apache County, Arizona and San Juan County, New Mexico. *U.S. Geol. Survey Circ.* 111, 5 p.

Strong, M. F. (1962) Pink marble in Arizona. *Gems and Minerals* 296: 26–27.

Strunz, H., and B. Contag (1965) Evenkite, idrialite, and refikite (abs.). *Amer. Min.* 50: 2109–2110.

Struthers, J. (1904) The production of the minor metals in 1903 — arsenic. *Engr. Mining Jour.* 77: 74.

Sun, Ming-Shan (1954) Titanclinohumite in kimberlitic tuff, Buell Park, Arizona (abs.). *Geol. Soc. Amer. Bull.* 65: 1311–1312.

———— (1961) Differential thermal analysis of shattuckite. *Amer. Min.* 46: 67–77.

———— (1963) The nature of chrysocolla from

Inspiration mine, Arizona. *Amer. Min.* 48: 649–658.

Taber, S., and W. T. Schaller (1930) Psittacinite from the Higgins mine, Bisbee, Arizona. *Amer. Min.* 15: 575–579.

Tainter, S. L. (1947a) Apex copper property, Coconino County, Arizona. *U.S. Bur. Mines Rept. Inv.* 4013.

—— (1947b) Amargosa (Esperanza) molybdenum-copper property, Pima County, Arizona. *U.S. Bur. Mines Rept. Inv.* 4016.

—— (1947c) Johnny Bull-Silver King lead-zinc property, Cerbat Mountains, Mohave County, Arizona. *U.S. Bur. Mines Rept. Inv.* 3998.

—— (1948) Christmas copper deposit, Gila County, Arizona. *U.S. Bur. Mines Rept. Inv.* 4293.

Talley, R. E. (1917) Mine fire methods employed by the United Verde Copper Co. *Trans. AIME* 55: 186–202.

Taylor, F. (1954) Jasper in Limestone Gulch. *Desert Magazine* 17: 18–21.

Taylor, O. J. (1960) Correlation of volcanic rocks in Santa Cruz County, Arizona. Univ. Arizona M.S. thesis, 59 p.

Tenney, J. B. (1936) Mineral survey of the State of Arizona. *Ariz. State Planning Board, Phoenix, Arizona.*

Thacpaw, S. C. (1960) Geology of the Ruby Star Ranch area, Twin Buttes mining district, Pima County, Arizona. Univ. Arizona M.S. thesis, 59 p.

Thoenen, J. R. (1941) Alunite resources of the United States. *U.S. Bur. Mines Rept. Inv.* 3561.

Thomas, B. E. (1949) Ore deposits of the Wallapai district, Arizona. *Econ. Geol.* 44: 663–705.

—— (1953) Geology of the Chloride quadrangle, Arizona. *Geol. Soc. Amer. Bull.* 64: 391–420.

Thomas, L. A. (1966) Geology of the San Manuel ore body, *in Geology of the porphyry copper deposits, southwestern North America*, S. R. Titley and C. L. Hicks (eds.). Univ. Arizona Press, Tucson: 133–142.

Thomas, W. L. (1931) Geology and ore deposits of the Rosemont area, Pima County, Arizona. Univ. Arizona M.S. thesis, 54 p.

Throop, A. H. (1970) The nature and origin of black chrysocolla at the Inspiration mine, Arizona. M.S. thesis, Arizona State University, 54 p.

——, and P. R. Buseck (1971) Nature and origin of black chrysocolla at the Inspiration mine, Arizona. *Econ. Geol.* 66: 1168–1175.

Titley, S. R., ed. (1968) Southern Arizona Guidebook III. *Ariz. Geol. Soc.* 354 p.

——, and C. L. Hicks (eds.) (1966) *Geology of the porphyry copper deposits, southwestern*

*North America.* Univ. Arizona Press, Tucson, 287 p.

Tolman, C. F., Jr. (1909) The geology of the vicinity of the Tumamoc Hills, *in* Spalding, V. M., *Distribution and movement of desert plants.* Carnegie Institution of Washington, Publ. 113: 67–82.

—— (1917) Observations on certain types of chalcocite and their characteristic etch patterns. *Trans. AIME* 54: 402–435.

Trebisky, T. J., and S. B. Keith (1975) Descloizite from the C and B Vanadium mine. *Mineral. Record* 6: 109.

Trischka, C. (1929) Diatomite in Arizona. *Eng. Mining Jour.* 127: 13–14.

——, O. N. Rove, and D. M. Barringer, Jr. (1929) Boxwork siderite. *Econ. Geol.* 24: 677–686.

Van Tassel, R. (1958) On carphosiderite. *Mineral. Mag.* 31: 818–819.

Vlisidis, A. C., and W. T. Schaller (1967) The formula of shattuckite. *Amer. Min.* 52: 782–786.

von Bernewitz, M. W. (1937) Occurrence and treatment of mercury ore at small mines. *U.S. Bur. Mines Inf. Circ.* 6966.

Walker, G. W., and J. W. Adams (1964) Mineralogy, internal structure and textural characteristics, and paragenesis of uranium-bearing veins in the conterminous United States. *U.S. Geol. Survey Prof. Paper* 455-D: 55–90.

Walker, L. W. (1957) The geodes of Kofa. *Nature* 50: 266–267.

Waller, H. E., Jr. (1960) The geology of the Paymaster and Olivette mining areas, Pima County, Arizona. Univ. Arizona M.S. thesis, 48 p.

Ward, G. W. (1931) A chemical and optical study of the black tourmalines. *Amer. Min.* 16: 145–190.

Wardwell, H. R. (1941) Geology of the Potts Canyon mining area near Superior, Arizona. Univ. Arizona M.S. thesis, 104 p.

Wargo, J. G. (1954) Geology of a portion of the Coyote-Quinlan complex, Pima County, Arizona. Univ. Arizona M.S. thesis, 67 p.

Warren, C. H. (1903) Native arsenic from Arizona. *Amer. Jour. Sci.* 16, *4th Series:* 337–340.

Warren, H. V., and R. W. Loofburrow (1932) The occurrence and distribution of the precious metals in the Montana and Idaho mines, Ruby, Arizona. *Econ. Geol.* 27: 578–585.

Watson, B. N. (1964) Structure and petrology of the eastern portion of the Silver Bell Mountains, Pima County, Arizona. Univ. Arizona Ph.D. dissertation, 168 p.

Watson, K. D., and D. M. Morton (1969) Eclogite inclusions in kimberlite pipes at Garnet Ridge,

Watson, K. D., and D. M. Morton (*cont'd*) northeastern Arizona. *Amer. Min.* 54: 267–285.

Webber, B. N. (1929) Marcasite in the contact metamorphic ore deposits of the Twin Buttes district, Pima County, Arizona. *Econ. Geol.* 24: 304–310.

Weber, R. H. (1950) The geology of the east-central portion of the Huachuca Mountains, Arizona. Univ. Arizona Ph.D. dissertation, 191 p.

Weeks, A. D., M. E. Thompson, and A. M. Sherwood (1954) Navajoite, a new vanadium oxide from Arizona. *Science* 119: 326.

———, ———, and ——— (1955) Navajoite, a new vanadium oxide from Arizona. *Amer. Min.* 40: 207–212.

Weight, H. O. (1949) Geodes and palms in the Kofa country. *Desert Magazine* 12: 17–22.

Weitz, J. H. (1942) High grade dolomite deposits in the United States. *U.S. Bur. Mines Inf. Circ.* 7226.

Wells, H. L., and S. L. Penfield (1885) Gerhardtite and artificial cupric nitrates. *Amer. Jour. Sci.* 30: 50–57.

Wells, R. C. (1913) A new occurrence of cuprodescloizite. *Amer. Jour. Sci.* 36, *4th Series:* 636–638.

——— (1937) Analysis of rocks and minerals from the laboratory of the USGS, 1914–36. *U.S. Geol. Survey Bull.* 878: 92–125.

Wells, R. L., and A. J. Rambosek (1954) Uranium occurrences in Wilson Creek area, Gila County, Arizona. *U.S. Atomic Energy Comm.* RME-2005 (rev.), 15 p. Oak Ridge, Tenn.

Wherry, E. T. (1915) Notes on wolframite, beraunite, and axinite. *U.S. Nat. Mus. Proc.* 47: 501–511.

Whitcomb, H. A. (1948) Geology of the Morgan mine area, Twin Buttes, Arizona. Univ. Arizona M.S. thesis, 83 p.

White, John S., Jr. (1974) What's new in minerals? *Mineral. Record,* 5: 233–236.

Wideman, F. L. (1957) A reconnaissance of sulfur resources in Wyoming, Colorado, Utah, New Mexico and Arizona. *U.S. Bur. Mines Inf. Circ.* 7770, 61 p.

Willden, R. (1960) Sedimentary iron-formation in the Devonian Martin formation, Christmas quadrangle, Arizona. *U.S. Geol. Survey Prof. Paper* 400-B: B21–23.

Williams, H. (1936) Pliocene volcanics of the Navajo-Hopi country. *Geol. Soc. Amer. Bull.* 47: 111–172.

Williams, S. A. (1960) A new occurrence of allanite in the Quijotoa Mountains, Pima County, Arizona. *Ariz. Geol. Soc. Digest* 3: 46–51.

——— (1961) Gerhardtite from the Daisy shaft, Mineral Hill mine, Pima County, Arizona. *Ariz. Geol. Soc. Digest* 4: 123.

——— (1962) The mineralogy of the Mildren and Steppe mining districts, Pima County, Arizona. Univ. Arizona Ph.D. dissertation, 145 p.

——— (1963) Oxidation of sulphide ores in the Mildren and Steppe mining districts, Pima County, Arizona. *Econ. Geol.* 58: 1119–1125.

——— (1966) The significance of habit and morphology of wulfenite. *Amer. Min.* 51: 1212–1217.

——— (1968) Wickenburgite, a new mineral from Arizona. *Amer. Min.* 53: 1433–1438.

——— (1970) Bideauxite, a new Arizona mineral. *Mineral. Magazine* 37: 637–640.

——— (1970) Tilasite from Bisbee, Arizona. *Mineral. Record* 1: 68–69.

——— (1976) Luetheite, $Cu_2Al_2(AsO_4)_2 (OH)_4 \cdot H_2O$, a new mineral from Arizona compared with chenevixite. *Mineral. Magazine,* Sept.

———, and J. W. Anthony (1968) Hemihedrite, a new mineral from Arizona (abs.). *Amer. Min.* 53: 1427.

———, and ——— (1970) Hemihedrite, a new mineral from Arizona. *Amer. Min.* 55: 1088–1102.

———, and R. A. Bideaux (1975) Creaseyite, $Cu_2Pb_2(Fe,Al)_2Si_5O_{17} \cdot 6H_2O$, a new mineral from Arizona and Sonora. *Mineral Mag.* 40: 227–231.

———, and B. Khin (1971) Chalcoalumite from Bisbee, Arizona. *Mineral. Record* 2: 126–127.

———, W. J. McLean, and J. W. Anthony (1970) A study of phoenicochroite — its structure and properties. *Amer. Min.* 55: 784–792.

———, and P. Matter III (1975) Graemite, a new Bisbee mineral. *Mineral. Record,* 6: 32–34.

Wilson, C. A. (1960) Ore controls of the San Xavier mine, Pima County, Arizona. Univ. Arizona M.S. thesis, 58 p.

Wilson, E. D. (1927) Arizona gold placers. *Ariz. Bur. Mines Bull.* 124, *Mineral Tech. Series* 30, 60 p.

——— (1928) Asbestos deposits of Arizona. *Ariz. Bur. Mines Bull.* 126, *Mineral Tech. Series* 31, 100 p.

——— (1929) An occurrence of dumortierite near Quartzsite, Arizona. *Amer. Min.* 14: 373–381.

——— (1933) Geology and mineral deposits of southern Yuma County, Arizona. *Ariz. Bur. Mines Bull.* 134, *Geol. Series* 7, 236 p.

——— (1941) Tungsten deposits of Arizona. *Ariz. Bur. Mines Bull.* 148, *Geol. Series* 14, 54 p.

——— (1944) Arizona nonmetallics. *Ariz. Bur. Mines Bull.* 152, *Min. Tech. Series* 41, 58 p.

——— (1961) Arizona gold placers, part 1, *in* Gold placers and placering in Arizona. *Ariz. Bur. Mines Bull.* 168: 1–86.

——— (1969) Mineral deposits of the Gila River Indian reservation, Arizona. *Ariz. Bur. Mines Bull.* 179, 34 p.

————, and G. M. Butler (1930) Manganese ore deposits in Arizona. *Ariz. Bur. Mines Bull.* 127, *Mineral Tech. Series* 32, 107 p.

————, J. B. Cunningham, and G. M. Butler (1934) Arizona lode gold mines and gold mining. *Ariz. Bur. Mines Bull.* 137, *Mineral Tech. Series* 37, 234 p.

————, et al., (1950) Arizona zinc and lead deposits, part I. *Ariz. Bur. Mines Bull.* 156, *Geol. Series* 18, 144 p.

————, et al., (1951) Arizona zinc and lead deposits, part II. *Ariz. Bur. Mines Bull.* 158, *Geol. Series* 19, 115 p.

————, and G. Roseveare (1949) Arizona nonmetallics. *Ariz. Bur. Mines Bull.* 155, *2nd ed., revised,* 60 p.

Wilson, R. L. (1956) Stratigraphy and economic geology of the Chinle Formation, Northeastern Arizona. Univ. Arizona Ph.D. dissertation, 255 p.

Wilson, W. E. (1971) Classic locality: the Apache mine. *Mineral. Record* 2: 252–258.

————, and D. K. Miller (1974) Minerals of the Rowley mine. *Mineral Record* 5: 10–30.

Winchell, R. E., and H. E. Wenden (1968) Synthesis and study of diaboleite. *Mineral. Magazine* 36: 933–939.

Wise, W. S., and R. W. Tschernich (1975) Cowlesite, a new ca-zeolite. *Amer. Min.* 60: 951–956.

Witkind, I. J. (1956) Uranium deposits at the base of the Shinarump conglomerate, Monument Valley, Arizona. *U.S. Geol. Survey Bull.* 1030-C: 99–130.

———— (1961) The uranium-vanadium ore deposit at the Monument No. 1-Mitten No. 2 mine, Monument Valley, Navajo County, Arizona. *U.S. Geol. Survey Bull.* 1107-C: 219–242.

————, and R. E. Thaden (1963) Geology and uranium-vanadium deposits of the Monument Valley area, Apache and Navajo Counties, Arizona. *U.S. Geol. Survey Bull.* 1103, 171 p.

Wood, M. M. (1963) Metamorphic effects of the Leatherwood Quartz Diorite, Santa Catalina Mountains, Pima County, Arizona. Univ. Arizona M.S. thesis, 68 p.

———— (1969) The crystal structures of ransomite, $CuFe_2(SO_4)_4 \cdot 6H_2O$, and roemerite, $Fe'' Fe'''_2 (SO_4)_4 \cdot 14H_2O$, and a proposed classification for the transition metal sulfate hydrates. Univ. Arizona Ph.D. dissertation, 59 p.

———— (1970) The crystal structure of ransomite. *Amer. Min.* 55: 729–734.

Woodbridge, D. E. (1906) Arizona and Sonora — V. — The Globe district. *Engr. Min. Jour.* 81: 1229.

Wright, R. J. (1955) Ore controls in sandstone uranium deposits of the Colorado Plateau. *Econ. Geol.* 50: 135–155.

Wrucke, C. T. (1961) Paleozoic and Cenozoic rocks in the Alpine-Nutrioso area, Apache County, Arizona. *U.S. Geol. Survey Bull.* 1121-H, 26 p.

Yagoda, H. (1945) The localization of copper and silver sulphide minerals in polished sections by the potassium cyanide etch pattern. *Amer. Min.* 30: 51–64.

Yildiz, M. (1961) Structure and petrography of Black Rock, Apache County, Arizona. Univ. Arizona M.S. thesis, 41 p.

Young, E. J., and P. K. Sims (1961) Petrography and origin of xenotime and monazite concentrations, Central City district, Colorado. *U.S. Geol. Survey Bull.* 1032-F: 273–297.

————, A. D. Weeks, and R. Meyrowitz (1966) Coconinoite, a new uranium mineral from Utah and Arizona. *Amer. Min.* 51: 651–663.

Young, R. G. (1964) Distribution of uranium deposits in the White Canyon-Monument Valley district, Utah-Arizona. *Econ. Geol.* 59: 850–873.

Zaslow, B., and L. M. Kellogg (1961) The analysis of metallic spheroids from Meteor Crater, Arizona. *Geochim. Cosmochim. Acta* 24: 315–316.

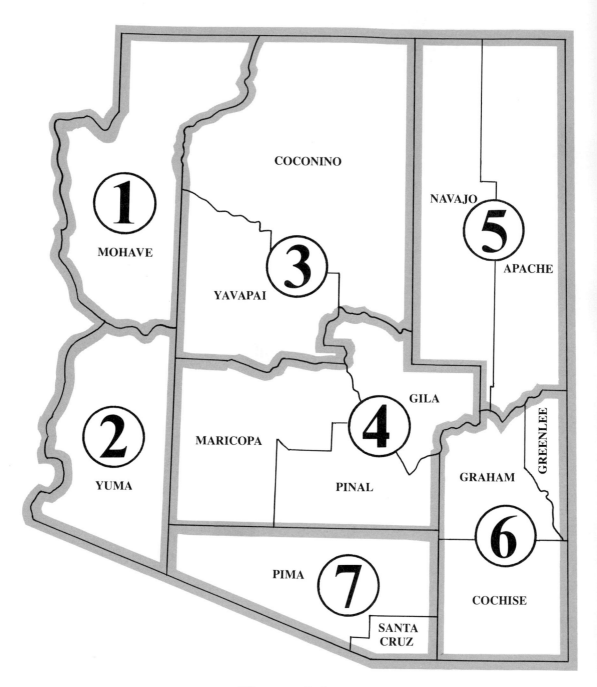

## Key to Maps

NO.

1 — Agua Dulce
2 — Agua Fria, Yavapai County
3 — Agua Fria, Maricopa County
   Aguila, see Big Horn (18)
4 — Ajo
5 — Alamo
6 — Amole (Tucson Mts.)
7 — Apache Iron (Chediski)
8 — Aquarius
9 — Aravaipa
10 — Arivaca
11 — Artillery Mts.
12 — Ash Creek
13 — Ash Peak (Duncan)
   Aztec, see Tyndall (222)
14 — Baboquivari
   Bagdad, see Eureka (72)
15 — Banner (Christmas, Troy)
   Bear Canyon, see Casador (37)
   Benson, see Whetstone (236)
16 — Bentley (Grand Gulch, Grand Wash Cliffs)
17 — Big Bug
18 — Big Horn (Aguila)
   Big Springs, see Twin Peaks (221)
   Bill Williams, see Santa Maria (194)
   Bisbee, see Warren (233)
19 — Black Canyon
20 — Black Hills
21 — Black Mesa
22 — Black Mt.
23 — Black Rock
24 — Blackwater (Sacaton Mts.)
25 — Bloody Basin
   Boundary Cone, see San Francisco (191)
26 — Blue Tank
27 — Bouse, also known as Plomosa (169)
28 — Bradshaw
29 — Bullard (Pierce)
30 — Bunker Hill (Copper Creek)
31 — Cababi (Comobabi)
32 — California (Chiricahua)
   Camp Creek, see Cave Creek (41)
33 — Camp Wood
34 — Camp Verde
35 — Canada del Oro
36 — Carrizo Mts.
37 — Casador (Bear Canyon)
   Casa Grande, see Silver Reef (202)
      and Vekol (225)
38 — Castle Creek
39 — Castle Dome
40 — Catalina (Santa Catalina)
41 — Cave Creek (Camp Creek)
42 — Cedar Valley
43 — Cerbat, also known as Wallapai (231)
44 — Cerro Colorado
45 — Chambers
   Chediski, see Apache Iron (7)
46 — Chemehuevis
47 — Cherry Creek
   Chiricahua, see California (32)

NO.

48 — Chloride, also known as Wallapai (231)
   Christmas, see Banner (15)
49 — Chrysotile
   Cibola, see Trigo Mts. (218)
50 — Cienega
51 — Cimarron Mts.
52 — Clark
53 — Cochise (Johnson)
   Comobabi, see Cababi (31)
   Congress, see Martinez (122)
   Control, see Oracle (145)
54 — Copper Basin
   Copper Creek, see Bunker Hill (30)
55 — Copper King Mt.
56 — Copper Mountain (Morenci)
   Cottonwood Cliffs, see Cottonwood,
      Mohave Co. (57)
57 — Cottonwood, Mohave County
      (Cottonwood Cliffs)
58 — Cottonwood, Pinal County (Crozier Pk.)
   Courtland, see Turquoise (220)
59 — Coyote
   Crown King, see Pine Grove (165)
   Crozier Pk., see Cottonwood,
      Pinal County (58)
60 — Cunningham Pass (Harcuvar, Ellsworth)
61 — Deer Creek
   Del Rio, see Granite Creek (86)
62 — Dome (Gila City)
63 — Dos Cabezas
64 — Douglas
65 — Dragoon (Golden Rule)
66 — Dripping Springs
   Duncan, see Ash Peak (13)
   Duquesne, see Patagonia (155)
67 — Eagle Tail
68 — Eldorado Pass
   Elfrida, see Swisshelm (210)
69 — Ellison
70 — Ellsworth (Harcuvar, Harquahala)
   Ellsworth, see also Cunningham Pass (60)
71 — Empire
72 — Eureka, Yavapai County (Bagdad)
   Eureka, Yuma County, see Silver (199)
73 — Flagstaff
74 — Florence
75 — Fluorine, also known as Sierra Ancha (198)
76 — Fortuna
77 — Francis
   Galiuro, see Gold Mountain (84)
78 — Garnet Mt.
   Gila Bend Mts., see Webb (235)
   Gila City, see Dome (62)
79 — Gila River
   Gleeson, see Turquoise (220)
80 — Globe, also known as Miami (127)
81 — Gold Basin
82 — Gold Bug
   Golden Rule, see Dragoon (65)
83 — Goldfields
   Gold Hill, see Nogales (143)

NO.

84 — Gold Mountain (Galiuro)
Gold Road, see San Francisco (191)
Grand Canyon, see Grand View (85)
Grand Gulch, see Bentley (16)
85 — Grandview (Grand Canyon)
Grand Wash Cliffs, see Bentley (16)
86 — Granite Creek (Del Rio)
87 — Greaterville
Greenlee, see Metcalf (126)
88 — Green Valley (Payson)
89 — Greenwood (Signal)
90 — Groom Creek
91 — Growler
92 — Gunsight (Meyer)
93 — Hackberry (Peacock)
94 — Hacks Canyon (Tucket)
Harcuvar, see Cunningham Pass (60)
and Ellsworth (70)
95 — Harquahala, see also Ellsworth (70)
Harrington, see Tiger, Yavapai Co. (214)
96 — Harshaw
97 — Hartford (Huachuca)
98 — Hassayampa (Prescott)
99 — Havasu Canyon
100 — Havasu Lake
101 — Heber
102 — Helvetia
103 — Horseshoe Basin
Huachuca, see Hartford (97)
Hualpai, see Maynard (124)
104 — Humbug
Indian Secret, see White Hill (237)
Jacob Canyon, see Warm Springs (232)
Jerome, see Verde (226)
Johnson, see Cochise (53)
Kaibito Plateau, see White Mesa (238)
105 — Katherine, also known as
San Francisco (191)
Kelvin, see Mineral Creek (129)
106 — Kimball (Peloncillo)
107 — Kirkland
108 — Kofa
109 — La Cholla
110 — La Paz (Weaver)
111 — La Posa (Wellton)
112 — Laguna (Las Flores)
Las Flores, see Laguna (112)
113 — Las Guijas
114 — Lone Star
115 — Long Valley
116 — Lost Basin
117 — Lukachukai
118 — Lynx Creek
119 — McConnico
McCraken, see Owens (149)
McMillen, see Richmond Basin (180)
Mac Morris, see Richmond Basin (180)
120 — Magazine
121 — Mammoth (Tiger, San Manuel, Old Hat)
122 — Martinez (Congress)
123 — Martinez Canyon
Mayflower, see Twin Peaks (221)

NO.

124 — Maynard (Hualpai)
125 — Mazatzal Mts.
126 — Metcalf (Greenlee)
Meyer, see Gunsight (92)
127 — Miami, also known as Globe (80)
Middle Camp, see Oro Fino (147)
128 — Midway
129 — Mineral Creek (Ray, Kelvin)
130 — Mineral Hill, Pinal County
Mineral Hill, Pima County, see Pima (162)
Mineral Hills Wash, see Santa Maria (194)
131 — Mineral Park, also known as Wallapai (231)
132 — Mineral Point
133 — Minnehaha
134 — Minnesota
Mocking Bird, see Virginia (227)
135 — Mohawk
136 — Montezuma (Puerto Blanco Mts.)
137 — Monument Valley
Moors, see New River (141)
Morenci, see Copper Mountain (56)
Morgan City, see Pikes Peak (160)
138 — Muggins
139 — Music Mountain
140 — Neversweat (Palomas Mts.)
141 — New River (Moors)
142 — New Water
143 — Nogales (Gold Hill)
Oatman, see San Francisco (191)
Octave, see Rich Hill (179)
144 — Old Baldy
Old Hat, see Mammoth (121), Oracle (145)
Olive, see Pima (162)
145 — Oracle (Control, Old Hat, Santa Catalina)
146 — Oro Blanco (Ruby)
147 — Oro Fino (Middle Camp)
148 — Osborn
149 — Owens (McCracken, Potts Mts.)
150 — Owl Head
151 — Pajarito
152 — Palmetto
Palomas Mts., see Neversweat (140)
153 — Papago (Sierrita)
154 — Parker Canyon
155 — Patagonia (Duquesne, Washington)
156 — Paul
Payson, see Green Valley (88)
Peacock, see Hackberry (93)
157 — Pearce
158 — Peck
Peloncillo, see Kimball (106)
159 — Picacho
Pierce, see Bullard (29)
160 — Pikes Peak (Morgan City)
161 — Pilgrim
162 — Pima (Olive, Mineral Hill, Twin Buttes)
163 — Pinal Mts. (Pioneer)
164 — Pinedale
165 — Pine Grove (Crown King)
166 — Pinto Creek (Pinto Valley)
Pinto Valley, see Pinto Creek (166)

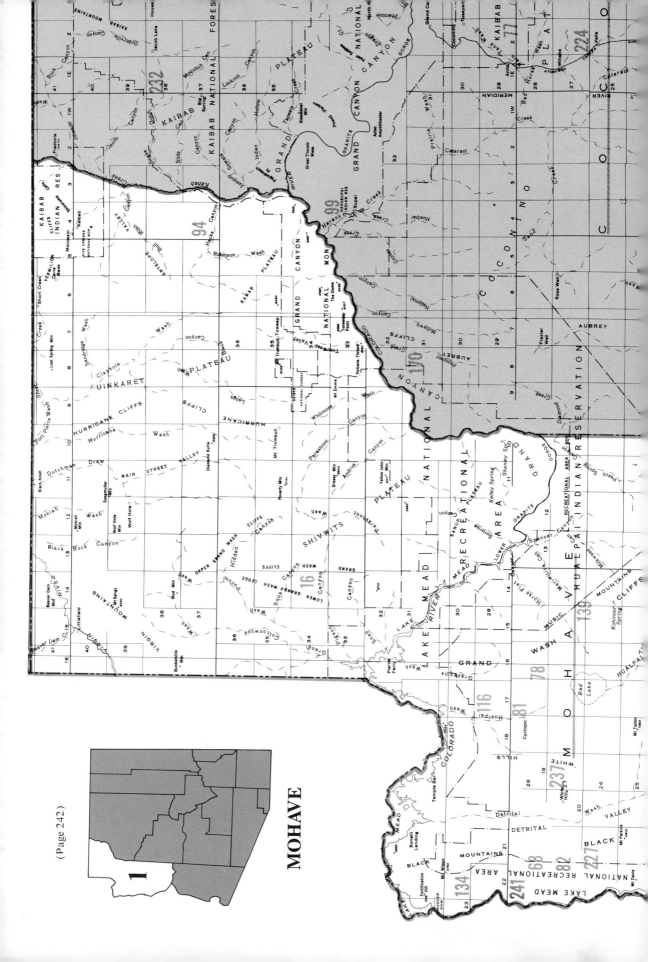

MOHAVE

(Page 242)

1

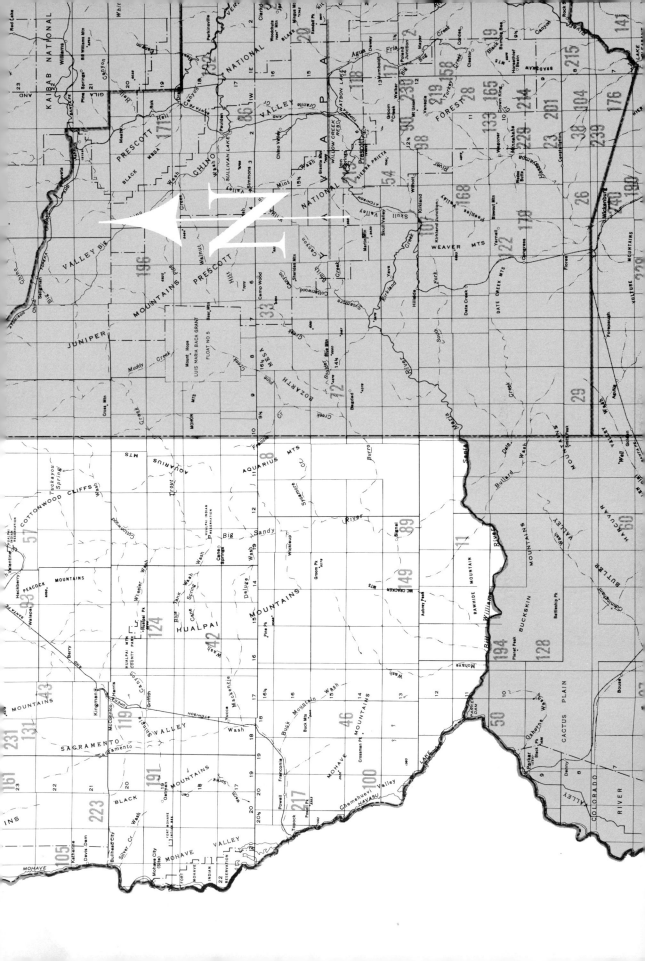

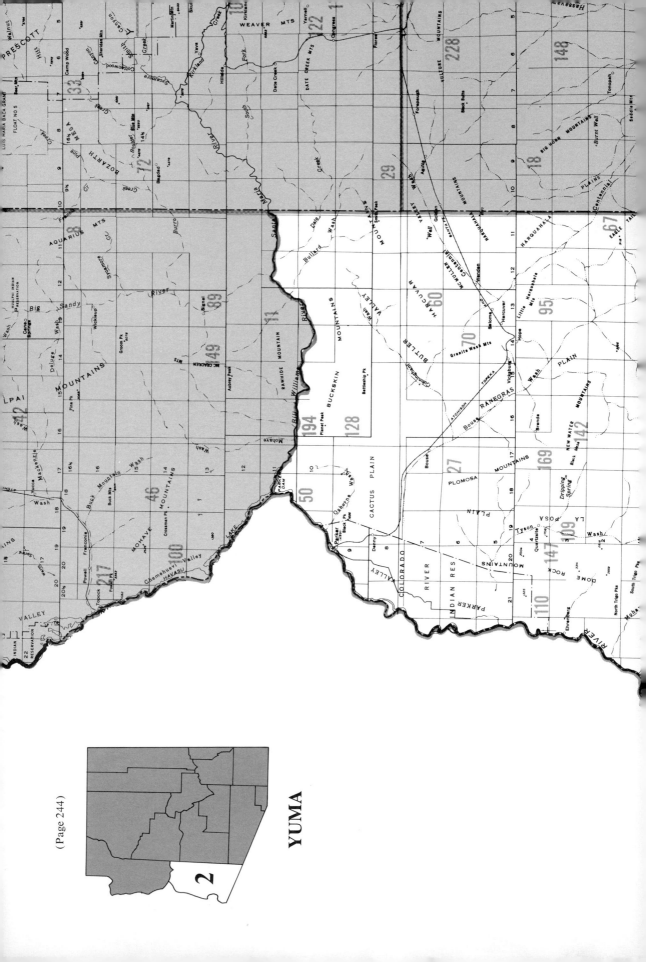

YUMA

2

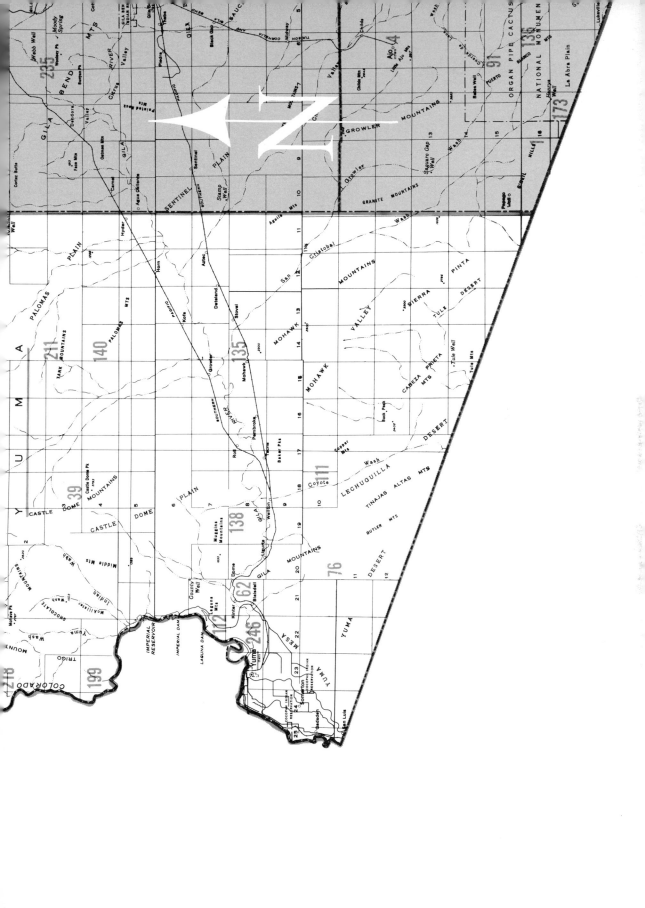

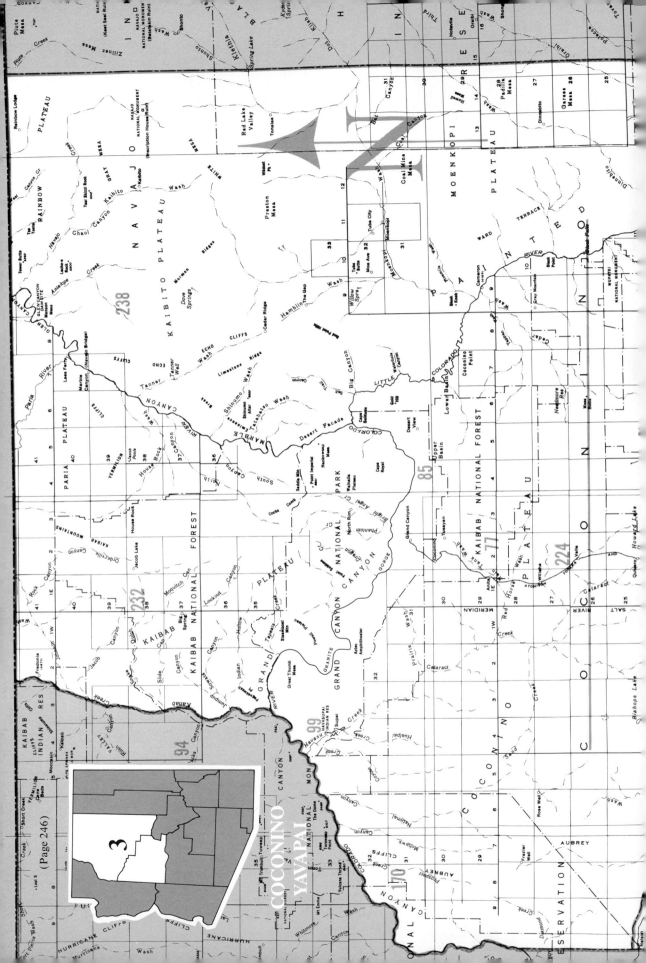

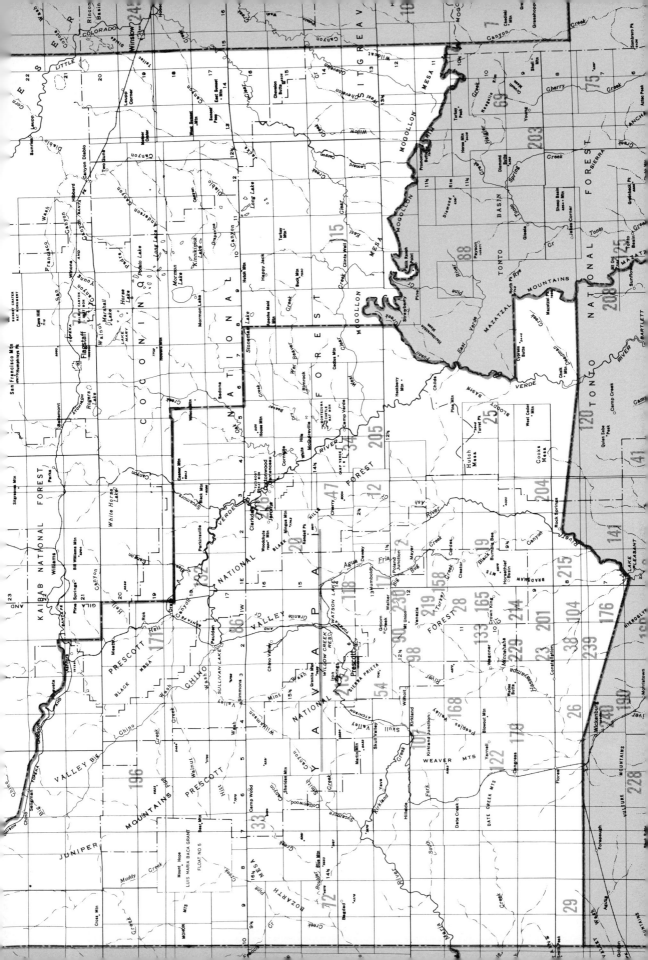

(Page 248)

MARICOPA
GILA
PINAL

4

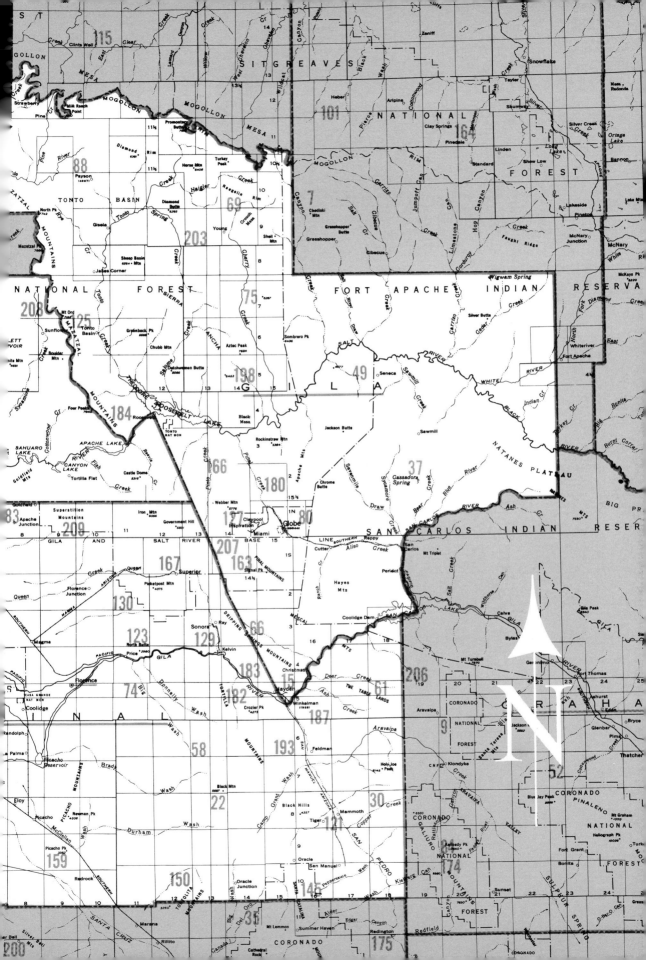

(Page 250)

# NAVAJO

# APACHE

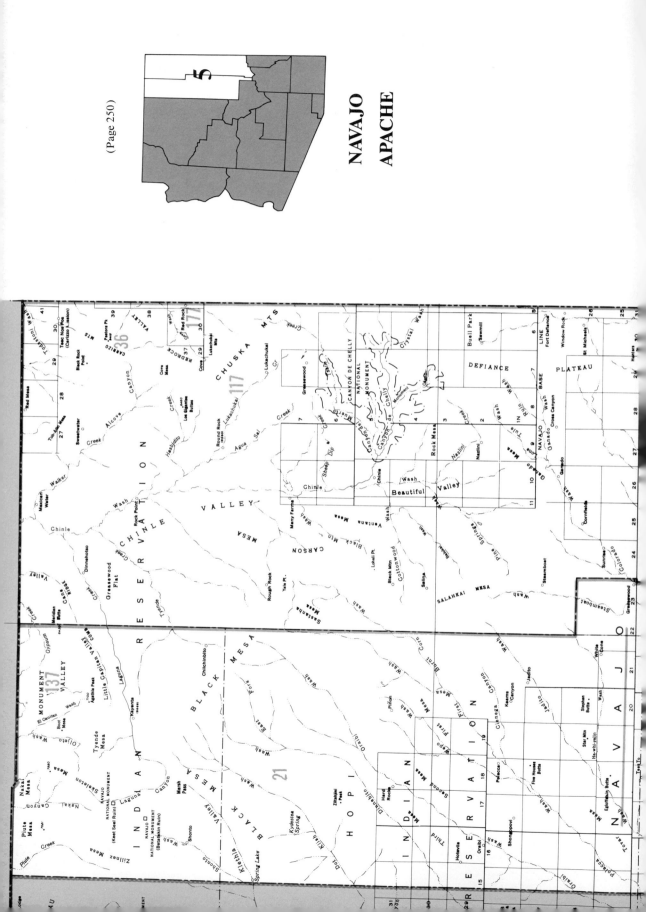

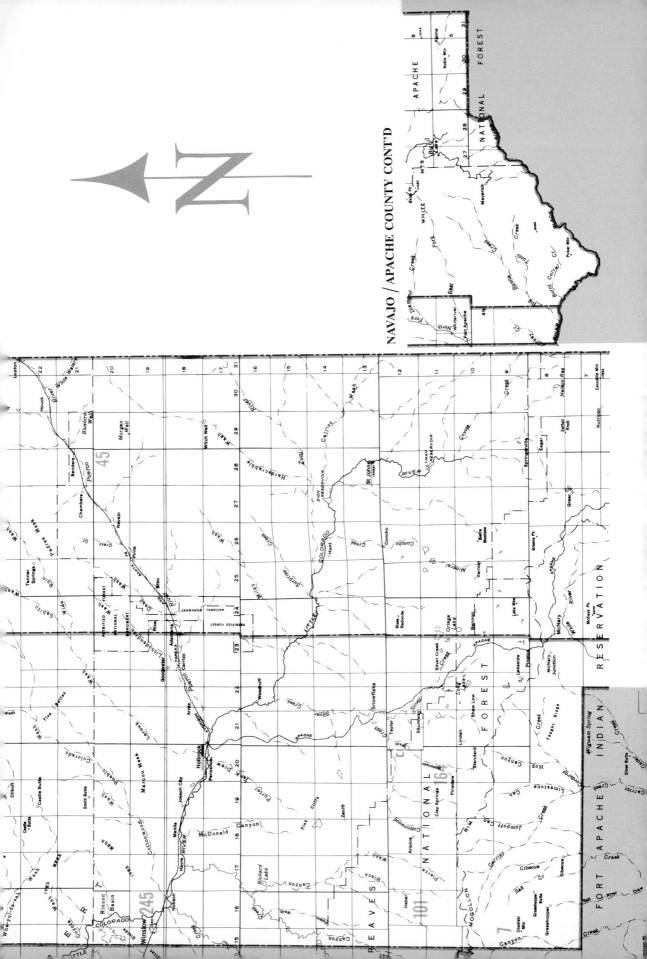

# GRAHAM
# GREENLEE
# COCHISE

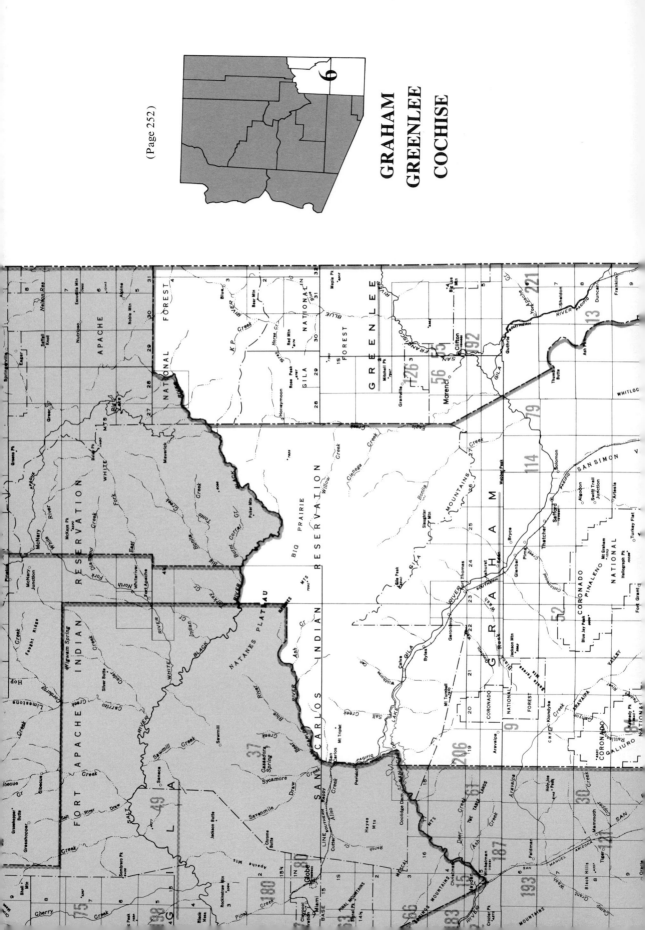

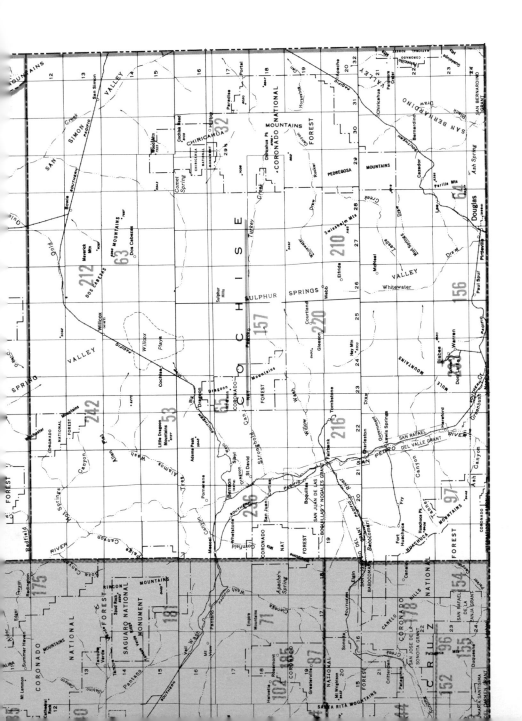

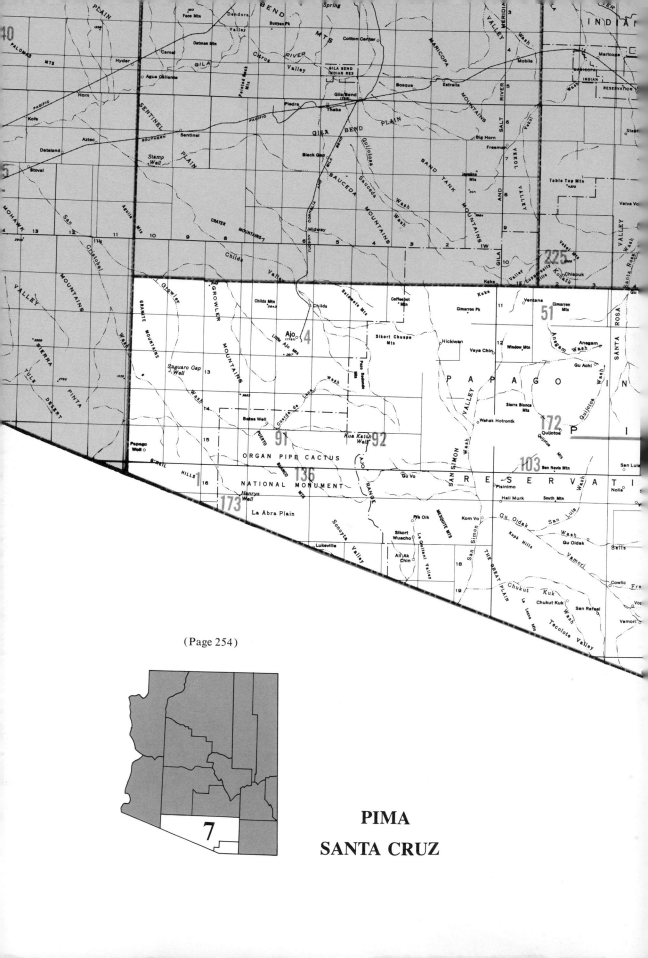

(Page 254)

PIMA

SANTA CRUZ